Environmental Risk-Based Analysis for Managers

Environmental Risk-Based Analysis for Managers

Edited by **Bernie Goldman**

SYRAWOOD
PUBLISHING HOUSE

New York

Published by Syrawood Publishing House,
750 Third Avenue, 9th Floor,
New York, NY 10017, USA
www.syrawoodpublishinghouse.com

Environmental Risk-Based Analysis for Managers
Edited by Bernie Goldman

International Standard Book Number: 978-1-68286-120-2 (Hardback)

Printed in the United States of America.

Contents

Permissions

List of Contributors

Preface

The recent exhaustion of natural resources and increase in pollution has given rise to the risks and negative impacts of such activities on the environment. This book elucidates the concepts and innovative models around prospective developments with respect to environmental risk based analysis. It presents detailed discussions on topics such as sustainable water resource management, pollution mitigation, environmental statistics, impacts and assessment of climate change, etc. The researches in this book cover the entire spectrum of environmental risk based analysis and will be an excellent reference material for professionals and academicians.

This book has been the outcome of endless efforts put in by authors and researchers on various issues and topics within the field. The book is a comprehensive collection of significant researches that are addressed in a variety of chapters. It will surely enhance the knowledge of the field among readers across the globe.

It gives us an immense pleasure to thank our researchers and authors for their efforts to submit their piece of writing before the deadlines. Finally in the end, I would like to thank my family and colleagues who have been a great source of inspiration and support.

Editor

Evaluating watershed management activities of campaign work in Southern nations, nationalities and peoples' regional state of Ethiopia

Kebede Wolka Wolancho

Abstract

Background: The Ethiopian government has been implementing watershed management mainly through public campaign work. However, its effects have not been evaluated in many micro-watersheds. This study evaluates watershed management activities and its socio-economic and biophysical role.

Results: Each kebele has institutional arrangements such as development teams comprising 3–35 households, and 5-person labor groups, which mobilize people and penalize absentees (if any). The survey indicated that common lands, subject to free resource exploitation such as grazing, were typically severely degraded. The majority of respondents wait for development agents and campaign work before repairing the conservation structures. Tree species selection was found to be appropriate in most areas. However, poor seedling survival (<5%) was observed in some micro-watersheds. In most micro-watersheds, structure selection, design, construction and spacing was appropriate.

Conclusions: Achievement in rehabilitating degraded lands was seen as excellent lessons for future efforts. The following issues need to be addressed in future watershed management campaign work: poor structure maintenance, low seedling survival, creating defined land user/owner for common land rehabilitated collectively, crop and cattle damage by wildlife residing in rehabilitated micro-watershed, incentivizing development agents, periodic auditing and repairing of built structures and seedling replacement.

Keywords: Conservation structures; Institutional arrangement; Property right; Seedling Survival; Watershed degradation

Background

Sustainable livelihood and increased food production in agricultural based developing countries require the availability of sufficient water and fertile land (Tesfaye 2011). In sub-Saharan Africa, unsustainable livelihoods often contribute to degradation of important watershed resources (Kerr 2002). Among the degrading watershed resources, fresh water and soil fertility take the lead in posing significant socio-economic, ecological, and environmental roles, especially for developing countries including Ethiopia where traditional agricultural-based economy is dominant. As a result of dependency of increasing population on traditional subsistence agriculture, most of the Ethiopian highlands are experiencing degradation of watershed resources.

Ever since people began manipulating land, various approaches and techniques were practiced to reduce degradation of watershed resources. However, the system thinking or modern watershed (generally a drainage area) management started in mid 20th century and adapted in most countries with the aim of controlling water pollution, sedimentation, soil erosion, flood, and discharge extremes. The watershed management effectively accounts multiple linkages between livelihood and natural resource management (Hope 2007; Tiwari et al. 2008). Vegetation, soil, and water resources can be protected more efficiently through this approach since whole ecosystems and people participation can significantly be considered (Kerr 2002; Srivastava et al. 2010; Price et al. 2011). This contributes for improvement of watershed resources and livelihood of the people (Pathak et al. 2013; Khajuria et al. 2014).

Correspondence: kebedewolka@gmail.com
Wondo Genet College of Forestry and Natural Resources, School of Natural Resources and Environmental Studies, Hawassa University, P.O. Box 128, Shashemene, Ethiopia

In Ethiopia, watershed management was initiated in the 1970s to tackle water-caused soil erosion impacts and water shortage in agricultural economy. Since this period, interest in the multiple environmental, economic and social benefits provided by watershed management has greatly increased and accordingly it has been recommended for achieving various purposes in different part of the country (Nigussie 2003; Woldeamlak 2003; Hengsdijk et al. 2004; Kefyalew 2004; Ludi 2004; Admasu 2005; Tamene et al. 2005; Ermias et al. 2006; German et al. 2006; Tamene et al. 2006; Tamene and Vlek 2007; Andualem 2008; Emiru 2009; Kebede 2012; HNCJ Hunger, Nutrition Climate Justice 2013).

There have been challenges (i.e. difficulty for rapid replication, engaging all land users in the watershed, implementation costs, etc.) to implement watershed management in different part of the country. However, as result of strong effort by government and community to overcome the challenges, exemplary successes at the mini-watershed (a drainage area covering 400–2000 ha) or micro-watershed (a drainage area covering less than 400 ha) scale has been recorded and globally appreciated HNCJ Hunger, Nutrition Climate Justice 2013). Evidence of success includes considerable improvement in water discharge levels in streams and springs, improved water table levels, and reduced sedimentation problems in the water harvesting ponds and reservoirs (Haregeweyn et al. 2005; Haregeweyn et al. 2008).

The current approach that much contributed for success is 'community based participatory integrated watershed management', which requires involvement and contribution of local people. The Ethiopian government understands the essence of this approach as evidences from successfully implemented pilot projects appear promising. Significant effort is occurring to replicate 'community based participatory integrated watershed management' activities in weredas[a] of most regions. As a component of this effort, in the last four years a nationwide 30 days public work campaign for watershed management has occurred. In Sidama, Kambata Tambaro, Wolayita and Dawro zone of the Southern Nations, Nationalities and Peoples Regional State (SNNPRS), such activities are ongoing. The 30 days watershed management labor contribution by farmers has been practiced in most weredas of these zones.

However, effect of such watershed management, effectiveness of institutional arrangement, influencing biophysical and socio-economic components, challenges and opportunities for replicating and sustaining the activities are rarely evaluated for most micro/mini-watersheds. Therefore, this study examines components of site specific packages of watershed management activities; evaluates the status of watershed management activities; analyzes effects of watershed management; and explores institutional, socio-economic and biophysical opportunities

and challenges in adopting and sustaining watershed management activities.

Methods
Site description

SNNPRS is one of the nine political regions of Ethiopia. The region borders Kenya to the south, South Sudan to the west, Gambela region to the northwest, Oromia region to the north and east. The region is divided in to 14 zones (Bench Maji, Dawro, Debub Omo, Gamo Gofa, Gedeo, Gurage, Hadiya, Kaffa, Kanbata Tambaro, Segen, Shaka, Sidama, Silte, Wolayita) and 4 special weredas (Alaba, Basketo, Konta and Yem) (Figure 1).

The 2007 census by Ethiopia's Central Statistical Agency (CSA) estimated that the SNNPRS region had a population of about 15 million (with annual growth rate of about 2.9%) from which 89.98% were rural inhabitants, making it Ethiopia's most rural region (Central Statistical Agency 2007). The region comprises an area of 105,887.18 km^2. The eastern, northern and central part of the region is densely populated whereas the southern and western part is sparsely populated.

The region has a diversity of agro-ecological zones, ranging from Berha (semi-arid) at south Omo to Wurch in Gamo Gofa zone. Accordingly, the lowland area (mainly part of south omo zone) of the region receive less than 600 mm annual rainfall and highlands of Shaka, Kaffa, Dawro, Wolayita, KambataTambaro, Sidama, Gamo Gofa zones receive more than 1200 mm per year (SNNPRS-BoFED 2004). Average temperature of lowland exceeds 20°c and about 28°c is reported for extremely hot area. Average temperature of 10–20°c is recorded for highland areas, for which less than 10°c also exists. Generally, different part of the region is characterized by climate categories of hot semi arid climate, tropical climate II, tropical climate III, warm temperature climate I, and warm temperature climate II (SNNPRS-BoFED 2004). Owing to its diverse agro-ecological zones, a range of plant species grow in the region. The region is rich in perennial crops such as *Ensete ventricosum, Coffee arabica*, and *Catha edulis*. Cereal crops such as *Triticum aestivum, Hordeum vulgare, Zea mays, Sorghum bicolor, Eragrostis tef, Phaseolus vulgaris, Pisum sativum, Vicia faba, Ipomoea batatas* and *Manihot esculenta are widely cultivated in the region.* Coffee based agroforestry is widely adopted. Planting fruit/tree species around homestead is commonly practiced. Soil types such as dystric nitosols, orthic acrisols, pellic vertisols, chromic luvisols, eutric fluvisols, eutric nitosols, mollic andosols, and chromic vertisols dominantly characterize different part of the region (SNNPRS-BoFED 2004). The three basins: Omo-Gibe, Baro-Akobo, and Rift valley lakes drain big area of the region.

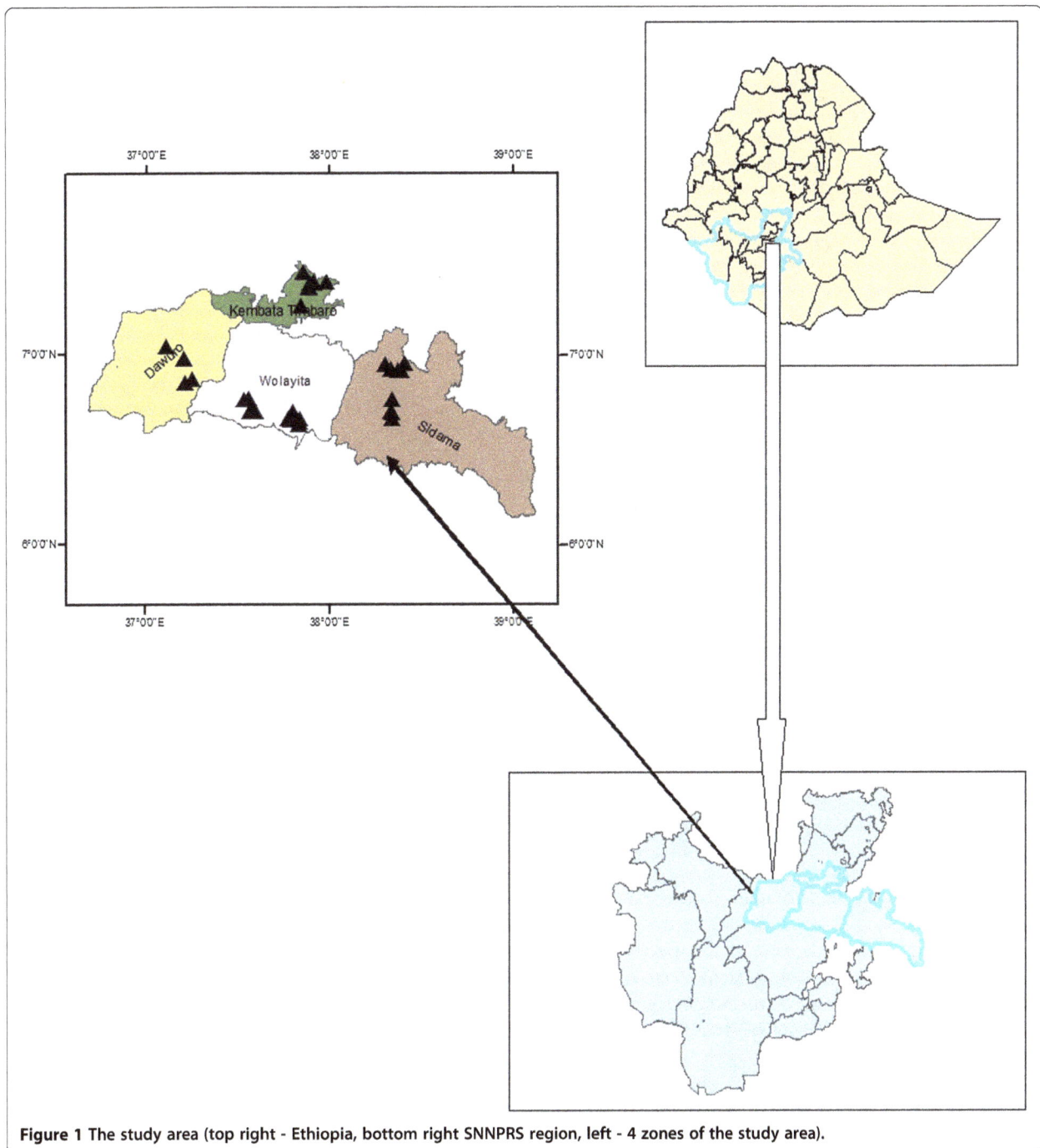

Figure 1 The study area (top right - Ethiopia, bottom right SNNPRS region, left - 4 zones of the study area).

The land area of eastern, northern and central part of the region is comparatively degraded due to increased pressure from dense populations and long-term cultivation.

Methodology

For this study, Sidama, KambataTambaro, Wolayita and Dawro zones were strategically selected. From each zone, two weredas were selected for accessibility with a four-wheel drive vehicle and the existence of a range of agro-ecological categories. In each wereda, all four-wheel accessible kebeles[b] were categorized into agro-ecological groups. After these considerations, a kebele was randomly selected from each agro-ecological category and all micro-watersheds within the kebele, where watershed management has been implemented by campaign work, were considered for evaluation (Table 1).

When the watershed management activities were implemented on common land, checklist guided focus group discussions were conducted with the farmers living around the impacted area. In the case where

Table 1 Sampled zone, wereda, kebele, micro-watershed

Zone	Wereda	Kebele	Micro-watersheds
Sidama	Borecha	Yiriba Dubancho	Garato, Lelanto
		Hanja Chafa	Chafa, Bantola
		Koran Goge	Harfato, Koncha
	Dale	Debub Kege	Golota, Begasine
		Megara	Alelecho, Manto-Danchame
		Kalete Senete	Gorbito,
Kambat-Tambaro	Damboya	Kota Kombola	Dilamo, Lantate, Eja
		Ha'amancho	Oliso
		Megere	Welecho
	Angacha	Zebecho	Ajacho, Koruwa, Halega
		Bucha	Kinham, lower Bucha, Ajacho
		Gede Genet	Bukuna, Bubuyesa, Farkasa
Wolayita	Humbo	Abela Faracho	Kole, Alata, Lali'ana
		Abela Sifa	Shafa, Basa, Loke
		Ela Kabala	Tewaye Hamasa, Tebo Hamasa
	Ofa	Yakima	Hoze, Halozia, Busho
		Sadoye	Sadoye
		Soresha	Bongota, Tuba, Awash
Dawro	Loma	Addis Bodare	Tida, Dulae
		Gendo Walcha	Toni
		Gedo Buna	Kuta
	Mareka	Mari Madara	Ali
		Gobo Shamena	Wuni

Table 2 Number of farmers participated in group discussion and interview in each considered kebele

Kebele	Number of participants in group discussion	Number of land users interviewed and farm field visited
Yiriba Dubancho	9	10
Hanja Chafa	15	18
Koran Goge	11	14
Debub Kege	6	-
Megara	5	19
Kalete Senate	8	-
Kota Kombola	7	11
Ha'amancho	9	10
Megere	7	18
Zebecho	-	15
Bucha	-	18
Gede Genet	5	18
Abela Faracho	5	14
Ela Kabala	5	8
Yakima	6	16
Sadoye	8	20
Soresha	7	10
Gendo Walcha	-	15
Gedo Buna	-	12
Mari Madara	-	8
Gobo Shamana	-	6

watershed management activities occurred on small-holders land, individual interviews were conducted. Five percent of the households were randomly selected for these interviews. Fifteen group discussions (each comprising 5–15 participants) were conducted in micro-watersheds. About 260 heads of household interviews were conducted (Table 2).

Discussions were conducted with each wereda watershed experts and kebele agricultural development agents. Secondary data, especially kebele achievement reports were reviewed.

When possible, field observations and expertise evaluations were conducted on watershed management activities in the micro-watersheds. The accomplished activities were evaluated using the following criteria:

- Compliance of the watershed management activity with the principles of watershed logic, especially commencing interventions such as constructing physical soil and water conservation structures from ridge or upper part of the micro-watershed and progressively proceeding to valley or lower part of the micro-watershed.
- Status of physical soil and water conservation structures:
  - Proportion of broken physical soil and water conservation structures.
  - The stabilization of physical soil and water conservation structures with biological measure (planting recommended grass/shrub/tree species on the structures).
  - Appropriateness of physical soil and water conservation structures selection (considering factors such as rainfall amount, soil texture, slope gradient, and existing land use).
  - Design of physical soil and water conservation structures by considering soil texture, rainfall amount, slope gradient and land use.
  - Compliance to recommended specification such as foundation for stone bund and channel depth, channel width, length, and berm for various other physical soil and water conservation structures.
- Existing management and maintenance of physical soil and water conservation structures.

- Seedlings planted and its management: appropriateness of species selection as affected by agro-ecology and related factors, survival percent of seedlings planted on common land in past two years, and on-going management.

An archive review was conducted at the kebele offices. The rainfall data for Yirba Dubancho, Yirgalem, Durame, Angecha, Humbo, Gesachare, and Gesuba stations was collected from Ethiopian Meteorological Agency,Hawassa Branch office. Micro-watershed elevation was measured using an altimeter. Survival of seedlings planted in the previous two years in common land was assessed by taking systematically distributed samples. Descriptive statistics was used to analyze the collected data.

Result and discussion

Due to an exploitive land use history, the study area's cultivated and grazing land has been degraded as characterized by: low fertility, gully and rills formation, moisture stress, declining productivity, etc. The objectives of watershed management in the studied micro-watersheds were related to these issues and mitigation activities focused on reducing further degradation and rehabilitating these impacted lands.

Role of the 'development team' in enhancing participation of local people

In public campaign work, people participate by contributing labor, farm tools, etc. in watershed management activities through a 'development team'. In each surveyed kebele, small 'development teams' have been organized. The number of members on a team depends on the local situation. The largest development team, with 35 members, existed in Sadoye kebele (Ofa wereda of Wolayita zone) and the smallest, with 3 members, was observed in Mari Madara kebele (Mareka wereda of Dawro zone). The team is responsible for developing and administering internal regulations that enforce, encourage, or punish absentees from campaign work. Since the implementation of watershed management campaign work is based on group consensus, the punishment on absentees is not particularly serious. For instance, in Zebecho, Soresha, Gede Genet, Ela Kabal, Abela Faracho and Adis Bodare kebele, some development teams only give advice and exercise no monetary or other penalties. The development teams in Gobo Shamenena, Zebecho, Megara, and Soresha kebele require that the absentees compensate missed work on another day(s) if they don't have convincing reasons for not participating. Some development teams in Bucha, Hanja Chafa and Gendo Walcha kebeles financially penalize the absentees and use the collected money for strengthening the development work (e.g. purchasing farm tools). Gendo Walcha kebele uses some of the

funds to partially pay guards to protect the exclosed site. The social and moral sanctions imposed by respective 'development team' members in a give locality seeks to encourage/enforce the members to participate in the watershed management work. Other study (Enwelu et al. 2014) also recognizes the importance of local rules and regulation in community participation. A study by Lullulangi et al. (2014) confirmed that community effort and norm in watershed resource management such as planting protective trees has positive contribution.

Work norm and supply of tools

In recent years and especially in 2013/14 campaign years, the watershed management activities in each kebele involved various community groups in the public work projects (Figure 2). The youth, women and head of households (adult male) participated in the various allocated tasks. In all the kebeles, the highest proportion of work has been achieved by the adult male group. For instance, in Debub Kege, Kota Kombola, Ha'amancho, Zebecho, Bucha and Sadoye kebeles respectively 45, 60, 60, 68, 83 and 95% of the work was accomplished by the adult male group. The remaining proportion of the watershed management activities were implemented by youth and women groups. In the respective order of these kebeles, 30 and 25, 15 and 25, 15 and 25, 12 and 20, 10 and 7, 2 and 3% of the activities were accomplished by youth and women respectively. The average soil bund construction is about 6 meters/day for male groups and 3 meters/day for women groups. In areas with hard or rocky soils these averages are not achieved.

Constructing physical soil and water conservation measures by those teams to achieve the intended norm requires various tools for lay out, digging, rock breaking, excavating, embanking etc. In Adis Bodare and Gedo

Figure 2 Public campaign workers building soil and water conservation measures in Hanja Chafa kebele, Chafa micro-watershed.

Buna, farmers use their own tools for such watershed management activities. Tools are provided by some weredas for the kebeles where equipment shortages exist as identified through preliminary assessments. This is part of the government contribution through the wereda agriculture office. Some kebeles (Hanja Chafa, Debub Kege, Kalete Senate, Kota Kombola, Zebecho, Abela Sifa, Mari Madera, and Gobo Shamena) confirmed they received one or more of the required tools through the wereda. Since these management activities are occurring simultaneously in all the kebeles in a given wereda, supplying sufficient tools is a challenge for kebeles, especially if they are not being supported by other organizations (e.g. Non Governmental Organizations, NGO).

Property rights and land degradation

Many of the micro-watersheds where management activities have been centered are characterized by gully formation, which indicates severe erosion of top soil and poor vegetation cover. Severely degraded micro-watersheds have common histories in land use including change in land ownership. The communal lands tend to be more susceptible to exploitative use and degradation problems.

In the Dale wereda (Golota micro-watershed of Debub Kege kebele), during the Imperial government (prior to 1974) the land was covered with forests and harbored abundant wildlife. Land was privately owned during that period. When the land was shifted from private to communal during the Derge government, the forest was removed by unregulated tree cutting and charcoal production. The land became eroded and degraded. A similar situation occurred with the Gorbito micro-watershed in Kalete Senete kebele.

A once severely degraded, but now recovering area is Dilamo micro-watershed of Kota Kombola kebele where privately owned land during Imperial government was converted to common land during the Derge government. The land was progressively degraded starting from the time it was used for common grazing and resource extraction and remained as barren wasteland until three years ago when watershed management intervention began.

Land in Oliso micro-watershed of Ha'amancho kebele has a similar history. Once privately owned and covered with thick forest, during the Derge regime the land was changed to common grazing land and soil erosion created serious degradation. About 60 years ago, there was small gully at lower part of the micro-watershed which later deepened and also elongated to the upper part of micro-watershed. As result of sever degradation for any economical use, farmers owning the land has ignored it and permitted land for common use such as grazing. This degradation created 'common land' now treated and appears promising resource for productive use in near future.

In Kota Kombola kebele, part of the Eja micro-watershed was used for intensive grazing and became severely degraded. Since 2011, watershed management activities have improved the land conditions (i.e. gullies filled, soil regenerating, grass growing) although the land is still considered common property. Similarly, in Welecho micro-watershed of Megere kebele, big gullies formed many years ago have been aggravated by intensive grazing. In Manto- Danchame micro-watershed, the land was naturally rock in lower watershed, but has now extended by an additional 50 meters uphill in past about 40 years. The land was owned by private called 'keberite'[c] during Imperial government but since Dergue government it has been communal land and the forest has been removed creating gully and land dissection consequences. In line with these observations, other studies also reported that land ownership has influence on soil conservation and its management (Gebremedhin and Swinton 2003; Bewket 2006). Land ownership and tenure influence motivation of land owner on watershed management (Rosenberg and Margerum 2008; Teshome et al. 2014).

Tree/shrub species selection, survival rate and management

A component of biological soil and water conservation, tree/shrub seedlings are often planted in watershed management areas. Species selection and survival rate for common lands were considered in this study whereas seedlings planted on individually owned land were not considered as this information is not easy to gather. In general, the species selection is vital as it determines future available products, seedling survival, and agro-ecological compatibility.

Species that can survive poor sites and also improve soil properties are generally preferred to rehabilitate degraded land through watershed management. In this regard, maintaining the Eucalyptus species in degraded communal land at Garato micro-watershed is not appropriate as this species restricts undergrowth and as result poorly contributes to soil conservation and land rehabilitation (Fikreyesus et al. 2011). Eucalyptus is also highly nutrient and water-consuming.

The micro-watershed areas that have been rehabilitated by retaining existing and regenerated trees and shrubs need technical support to continue sound management and utilization. In Ela Kabala kebele, the lower watershed of 2012 and 2014 intervention is in commonly owned woodland. The objective of this micro-watershed management is to enhance grass growth for livestock, which currently has poor grass availability. Thus, the shrubs should be harvested under technical guidance and implemented to meet the farmers' interests. Retaining the shrub component will probably not contribute to the intended goals.

In Tuba micro-watershed of Soresha kebele, development agent reported that none of the seedling planted in

2012 survived due to late planting in reference to the rainy season, and seedlings were planted without polyethylene tubes. In addition, water logging properties of soils influenced the survival rate. On other hand, the poor survival of seedlings in micro-watersheds of some kebeles such as Soresha, Yirba Dubancho and Koran Goge (Table 3) is related to land degradation problem. Since the degraded land has poor soil fertility and water holding capacity, seedling survival is negatively affected.

When the ecology of the micro-watersheds are surveyed for their suitability for dominantly planted tree and shrub species most of the micro-watersheds were found to be appropriate (Azene and Birnie 1993) (Table 3).

Soil and water conservation structures: specification and selection

Depending on slope, rainfall and workable soil depth, each soil and water conservation structure has specific standards. Research findings suggest that structures built below these standards are less effective in controlling erosion. Sometimes a poorly constructed structure becomes dangerous as it aggravates erosion by collecting the surface runoff and enhancing collective high volume flow. Therefore, maintaining the construction standards is obligatory. In Alelecho micro-watershed of Hanja Chafa kebele (Figure 3), some of the structures built in 2014 are in need of improvement. For example the trench depth, which is less than the 50 cm requirement, needs improvement. The *fanya juus* and trenches have either no, or a narrow berm, to reduce back flow of soil into the channel. Similarly, a narrow berm was observed in the *fanya juus* built in 2014 at the Welecho micro-watershed (Megere kebele). Mounding and compacting of embankment soil was poorly managed. Stone bunds can be built in areas where sufficient stone can be collected. The specifications for these structures need to be followed to reduce their dismantling and increase their soil

conservation effectiveness. Among the specifications, a properly installed foundation (to the depth not less than 20 cm) is required. However, a weak or no foundation was observed in most stone bunds built in micro-watersheds of Alelecho, Bongota, Tuba, Sadoye, Tewaye Hamasa and Alata.

Soil and water conservation structures built on cultivated land need to be suitable for farming activities, including easy travel across a farmland. To achieve this, the recommended maximum length of a structure should be 40–50 meters CFSCDD Community Forests and Soil Conservation Development Department (1986) with a gap of 5–10 meters before restarting (if required). In Welecho micro-watershed of Megere kebele and Teba Hamasa micro-watersheds of Ela Kabala, soil bunds over 100 meters long were observed.

According to CFSCDD (Community Forests and Soil Conservation Development Department) (1986) and Lakew et al. (2005), most of implemented structures are appropriate for the sites (Table 4). However, in micro-watersheds of some kebele such as Yakima, Mari Madara, and Zebecho, soil bunds and *fanya juus* were constructed on crop lands with steep slopes (>40%). In principle, such steep sloped lands are less feasible for crop cultivation. In practice, due to land shortage, it becomes necessary to cultivate these lands. These structures can be suitable for such slopes with cautious design and management CFSCDD Community Forests and Soil Conservation Development Department (1986). In comparison, in micro watersheds of Kalete Senete, Megere, Abela Faracho, and Abela Sifa kebele, soil bunds and/or *fanya juus* has been built on the land with slopes less than 3% (Table 4). Soil erosion for such slopes can be managed by less costly activities such as grass strips, unless the area has critical water shortage to establish the grass. Practically, conservation of the water, itself, in water scarce lands, may demand the structures.

Table 3 Planted species, seedling survival and their appropriateness

Kebele	Considered micro-watersheds	Elevation range (m)	Mean annual Rainfall (mm)	Dominantly planted species on common land	Survival %	Comment on planted species
Yiriba Dubancho	1	1980-2020	1070.6	*Acacia saligna*	5	Appropriate
Hanja Chafa	2	1928-1945	1070.6	*A. saligna, Grevillea robusta*	25	Appropriate
Koran Goge	1	1925-1940	1070.6	*A. saligna*	7.5	Appropriate
Debub Kege	1	1730-1734	1209.1	*A. saligna, Gravilea*	65	Appropriate
Megara	2	1765-1792	1209.1	*Olea, Cordia, G. robusta, A. saligna*	70	Appropriate
Kalete Senete	1	1970	1209.1	*A. saligna, G. robusta*	50	Appropriate
Kota Kombola	3	2000-2110	1156.5	*G. robusta, Acacia decurrens, A. saligna*	43	Appropriate
Ha'amancho	1	2100-2200	1156.5	*A. saligna, A. decurrens, Cupressus lusitanica*	60	Appropriate
Ela Kabala	1	1520-1580	1132.3	*G. robusta*	85	Appropriate
Soresha	1	1570-1673	1056.1	*Jacaranda mimosifolia, G. robusta*	2	Partially appropriate

Figure 3 Examples of bund with insufficient berm in Hunja Chafa kebele (the mounded soil at the side of channel in both pictures shows poor berm and embanking, apparently without berm (a); poor berm (b)).

The principle of watershed management requires starting from the upslope/ridge and progressively intervening toward the lower watershed. The intervened micro-watersheds in public campaign work heed administrative boundaries as it is confined within a kebele or wereda. However, the micro-watershed boundary may not fit the administrative boundary making the structures ineffective. In Bantola micro-watershed, intervention of the upper watershed commenced three years after the treatment of the area below it. This occurred because the upper watershed (even though a comparatively small area) belongs to another kebele. A similar case was observed in Welencho micro-watershed of Megere kebele. In Garato micro-watershed of Yirba Dubancho kebele, effective intervention was undertaken to rehabilitate degraded lands even though the lower watershed that is being protected from the severe water flows is in another wereda. Even though only a few fragmentations in the intervention process were observed, from the perspective of making watershed management effective, when fragmentation does occur, there needs to be collaboration in planning and management activities.

Challenges in practicing activities and in sustaining outcomes

Key informants and development agents indicated various site specific challenges with watershed management activities. The key informant in Bucha kebele explained that soil is too hard to dig and it is difficult to achieve the work norm stated in Work norm and supply of tools. In the lower watershed of this kebele, planting grass on soil bunds/fanya juu is poorly practiced compared to other parts of the micro-watershed, which may be due to the large land area per household, as the area's population density appears to be low. In contrary, in many areas land shortage challenges the size of channel and embankments. In various kebeles, farmers reduced conservation structure embankment widths in order to have more land for crops. This resulted in many soil and water conservation structures being overtopped by flooding.

Sustaining community participation in structure construction and management is other challenge in Abela Faracho and Abela Sifa kebeles. Participant numbers are small because it is a democracy and during the off season, cattle graze on farm land and disrupt the stability of structures. In addition, farmers do not like having stone bunds built close to their houses as they tend to be good snake habitat. In Abela Sifa kebele, since there was payment for constructing structures by NGO and safety net, the farmer hesitates to accept campaign work. In this kebele's Shafa micro-watershed, the surface flow resulting from few days' rainfall overtopped and significantly damaged the structures due to poor design and management.

In Ela Kabala kebele, experts and farmers explained that monkeys residing in the watershed management - woodland area and damage their crops. In Sadoye and Megere kebele (Welecho micro-watershed), beginning in 2011, hyenas have moved into the managed watershed and have hurt cattle and people. This has had a negative effect on the community's perception about the management activities. The progressive decrease in number of participants after a few days of campaign work is another challenge in Ela Kabala kebele. In Tuba micro-watershed of Soresha kebele, water accumulating in the soil bund increased land sliding at the upper end of gully. In addition, water logging inhibited seedlings survival.

Good practices that help maintain or improve the micro-watersheds' functioning were observed in some of the kebeles. In Angacha and Borecha wereda, 78 and 76% of the respondents (respectively) practice a tie and graze system that help to regulate degradation of the micro-watershed by livestock. The remaining 22 and 24% of the respondents in the respective weredas practice grazing openly on common land. In most kebeles of Kambata Tambaro zone, there is an agreement to implement zero grazing on cultivated land, which supports sustainability of the built structures and maintains soil fertility.

Table 4 Commonly implemented soil and water conservation structures in respective micro-watersheds and their appropriateness

Kebele	No of evaluated Micro-watersheds	Elevation (m a s l)	Average annual Rainfall (mm)	Slope (%)	Commonly Implemented physical soil and water conservation structures	Appropriateness of structures selection
Yiriba Dubancho	2	1980-2020	1070.6	13-30	soil bund, fanyajuu, trenches	Appropriate
Hanja chafa	2	1928-1945	1070.6	10-30	soil bund, fanyajuu, trenches, micro- basin, stone bund	Appropriate
Koran Goge	2	1925-1940	1070.6	5-21	trenches, micro- basin, eye brow	Appropriate
Debub Kege	2	1730-1734	1209.1	5-15	Trench, soil bund, fanya juu	Appropriate
Megara	2	1765-1792	1209.1	3-6	Fanya juu, stone bund, trench	Appropriate
Kalete Senete	1	1970	1209.1	1.5	Soil bund, cutoff drain, eye brow, trench	Appropriate from the point as it formed depressed land due to erosion
Kota kombola	3	2000-2110	1156.5	10-40	Soil bund, fanya juu, trenches, micro-basin	Appropriate
Ha'amancho	1	2100-2200	1156.5	26.5	Soil Bund, Fanya juu, trench, Micro-basin	Appropriate
Megere	1	2040-2160	1156.5	1.5-30	Soil Bund, Fanya juu, trench, Micro-basin	Partially appropriate
Zebecho	3	2200-2729	1565	20-45	Soil Bund, Fanya juu, trench, Micro-basin	Partially appropriate
Bucha	3	2060-2170	1565	5-20	Soil bund, fanya juu	Appropriate
Gede Genet	3	2100-2180	1565	5-20	Soil bund, fanya juu	Appropriate
Abela Faracho	3	1386-1417	1132.3	1.5-3	Soil bund fanya fuu	Partially appropriate
Abala Sifa	3	1440-1449	1132.3	1.5-3	Soil bund fanya fuu	Partially appropriate
Ela kabala	2	1520-1580	1132.3	3-17	Soil bund fanya fuu, trenches	Appropriate
Yakima	3	1920-2200	1056.1	10-55	Stone bund, soil bund, fanya juu	Partially appropriate
Sadoye	2	1700-1780	1056.1		Soil bund, fanyajuu	Appropriate
Soresha	3	1570-1673	1056.1	3-10	Stone bund, soil bund, fanya juu, micro-basin, trench, check dam	Appropriate
Addis Bodare	2	1351-1399		5-35	Stone bund, soil bund	Appropriate
Gendo Walcha	1	1970-1870	1731.1	3-25	Soil bund	Appropriate
Gedo Buna	1	2598-2610	1731.1	10-30	Soil bund, fanya juu	Appropriate
Mari Madara	1	2400	1731.1	10-67	Soil bund	Partially appropriate
Gobo Shamana	1	1900-2200	1731.1	10-30	Soil bund fanya juu	Appropriate

In Garato micro-watershed, the common land is used for grazing and is poorly guarded. Thus, the effect of watershed management is minimal. In Garato micro-watershed, all the interviewed farmers expressed appreciation for the good effects created by the soil conserving structures. However, some farmers demolished the campaign-built structures and/or poorly maintained sections broken by flooding or livestock.

In Sadoye kebele and some other micro-watersheds (i. Golota, Alelecho, Manto-Danchame, Gorbito, and Awash) on common lands, exclosures are created along with various interventions, such as planting seedlings, and these areas are guarded by employed personnel. Payment for

people to guard the management areas is arranged by the woreda agriculture office, either from their normal budget or through a safety net program (if the kebele is a beneficiary of the program). Even though the community understands they have invested their labor to improve the land and thus have a stake in the project's success, they tend to expect the guards to protect the area. In practice, such guarding is difficult to sustain because of uncertainty of wages from temporary sources and the changing interest of the community towards the land use.

In Alelecho, Bantola, and Golota micro-watersheds and Soresha kebele, there is a plan to handover the rehabilitated common land to a jobless youth association who will sustainably manage the land. This intention appears hopeful provided that there is sufficient discussion with the community, a clear plan is developed, regulations on resource utilization and management are agreed upon, and relevant technical support implemented.

Intensive rainfall, livestock, and land use processes can damage the physical soil and water conservation structures such as soil bunds, stone bunds, *fanya juus*, trenches, and micro-basins. Therefore, it requires frequent removal of sediment from channel and other maintenance activities. In micro-watersheds of Soresha and Sadoye kebeles where management activities occur on common land, the maintenance work is undertaken by the community through labor investment or safety net program support. On individually owned land, structures built by the public campaign are expected to be maintained by the owners. Through field observations and discussions with key informants, it was determined that most of the land owners in Kota Kombola, Ha'amancho, Megere and Zobecho kebeles are both undertaking the normal maintenance work and also strengthening the structures (soil bunds and *fanya juus*) with 'desho' grass (*Pennisetum pedicellatum)*. In Angacha and Damboye wereda, good interest for repairing the structures was observed with 86% of respondents willing to maintain the conservation practices by investing family labor and the remaining14% preferring to wait for external support.

Even though some efforts exist, the trend of repairing structures built by public campaign work, and which land users are expected to maintain, is poor in the assessed micro-watersheds of 19 kebeles (excluding the above 4 kebeles). Generally, they tend to wait for public campaign work and development agents to take care of the maintenance. For instance, about 78% of respondents in Borecha wereda wait for campaign work and 22% attempt to repair by using family labor. Some farmers ignore the structure after they have been damaged by rain and livestock and others remove the improvements during the tillage process. Reasons such as lack of awareness and household labor shortage can be considered as excuses for these behaviors, but the existing situation hampers the long-term defensive function of these structure and challenges productive land sustainability.

Another challenge of watershed management public campaign work is the poor program adoption and replication. The rate of replication and adoption of the introduced technology by land users is important for sustainability. The ongoing public campaign is actively promoted for only a month and less active for 11 months of the year. In this survey of 23 kebeles, efforts were made to determine replicated and adopted farmers or farmers who built a significant number and length of structures by involving the family labor. Except for the traditional stone bunds in Adis Bodare kebele, no significant replication or adoption effort

occurred. In previous studies many site specific socioeconomical and biophysical situations influenced the replication and maintenance of structures (Amsalu and de Graaff 2006; Bewket 2006; Nyangena 2007; de Graaff et al. 2008; Kebede 2014).

In Kota Kombola kebele, there is significantly big degraded and abandoned area (Figure 4), which requires huge investment than the local labor. This challenges the community to achieve watershed management for entire kebele within few years. In all considered kebeles, the achievement data obtained from records/reports did not fit the amount of activity audited in many micro-watersheds. This also needs serious attention and discussion with the concerned institutions.

Lessons learned from watershed management practices

The watershed management activities have acted as practical models for understanding environmental issues and rehabilitating degraded land. Participatory integrated watershed management has been implemented in degraded micro-watersheds for four consecutive years by public campaign approach. The effective and quickly responding impacts of implemented practices in 'severely degraded' micro-watersheds demonstrate the real possibilities of rehabilitating degraded areas. Related effort and significant achievement was reported for Nepal (Tiwari et al. 2008). In India, such watershed management improved ecological and environmental status of the resource and positively enhanced socio-economic situation of the community in the watershed (Pathak et al. 2013).

In Hanja Chafa kebele, farmers were not interested in participating in management work because the land was badly degraded and it appeared it would be hopeless to restore. However, after structures were built, grass began to regenerate/grow, tree species appeared and flooding was reduced in the restored area (Figure 5d). In semi-arid area of India, such watershed management improved ground and surface water resources (Khajuria et al. 2014). Another study in Millsboro indicated that watershed management activities such as improved nutrient and land cover management reduced nutrient load in water (Sood and Ritter 2010). In Blue Nile basin watershed management also significantly reduced soil erosion (Amdihun et al. 2014).

In Alelecho micro-watershed, the severely degraded land appeared hopeless to try and restore. Following management work, the farmers saw a surprising change. The rainwater was controlled, grass began to grow, and seedling survived. The farmers learned that the watershed improvement activities were a good investment toward the future. Similarly, in Dilamo micro-watershed, Kota Kombola kebele, the previously degraded area containing a network of gullies and exposed parent materials, was hardly suitable for humans even to cross. Intensively managed

Figure 4 Partial view of abandoned degraded land in Kota Kombola kebele, waiting for watershed management.

since 2010, significant rehabilitation has occurred characterized by abundant grass, controlled water runoff, suitable micro-climates, and improved soil fertility. A farmer reported 'the fertility and land cover is improved in such a way that people who knew this area previously, but have been gone for a while, may hesitate/get confused on accepting the improved situation'. Areas like this help to develop awareness and serve as a real-live demonstration.

The improved survival and vigorous growth of seedling in Golota micro-watershed of Debub Kege kebele

Figure 5 (a) Introduced fodder grass in zobcho kebele (b) *cajanus cajan* in Gendo Walcha kebele (c) typical soil bund modified to reduce the land occupied by it in Gede Genet kebele (d) exemplary rehabilitated land by watershed management in Hanja Chafa kebele.

has been achieved by planting seedlings in pits filled with forest soil and imported compost. The area serves as a model for tree seedling planting on poor land. In Gede Genet, Bucha and other kebeles the farmers modified the soil bunds to reduce the area of land occupied by the structure (Figure 5c). The effectiveness of this new design is still being tested. In most micro-watersheds, farmers attempt to make the embankment of structures productive by growing fodder grass (Figure 5a) and *cajnus cajan* (Figure 5b).

Even though landslides are not frequently observed as result of building physical water impound structures, they were seen in soil bunds built in Soresha kebele in Tuba micro-watershed. This land slide sensitivity needs to be considered for future management planning as it is important lesson. This risk of land slide might be due to poor subsurface water flow due to local geology.

Conclusion and recommendation

In most of the surveyed micro-watersheds, even though improper land use affected soil productivity on individually owned lands, common use lands were universally severely degraded and will demand intensive investment to restore. All watershed management by public campaign work had the objective to rehabilitate such degraded lands. To alleviate these problems, the government attempted to mobilize communities by using preferred conservation structures. The survey revealed that in each studied kebele there is strong organizational structure that helps to inform and mobilize people for intensive watershed management work. The kebele development teams have authority and responsibility to motivate participation, and to punish absentees in campaign work. It was this team that significantly contributed to the achievement of observed results. These results, such as well established tree/shrub plantations, rehabilitated lands, fodder grass established on bunds, and the planting of soil fertility improving species such as *Cajanus cajan* can be considered as excellent land management lessons which can motivate the public to participate in such labor-intensive tasks.

In the studied micro-watersheds, most of the species planted on these degraded lands were suited to the agro-ecology and poor site conditions. However, the poor survival observed in some micro-watersheds needs technical and planting time consideration. In most micro-watersheds, structure selection, design, construction and spacing were appropriate. However, in some micro-watersheds errors need correcting such as poor stone bund foundations, bunds with narrow berms, shallow channel depth, and too long bunds without space for land users to move across farmland. In addition, the effort to repair the broken/sediment filled structures is poor in many area and needs attention, which influence the long-term fate of these structures.

Micro-watersheds, where the habitat and vegetation has improved, often see an increase in wildlife use. Farmers complained that these animals damage their crop and/or attack their livestock. This might create a negative perception towards the management work. Ways to effectively control the impacts caused by wildlife needs to be investigated.

Experts in many kebele were optimistic on finishing the first round of campaign work, wanting to complete the entire kebele within a few years. However, in some kebeles, such as Kota Kombola there are large areas, with serious degradation problems that will require long-term intervention and probably external support (Figure 4). In all kebeles, the actual work needs to match the reported accomplishments. Managed watersheds, especially common lands, need arrangements to assure future sustainable management and utilization.

Endnotes

[a]weredas is roughly equivalent to district and is the next higher administrative division to kebele in the country.

[b]kebeles is the lowest government structure in the country.

[c]keberite' The local leader who own comparatively wide rural land and rent for other landless farmers during regime of emperor Hailesilase.

Abbreviations

BoFED: Bureau of Finance and Economic Development; CFSCDD: Community Forests and Soil Conservation Development Department; CSA: Central Statistical Authority; NGO: Non Governmental Organization; SNNPRS: Southern Nations, Nationalities and Peoples Regional State.

Competing interests

The author declares that he has no competing interests.

Author's contributions

KW: designed the research idea, field method, collected data, analyzed data, intrepretated data and produced report.

Author's information

Kebede Wolka: Assistant professor at Hawassa University, Wondo Genet College of Forestry and Natural Resources. He mainly teaches and undertakes research on soil, soil erosion, soil and water conservation and watershed management. He published 8 articles mainly in international journals and few other are in peer review process.

Acknowledgements

This research was financially supported by Hawassa University, Wondo Genet College of Forestry. The author would like to thank Bob Sturtevant, Colorado State University, for language editing and for the input from the anonymous reviewers.

References

Admasu A (2005) Study of Sediment Yield from the Watershed of Angereb Reservoir. Master Thesis, Department of Agricultural Engineering, Alemaya University, Ethiopia

Amdihun A, Gebremariam E, Rebelo R, Zeleke G (2014) Suitability and scenario modeling to support soil and water conservation interventions in the Blue Nile Basin. Ethiopia Environ Syst Res 2014(3):23–25

Amsalu A, de Graaff J (2006) Determinants of adoption and continued use of stone terraces for soil and water conservation in an Ethiopian highland watershed. Ecol Econ 61:294–302

Andualem G (2008) Prediction of Sediment Inflow for Legedadi Reservoir (using SWAT Watershed and CCHE1D Sediment Transport Models). Masters thesis, Faculty of Technology Addis Ababa University, Ethiopia

Azene BT, Birnie A (1993) Useful Trees and Shrubs for Ethiopia: Identification, Propagation and Management for Agricultural and Pastoral Communities, Regional Soil Conservation Unit. Swedish International Development Authority, Nairobi, Kenya

Bewket W (2006) Soil and water conservation intervention with conventional technologies in northwestern highlands of Ethiopia: acceptance and adoption by farmers. Land Use Policy 24:404–416

CSA Central Statistical Agency (2007) Population and housing census report. Addis Ababa, Ethiopia

CFSCDD (Community Forests and Soil Conservation Development Department) (1986) Guidelines for Development Agents on Soil Conservation in Ethiopia. Ministry of Agriculture, Addis Ababa, Ethiopia

De Graaff AA, Bodna RF, Kessler A, Posthumus H, Tenge A (2008) Factors influencing adoption and continued use of long-term soil and water conservation measures in five developing countries. Appl Geogr 28:271–280

Emiru WK (2009) Hydrological Responses to Land Cover Changes in Gilgel Abbay Catchment. Masters thesis, International Institute for Geo-Information Science and Earth Observation, Enshede, the Netherlands

Enwelu A, Agwu AE, Igbokwe EM (2014) Challenges of participatory approach to watershed management in rural communities of Enugu state. J Agric Ext 14(1):69–79

Ermias A, Solomon A, Alemu E (2006) Small-scale reservoir sedimentation rate analysis for a reliable estimation of irrigation schemes economic lifetime: a case study of Adigudom area, Tigray, Northern Ethiopia. Html: http://www.zef.de/module/register/media/6dd3_Siltation_Tigray_Ethiopia_Ermias.pdf. Accessed on 26/02/2010

Fikreyesus S, Kebebew Z, Nebiyu A, Zeleke N, Bogale S (2011) Allelopathic effects of *Eucalyptus camaldulensis* dehnh. on germination and growth of tomato. Am-Eurasian J Agric Environ Sci 11(5):600–608

Gebremedhin B, Swinton SM (2003) Investment in soil conservation in northern Ethiopia: the role of land tenure security and public programs. Agric Econ 29:69–84

German L, Mansoor H, Getachew A, Wagengia W, Amede T, Stroud A (2006) Participatory integrated watershed management: evolution of concepts and methods in an ecoregional program of the eastern African highlands. Agric Syst 94:189–204

Haregeweyn N, Poesen J, Nyssen J, De Wit J, Haile M, Govers G, Deckers S (2005) Reservoirs in Tigray (northern Ethiopia): characteristics and sediment deposition problems. Land Degrad Dev 17(2):211–230

Haregeweyn N, Poesen J, Nyssen J, Govers G, Verstraeten G, Vente J, Deckers J, Moeyersons J, Haile M (2008) Sediment yield variability in Northern Ethiopia: a quantitative analysis of its controlling factors. Catena 75(1):65–76

Hengsdijk H, Meijerink GW, Mosugu ME (2004) Modeling the effect of three soil and water conservation practices in Tigray, Ethiopia. Agric Ecosyst Environ 105(1–2):29–40

HNCJ Hunger, Nutrition Climate Justice (2013) Scaling up an Integrated Watershed Management approach through Social Protection Programmes in Ethiopia. The MERET and PSNP schemes: case studies, policy response, Dublin, Ireland

Hope RA (2007) Evaluating social impacts of watershed development in India. World Dev 35:1436–1449

Kebede W (2012) Watershed management: an option to maintain dam and reservoir function in Ethiopia. J Environ Sci Technol 5(5):263–272

Kebede W (2014) Effects of soil and water conservation measures and challenges for its adoption: Ethiopia in focus. J Environ Sci Technol 7(4):185–199

Kefyalew A (2004) Integrated Flood Management WMO and Global Water Partnership, Associated Programme on Flood Management. Editor, technical unit, Ethiopia

Kerr J (2002) Watershed development, environmental services, and poverty alleviation in India. World Dev 30(8):1387–1400

Khajuria A, Yoshikawa S, Kanae S (2014) Adaptation technology: benefits of hydrological services: watershed management in semi-arid region of India. J Water Resour Prot 6:565–570

Lakew D, Carucci V, Asrat W, Yitayew A (2005) Community Based Participatory Watershed Development. A guide, Ministry of Agriculture and Rural development, Addis Ababa

Ludi E (2004) Economic Analysis of Soil Conservation: Case Studies from the Highlands of Amhara Region, African Studies Series A18. Bernee: Geographical Bernensia, Ethiopia

Lullulangi M, Ardi M, Pertiwi N, Bakhrani R, Dirawan G (2014) Subjective norms adopted by the local community in preserving environment of settlement in the watershed, Mamasa. J Environ Sci Technol 7(5):305–313

Nigussie H A (2003) Sediment deposition in reservoirs in Tigray (Northern Ethiopia): Modeling rates, sources and target areas for intervention. http://www.kuleuven.be/geography/frg/staff/41853/index.php. Accessed on 26/02/2010

Nyangena W (2007) Social determinants of soil and water conservation in rural Kenya. Environ Dev Sustain 10(6):745–767

Pathak P, Chourasia AK, Wani SP, Sudi R (2013) Multiple impact of integrated watershed management in low rainfall semi-arid region: a case study from eastern Rajasthan, India. J Water Resour Prot 3(5):27–36

Price K, Jackson RC, Parker JA, Reitan T, Dowd J, Cyterski M (2011) Effects of watershed land use and geomorphology on stream low flows during severe drought conditions in the southern Blue Ridge Mountains, Georgia and North Carolina, United States. Water Resour Res 47(W02516):1–19

Rosenberg S, Margerum RD (2008) Landowner motivations for watershed restoration: lessons from five watersheds. J Environ Plan Manag 51(4):477–496

SNNPRS-BoFED (2004) Regional Atlas. Southern Nation, Nationalities and Peoples Regional State, Bureau of Finance and Economic Development, Bureau of Statistics and population. Awassa, Ethiopia

Sood A, Ritter WF (2010) Evaluation of best management practices in Millsboro pond watershed using soil and water assessment tool (SWAT). Model J Water Resour Prot 2:403–412

Srivastava RK, Sharma HC, Raina AK (2010) Suitability of soil and water conservation measures for watershed management using geographic information system. J Soil Water Conserv 9(3):148–153

Tamene L, Vlek PLG (2007) Assessing the potential of changing land use for reducing soil erosion and sediment yield of catchments: a case study in the highlands of northern Ethiopia. Soil Use Manag 23(1):82–91

Tamene L, Park SJ, Dikau R, Vlek PLG (2005) Analysis of factors determining sediment yield variability in the highlands of Northern Ethiopia. Geomorphology 76:76–91

Tamene L, Park SJ, Dikau R, Vlek PLG (2006) Reservoir siltation in the semi-arid highlands of northern Ethiopia: sediment yield–catchment area relationship and a semi-quantitative approach for predicting sediment yield. Earth Surf Process Landf 31(11):1364–1383

Tesfaye H (2011) Assessment of sustainable watershed management approach case study Lenche Dima, Tsegur Eyesus and Dijjil watershed. Master thesis, Faculty of the Graduate School of Cornell University

Teshome A, de Graaff J, Coen Ritsema C, Kassie M (2014) farmers' perceptions about the influence of land quality, land fragmentation and tenure systems on sustainable land management in the north western Ethiopian highlands. Land Degradation and Development. doi: 10.1002/ldr.2298

Tiwari KR, Bajracharya RM, Sitaula BK (2008) Natural resource and watershed management in south Asia: a comparative evaluation with special references to Nepal. J Agric Environ 9:72–89

Woldeamlak B (2003) Towards integrated watershed management in highland Ethiopia: the Chemoga watershed case study, PhD dissertation, Wageningen Universty

2

Global existence and convergence rates for the smooth solutions to the compressible magnetohydrodynamic equations in the half space

Qing Chen[1*] and Zhong Tan[2]

Abstract

Background: With the characteristics of low pollution and low energy consumption, the magnetohydrodynamics has made widely attention. This paper provides the standard energy method to solve the magnetohydrodynamic equations (MHD) in the half space $\mathbb{R}^3_+$. It proves the global existence for the compressible (MHD) by combining the careful a priori estimates and the local existence result. This study also considers the large time behaviors of the solutions.

Results: The interactions between the viscous, compressible fluid motion and the magnetic field are modeled by the magnetohydrodynamic system which describes the coupling between the compressible Navier-Stokes equations and the magnetic equations. This study has applied the analytical method to obtain the solutions to (MHD) in $\mathbb{R}^3_+$. It proves that under the assumption that the initial data are close to the constant state, the global existence of smooth solutions can be established. Moreover, the various decay rates of such solutions in L^p-norm with $2 \leq p \leq +\infty$ and their derivatives in L^2-norm can also be derived from combining the decay estimates of the linearized system and the energy method.

Conclusions: This study demonstrates that the global existence and the decay rates for the compressible (MHD) can be established under the similar initial assumptions as for the compressible Navier-Stokes equations. Especially, the results suggest that if the initial velocity is small, the velocity decays at a certain rate. This implies that only under the initial assumption that the data are large, it may reach the requirements of (MHD) power generation, which can be used to achieve the value of industrial application and environmental protection.

Keywords: Magnetohydrodynamics; Low pollution; Global existence; Convergence rates

Background

Magnetohydrodynamics, which combines the environmental fluid mechanics and electrodynamics theories to study the interaction discipline between the conduction fluid and electromagnetic, is the theory of the macroscopic, and it has spanned a very large range of applications (Gerebeau et al. 2006). Due to the lower environmental pollution, especially in energy industry, magnetohydrodynamic power generation is used to conserve

energy and mitigate pollution in order to protect the environment. In virtue of the industrial importance and theoretical challenges, the study on (MHD) has attracted many scientists. In the present paper, we are interested in the well-posedness theory of (MHD). Many results concerning the existence and uniqueness of (weak, strong or smooth) solutions in one dimension can be found in (Chen and Wang 2002, 2003; Kawashima and Okada 1982) and the references cited therein. In multidimensional case the global existence of weak solutions for the bounded domains has been established recently in (Ducomet and Feireisl 2006; Tan and Wang 2009). The local unique strong solution has been obtained in

*Correspondence: chenqing@xmut.edu.cn
[1] School of Applied Mathematics, Xiamen University of Technology, Ligong Road, 361024 Xiamen, China
Full list of author information is available at the end of the article

(Fan and Yu 2009). In (Chen and Tan 2010, 2012) we established the global existence and decay rates of the smooth solutions for the Cauchy problem. However, many fundamental problems for the compressible (MHD) in the half space are still open. In this paper, we will extend our results (Chen and Tan 2010) to the initial boundary problem in the half space.

Results

In this paper, we will consider the initial boundary value problem for the compressible magnetohydrodynamic equations (MHD) in the half space $\mathbb{R}^3_+ = \left\{ x = (x', x_3) : x' \in \mathbb{R}^2, x_3 > 0 \right\}$ (cf. (Gerebeau et al. 2006)):

$$
\begin{cases}
\rho_t + \text{div}(\rho \mathbf{u}) = 0, \\
(\rho \mathbf{u})_t + \text{div}(\rho \mathbf{u} \otimes \mathbf{u} - \mathbf{P}) = \mu_0 \text{curl} \mathbf{H} \times \mathbf{H}, \\
\mathbf{H}_t - \text{curl}(\mathbf{u} \times \mathbf{H}) + \dfrac{1}{\sigma \mu_0} \text{curlcurl} \mathbf{H} = 0, \ \text{div} \mathbf{H} = 0.
\end{cases}
\tag{1}
$$

Here $\rho, \mathbf{u} = \left(u^1, u^2, u^3 \right), \mathbf{H} = \left(H^1, H^2, H^3 \right)$ represent the density, velocity of the fluid and the magnetic field respectively. $\mu_0 > 0$ stands for permeability of free space, and $\sigma > 0$ is the electric conductivity. The stress tensor $\mathbf{P}$ is given by

$$
\mathbf{P} = -p \mathbf{I} + \mu \left(\nabla \mathbf{u} + \nabla \mathbf{u}^T \right) + \lambda \text{div} \mathbf{u} \mathbf{I},
$$

where $p = p(\rho)$ is the pressure and the viscosity coefficients λ, μ satisfy

$$
2\mu + 3\lambda > 0 \text{ and } \mu > 0.
$$

For convenience, we reformulate the system (1) as

$$
\begin{cases}
\rho_t + \text{div}(\rho \mathbf{u}) = 0, \\
\rho \mathbf{u}_t + \rho \mathbf{u} \cdot \nabla \mathbf{u} + \nabla p = \Delta \mathbf{u} + \nabla \text{div} \mathbf{u} + \text{curl} \mathbf{H} \times \mathbf{H}, \\
\mathbf{H}_t + \text{curl}(\mathbf{u} \times \mathbf{H}) - \Delta \mathbf{H} = 0, \ \text{div} \mathbf{H} = 0,
\end{cases}
\tag{2}
$$

in $(0, \infty) \times \mathbb{R}^3_+$. Notice that we have normalized some physical constants to be unit but without reducing any essential difficulties for our analysis. We complement (2) the initial condition

$$
(\rho, \mathbf{u}, \mathbf{H})(0, x) = (\rho_0(x), \mathbf{u}_0(x), \mathbf{H}_0(x)), \quad x \in \mathbb{R}^3_+, \tag{3}
$$

and the following boundary conditions

$$
\mathbf{u}|_{\{x_3=0\}} = \mathbf{0}, \ \mathbf{H}|_{\{x_3=0\}} = \mathbf{0}, \tag{4}
$$

or

$$
\mathbf{u}|_{\{x_3=0\}} = \mathbf{0}, \ \mathbf{H} \cdot \mathbf{n}|_{\{x_3=0\}} = 0, \ \text{curl} \mathbf{H} \times \mathbf{n}|_{\{x_3=0\}} = \mathbf{0}, \tag{5}
$$

where $\mathbf{n} = (0, 0, -1)$ is the normal vector of $\mathbb{R}^3_+$. We assume that throughout the paper the initial boundary data satisfy certain compatibility conditions as usual in (Matsumura and Nishida 1983).

Before stating out our results, we shall introduce some standard notations.

Notations. We denote by L^p, $W^{m,p}$ the usual Lebesgue and Sobolev spaces on $\mathbb{R}^3_+$ and $H^m = W^{m,2}$, with norms $|\cdot|_{L^p}, |\cdot|_{W^{m,p}}, |\cdot|_{H^m}$ respectively. For the sake of conciseness, we do not precise in functional space names when they are concerned with scalar-valued or vector-valued functions. We denote $\nabla = \partial_x = (\partial_1, \partial_2, \partial_3)^t$, where $\partial_i = \partial_{x_i}$, and put $\partial_x^l f = \nabla^l f = \nabla \left(\nabla^{l-1} f \right)$ for $l = 1, 2, 3, \cdots$. We assume that C be a positive generic constant throughout this paper that may vary at different places and the integration domain $\mathbb{R}^3_+$ will be always omitted without any ambiguity. Now our main results can be formulated as the following theorems. Firstly we state the results on the global existence and uniqueness of smooth solutions as:

Theorem 1. Assume that the initial data are close enough to the constant state $(\bar{\rho}, \mathbf{0}, \mathbf{0})$ with $\bar{\rho} > 0$, i.e., there exists a constant δ_0 such that

$$
| (\rho_0 - \bar{\rho}, \mathbf{u}_0, \mathbf{H}_0) |_{H^3} \leq \delta_0. \tag{6}
$$

Then there exists a unique globally smooth solution $(\rho, \mathbf{u}, \mathbf{H})$ of the initial boundary problem (2)–(4) or (2), (3) and (5) such that for any $t \in [0, \infty)$, it holds

$$
|(\rho - \bar{\rho}, \mathbf{u}, \mathbf{H})(\cdot, t)|^2_{H^3} + \int_0^t |\partial_x \rho(\cdot, s)|^2_{H^2} + |(\partial_x \mathbf{u}, \partial_x \mathbf{H})(\cdot, s)|^2_{H^3} ds
$$

$$
\leq C |(\rho_0 - \bar{\rho}, \mathbf{u}_0, \mathbf{H}_0)|^2_{H^3}. \tag{7}
$$

Remark 1. We will only prove Theorem 1 under the boundary condition (4). Due to the special geometry of the boundary of the half space, a simple calculation shows that the boundary conditions on $\mathbf{H}$ in (5) are equivalent to the following Dirichlet-Neumann boundary conditions:

$$
\partial_3 H^1 |_{\{x_3=0\}} = \partial_3 H^2 |_{\{x_3=0\}} = 0, \quad H^3 |_{\{x_3=0\}} = 0.
$$

Hence to treat H^1, H^2 as in (Matsumura and Nishida 1983), we can also prove Theorem 1 under the boundary conditions (5) in the similar way as we will proceed.

By imposing some additional conditions on the initial data we will establish the following various decay rates of the solutions obtained in Theorem 1:

Theorem 2. Let $(\rho, \mathbf{u}, \mathbf{H})$ be the solution obtained in Theorem 1 and assume in addition that the initial data $(\rho_0 - \bar{\rho}, \mathbf{u}_0, \mathbf{H}_0) \in L^1 \left(\mathbb{R}^3_+ \right)$ and there exists $\delta_1 > 0$ such that

$$
|(\rho_0 - \bar{\rho}, \mathbf{u}_0, \mathbf{H}_0)|_{L^1} + |(\rho_0 - \bar{\rho}, \mathbf{u}_0, \mathbf{H}_0)|_{H^3} < \delta_1, \tag{8}
$$

then for all $t \geq 0$, it holds that

$$|(\rho - \bar{\rho}, \mathbf{u}, \mathbf{H})(t)|_{L^p} = O\left((1+t)^{-\frac{3}{2}\left(1-\frac{1}{p}\right)}\right), \ \forall \, p \in [2, \infty],$$
(9)

$$|\partial_x(\rho - \bar{\rho}, \mathbf{u}, \mathbf{H})(t)|_{L^2} = O\left((1+t)^{-\frac{5}{4}}\right),$$
(10)

and

$$|(\rho - \bar{\rho}, \mathbf{u}, \mathbf{H})(t)|_{H^3} \leq C(\delta_1)\left((1+t)^{-\left(\frac{3}{4}-\varepsilon_1\right)}\right), \forall \, 0 < \varepsilon_1 < \widetilde{\varepsilon},$$
(11)

where $\widetilde{\varepsilon}$ is some positive number.

Moreover, if the data satisfy $(\rho_0 - \bar{\rho}, \mathbf{u}_0, \mathbf{H}_0) \in H^4\left(\mathbb{R}^3_+\right)$ and there exists $\delta_2 > 0$ such that

$$|(\rho_0 - \bar{\rho}, \mathbf{u}_0, \mathbf{H}_0)|_{L^1} + |(\rho_0 - \bar{\rho}, \mathbf{u}_0, \mathbf{H}_0)|_{H^4} < \delta_2,$$
(12)

then

$$|\partial_x(\rho - \bar{\rho}, \mathbf{u}, \mathbf{H})(t)|_{H^3} \leq C(\delta_2)(1+t)^{-\left(\frac{5}{4}-\varepsilon_2\right)}, \ \forall \, 0 < \varepsilon_2 < \widehat{\varepsilon},$$
(13)

for all $t \geq 0$, where $\widehat{\varepsilon} < \widetilde{\varepsilon}$ is a positive number. In fact, it holds

$$\left|\left(\partial_x^2\rho, \partial_x^2\mathbf{u}, \partial_x^2\mathbf{H}, \partial_x^3\mathbf{H}\right)(t)\right|_{L^2} \leq C(\delta_2)(1+t)^{-\frac{5}{4}},$$
(14)

for all $t \geq 0$.

We will prove the global existence of smooth solutions by the standard energy method in spirit of (Matsumura and Nishida 1983, 1979, 1980). And we remark that we can also obtain the global existence of strong solutions for the small initial data in the H^2-framework, which can be proved in the similar way. On the other hand, the L^2-L^∞ decay rates for the smooth solutions and the L^2 decay rates for the derivatives of first order are optimal since (9)–(10) concerning $(\rho - \bar{\rho}, \mathbf{u})$ and $\mathbf{H}$ are the same as the optimal decay rates for the compressible Navier-Stokes equations (Kagei and Kobayashi 2005) and the heat equation respectively. Related convergence rates of solutions for the Navier-Stokes equations on the unbounded domain can be found in (Kobayashi 2002; Kobayashi and Shibata 1999; Matsumura and Nishida 1979) and the references cited therein. Although our proofs are in spirit of those for the Navier-Stokes equations, (Kagei and Kobayashi 2005; Kobayashi 2002; Kobayashi and Shibata 1999; Matsumura and Nishida 1979, 1980), we should derive the new estimates arising from the presence of the magnetic field and overcome the strong coupling between the mass, momentum equations and magnetic equation. However, it is easy to obtain the optimal decay rates of $\mathbf{H}$ and its first derivatives by the properties of heat kernel.

Indeed, we can rewrite the equation $(19)_3$ analogously to the form of $(19)_1$–$(19)_2$, i.e.,

$$\begin{cases} 0_t + \text{div}\mathbf{H} = 0, \\ \mathbf{H}_t + \nabla 0 - \Delta\mathbf{H} - \nabla\text{div}\mathbf{H} = S_3. \end{cases}$$

Thus, we can get the estimates of $(0, \mathbf{H})$ which are similar to $(\varrho, \mathbf{v})$ in the system $(19)_1$–$(19)_2$. Moreover, we will get the better decay rate of the magnetic field by the elliptic system.

Discussion

As well known, the heavy emissions of Greenhouse gases, such as CO_2, CH_4, N_2O, SF_6 cause global warming, and also result in a great deal of harm to the environment. It has been a hot topic and widespread concern to study on how to strictly control the greenhouse gases emissions. In order to profoundly reduce the environment pollution, we must focus on energy structure adjustment. Without any course of mechanical motion, Magnetohydrodynamics (MHD) power generation technology, also called plasma power generation technology, transforms thermal energy and kinetic energy directly into electricity. Thus by applying (MHD) power generation technology, we can realize the desulphuriz and reduce the production of NO_x effectively, so as to achieve the effect of high efficiency and low pollution.

To complete the (MHD) generation process, which is of high industrial application value, a conductive gas (plasma) will be directed through a magnetic field with a large velocity, under a high temperature condition. In this situation, how to control the initial velocity of the conductive gas has to be considered. From the results in Section Results, we can conclude that if we assume that the initial data are close enough to the constant state, then there exists a unique globally solution to the (MHD) system and the solution decays at some rates. This indicates that if the initial velocity is sufficiently small, although the solution to the (MHD) system exists globally, then the velocity will decays and never be large, which implies that it may never reach the requirements of (MHD) power generation. However, the problem of the global existence of the solutions with the large initial data is still open.

Conclusions

In this paper, we demonstrate that the global existence and the decay rates for the compressible (MHD) in $\mathbb{R}^3_+$ can be established under the similar initial assumptions as for the compressible Navier-Stokes equations which can be seen in (Matsumura and Nishida 1983). It implies that the magnetic field does not affect the decay rates of the velocity. Indeed, the results (9)–(13) in Theorem 2 suggest that the decay rates for the derivatives of the magnetic field are the same as the velocity's. And in (14), we cannot get the estimates for $\partial_x^3 \mathbf{u}$ but $\partial_x^3 \mathbf{H}$. Furthermore, the results suggest that if the initial velocity is small, the velocity decays at the optimal rate. This implies that it may never reach the requirements of (MHD) power generation unless giving the gas an large initial velocity.

Methods

Proof of theorem 1

In this section, we will prove the existence part of Theorem 1 and the uniqueness is standard so it will be omitted.

Some elementary inequalities

The first bright idea to reduce many complicated computations lies in that we just need to do the lowest-order and highest-order energy estimates for the solutions. This is motivated by the following observation:

$$|f|_{H^k}^2 \le C|\left(f, \partial_x^k f\right)|_{L^2}^2, \ \forall f \in H^k\left(\mathbb{R}_+^3\right).$$

$$(15)$$

The inequality (15) can be easily proved by combing Young's inequality and Gagliardo-Nirenberg's inequality

$$|\partial_x^i f|_{L^p} \le C(p)|f|_{L^q}^\alpha |\partial_x^k f|_{L^r}^{(1-\alpha)}, \ \forall f \in H^k\left(\mathbb{R}_+^3\right)$$

$$(16)$$

where $\frac{1}{p} - \frac{i}{3} = \frac{1}{q}\alpha + \left(\frac{1}{r} - \frac{k}{3}\right)(1-\alpha)$ with $i \le k$. Indeed (16) can be proved by the extension technique together with Gagliardo-Nirenberg's inequality in the whole space. We can obtain the following useful inequality by Hölder inequality and (16):

$$|\partial_x^k(fg)|_{L^2} \le C\left(|f|_{L^\infty}|\partial_x^k g|_{L^2} + |\partial_x^k f|_{L^2}|g|_{L^\infty}\right),$$

$$\forall f, g \in H^k\left(\mathbb{R}_+^3\right), \ k \ge 2$$

$$(17)$$

and the general form of (3) can be deduced directly in the following:

$$|f_1 f_2 \cdots f_s|_k \le C\sum_{j=1}^s |f_1|_{L^\infty} \cdots |f_{j-1}|_{L^\infty}|f_j|_k|f_{j+1}|_{L^\infty} \cdots |f_s|_{L^\infty},$$

$$\forall f_j \in H^k\left(\mathbb{R}_+^3\right), \ k \ge 2, \text{for } j = 1, 2, \cdots, s.$$

The linearized system

We will linearize the problem (2)–(4) as follows. Setting $\gamma = \sqrt{P'(\bar\rho)}$, $\mu = 1/\bar\rho$ and introducing new variables by

$$\varrho = \rho - \bar\rho, \ \mathbf{v} = \frac{1}{\mu\gamma}\mathbf{u}, \ \mathbf{H} = \mathbf{H}.$$

$$(18)$$

Hence the initial boundary value problem (2)–(4) can be reformulated as

$$\begin{cases} \varrho_t + \gamma \operatorname{div}\mathbf{v} = S_1, \\ \mathbf{v}_t + \gamma\nabla\varrho - \mu\Delta\mathbf{v} - \mu\nabla\operatorname{div}\mathbf{v} = S_2, \\ \mathbf{H}_t - \Delta\mathbf{H} = S_3, \ \operatorname{div}\mathbf{H} = 0, \\ (\varrho, \mathbf{v}, \mathbf{H})(x, 0) = (\varrho_0, \mathbf{v}_0, \mathbf{H}_0)(x) = (\rho_0 - \bar\rho, \mathbf{v}_0, \mathbf{H}_0), \\ \mathbf{v}|_{\{x_3=0\}} = \mathbf{H}|_{\{x_3=0\}} = \mathbf{0}, \end{cases}$$

$$(19)$$

where

$$S_1 = -\mu\gamma\operatorname{div}(\varrho\mathbf{v}),$$

$$(20)$$

$$S_2 = \frac{1}{\mu\gamma\rho}\operatorname{curl}\mathbf{H} \times \mathbf{H} + \left(\frac{1}{\rho} - \frac{1}{\bar\rho}\right)\Delta\mathbf{v} + \left(\frac{1}{\rho} - \frac{1}{\bar\rho}\right)\nabla\operatorname{div}\mathbf{v}$$

$$- \mu\gamma\mathbf{v}\cdot\nabla\mathbf{v} - \frac{1}{\mu\gamma}\left[\frac{P'(\rho)}{\rho} - \frac{P'(\bar\rho)}{\bar\rho}\right]\nabla\varrho,$$

$$(21)$$

and

$$S_3 = -\mu\gamma\operatorname{curl}(\mathbf{v} \times \mathbf{H}).$$

$$(22)$$

In order to state our results more concisely, we define an energy functional as:

$$N(t_1, t_2) = \left\{ \sup_{t_1 \le t \le t_2} |(\varrho, \mathbf{v}, \mathbf{H})(t)|_3^2 + \int_{t_1}^{t_2} |\partial_x\varrho(s)|_2^2 \right. $$

$$\left. + |(\partial_x\mathbf{v}, \partial_x\mathbf{H})(s)|_3^2 ds \right\}^{\frac{1}{2}},$$

and change the condition of the initial data (6) as

$$|(\varrho_0, \mathbf{v}_0, \mathbf{H}_0)|_{H^3} \le \delta_0' = \max\left(\frac{1}{2\mu\gamma}, 1\right)\delta_0.$$

$$(23)$$

Local and global existence

We will finish the proof of Theorem 1 in this subsection. First we state out the local existence without proof, since it can be proved in a standard way (Matsumura and Nishida 1980) or can be found in (Ströhmer 1990, Vol'pert and Hudjaev 1972):

Theorem 3. (local existence) Under the assumption (23), there exists a positive constant T such that the initial boundary value problem (19) has a unique solution $(\varrho, \mathbf{v}, \mathbf{H})$ which is continuous in $[0, T] \times \mathbb{R}^3$ together with its derivatives of first order in t and of second order in x. Moreover, there exists a constant $C_1 > 1$ such that it holds $N(0, t) \le C_1 N(0, 0)$, for any $t \in [0, T]$.

We will prove in this subsection the following a priori estimate:

Theorem 4. (a priori estimate) There exists a constant $\delta \ll 1$ such that if $N(0, T) \leq \delta$, then there exists a constant $C_2 > 1$ such that $N(0, T) \leq C_2 N(0, 0)$.

The global existence of smooth solutions will be proved via a continued argument by combining the local existence theorem and the a priori estimate theorem. We shall state the global existence of smooth solutions to the linearized problem (19) as follows.

Proposition 3.1. (global existence) Under the assumptions of Theorem 1, the initial boundary value problem (19) has a unique global solution such that for $t \in [0, \infty)$, it holds $N(0, t) \leq CN(0, 0)$. Thus $(\rho, \mathbf{u}, \mathbf{H})$ which satisfies (4) uniquely solves the initial boundary value problem (2)–(4) for all time.

Proof. See in (Chen and Tan 2010). □

A priori estimates

We observe that the a priori assumption in Theorem 4 and the embedding inequality together with the continuity equation $(19)_1$ imply

$$\sup_{0 \leq t \leq T} |(\varrho, \varrho_t, \partial_x \varrho, \mathbf{v}, \partial_x \mathbf{v}, \mathbf{H}, \partial_x \mathbf{H})(t)|$$
$$\leq C \sup_{0 \leq t \leq T} |(\rho - \bar{\rho}, \mathbf{v}, \mathbf{H})(\cdot, t)|_{H^3} \leq C\delta. \tag{24}$$

In particular

$$\frac{\bar{\rho}}{2} \leq \rho = \varrho + \bar{\rho} \leq 2\bar{\rho}. \tag{25}$$

In the sequel, we will always use the smallness assumption of δ and (24)–(25).

Next we shall do some preparatory work from Lemma 3.1 to Lemma 3.7. Firstly, we regard the equations $(19)_2$–$(19)_3$ as the elliptic system with respect to x variables, i.e.,

$$\begin{cases} \Delta \mathbf{v} + \nabla \operatorname{div} \mathbf{v} = \dfrac{1}{\mu}\mathbf{v}_t + \dfrac{\gamma}{\mu}\nabla\rho - \dfrac{1}{\mu}S_2, \\ \Delta \mathbf{H} = \mathbf{H}_t - S_3, \\ \mathbf{v}|_{\{x_3=0\}} = 0, \ \mathbf{H}|_{\{x_3=0\}} = 0. \end{cases}$$

Thus we have the following estimates which we can found in (Cho et al. 2004):

Lemma 3.1. Under the assumptions of Theorem 4, we have for $k = 2, 3, 4$ that

$$|\partial_x^k \mathbf{v}|_{L^2} \leq C\left\{|\mathbf{v}_t|_{k-2} + |\partial_x\varrho|_{k-2} + |S_2|_{k-2} + |\mathbf{v}|_{L^2}\right\}, \tag{26}$$

and

$$|\partial_x^k \mathbf{H}|_{L^2} \leq C\left\{|\mathbf{H}_t|_{k-2} + |S_3|_{k-2} + |\partial_x\mathbf{H}|_{L^2}\right\}. \tag{27}$$

Next we derive the following stokes equation from the equations $(19)_1$–$(19)_2$:

$$\begin{cases} \gamma \operatorname{div} \mathbf{v} = g = -\dfrac{d\rho}{dt} - \mu\gamma\varrho \operatorname{div}\mathbf{v}, \\ -\mu\Delta\mathbf{v} + \gamma\nabla\rho = \mathbf{h} = -\mathbf{v}_t - \dfrac{\mu}{\gamma}\nabla g + S_2, \\ \mathbf{v}|_{\{x_3=0\}} = 0. \end{cases} \tag{28}$$

We have the following estimates which can be found in (Galdi et al. 1994):

Lemma 3.2. Under the assumptions of Theorem 4, we have for $k = 2, 3, 4$ that

$$|\partial_x^k \mathbf{v}|_{L^2}^2 + |\partial_x^{k-1}\rho|_{L^2}^2 \leq C\left\{|g|_{k-1}^2 + |h|_{k-2}^2\right\}. \tag{29}$$

Next we shall do the estimates for the terms contained in $N(0, t)$.

Lemma 3.3. Under the assumptions of Theorem 4, we have that

$$|(\varrho, \mathbf{v}, \mathbf{H})|_{L^2}^2 + \int_0^t |\partial_x(\mathbf{v}, \mathbf{H})|_{L^2}^2 ds \leq CN(0,0)^2 + C\delta N(0,t)^2, \tag{30}$$

and

$$|(\varrho_t, \mathbf{v}_t, \mathbf{H}_t)|_{L^2}^2 + \int_0^t |\partial_x(\mathbf{v}_t, \mathbf{H}_t)|_{L^2}^2 ds \leq CN(0,0)^2 + C\delta N(0,t)^2. \tag{31}$$

Proof. By multiplying the equations $(19)_1$–$(19)_3$ by ϱ, $\mathbf{v}$ and $\mathbf{H}$ respectively, integrating over $\mathbb{R}^3_+$ and adding the resulting equations, we have

$$\frac{1}{2}\frac{d}{dt}|(\varrho, \mathbf{v}, \mathbf{H})|_{L^2}^2 + \mu|(\operatorname{div}, \partial_x\mathbf{v})|_{L^2}^2 + |\partial_x\mathbf{H}|_{L^2}$$
$$= \langle S_1, \varrho \rangle + \langle S_2, \mathbf{v} \rangle + \langle S_3, \mathbf{H} \rangle.$$

We shall estimate the terms in the right-hand side term by term as:

$$\langle S_1, \varrho \rangle = \int \gamma\mu(\varrho\mathbf{v}) \cdot \partial_x\varrho\, dx \leq C|\varrho|_{L^3}|\mathbf{v}|_{L^6}|\partial_x\varrho|_{L^2}$$
$$\leq C|\varrho|_{H^1}|(\partial_x\mathbf{v}, \partial_x\varrho)|_{L^2}^2 \leq C\delta|(\partial_x\mathbf{v}, \partial_x\varrho)|_{L^2}^2,$$
$$\langle S_2, \mathbf{v} \rangle \leq C|(\varrho, \mathbf{v}, \mathbf{H})|_{L^3}|(\partial_x\varrho, \partial_x\mathbf{v}, \partial_x^2\mathbf{v}, \partial_x\mathbf{H})|_{L^2}|\mathbf{v}|_{L^6}$$
$$\leq C|(\varrho, \mathbf{v}, \mathbf{H})|_{H^1}|(\partial_x\varrho, \partial_x\mathbf{v}, \partial_x^2\mathbf{v}, \partial_x\mathbf{H})|_{L^2}^2$$
$$\leq C\delta|(\partial_x\varrho, \partial_x\mathbf{v}, \partial_x^2\mathbf{v}, \partial_x\mathbf{H})|_{L^2}^2,$$
$$\langle S_3, \mathbf{H} \rangle \leq C|(\mathbf{v}, \mathbf{H})|_{L^3}|(\partial_x\mathbf{v}, \partial_x\mathbf{H})|_{L^6}|\mathbf{H}|_{L^2}$$
$$\leq C|(\mathbf{v}, \mathbf{H})|_{H^1}|(\partial_x\mathbf{v}, \partial_x\mathbf{H})|_{L^6}|\mathbf{H}|_{L^2}$$
$$\leq C\delta|(\partial_x\mathbf{v}, \partial_x\mathbf{H})|_{L^2}^2,$$

where Hölder's inequality and Sobolev's inequality are used. Thus by integrating the above inequality in time and the definition of $N(0, t)$, we have got (30).

Similarly, by taking ∂_t to $(19)_1$–$(19)_3$, multiplying by ϱ_t, $\mathbf{v}_t$ and $\mathbf{H}_t$ respectively and integrating the resulting inequalities, we obtain

$$|(\varrho_t, \mathbf{v}_t, \mathbf{H}_t)|^2_{L^2} + \int_0^t |\partial_x(\mathbf{v}_t, \mathbf{H}_t)|^2_{L^2} ds$$

$$\leq CN(0,0)^2 + C \left| \int_0^t \langle \partial_t S_1, \varrho_t \rangle + \langle \partial_t S_2, \mathbf{v}_t \rangle + \langle \partial_t S_3, \mathbf{H}_t \rangle ds \right|.$$

Thus we have to estimate the terms in the right-hand side of the above inequality.

$$\langle \partial_x S_1, \varrho_t \rangle = \int \gamma \mu \partial_t (\varrho \operatorname{div} \mathbf{v} + \nabla \varrho \cdot \mathbf{v}) \partial_t \varrho \, dx$$

$$\leq C|(\varrho, \partial_x \varrho, \mathbf{v}, \partial_x \mathbf{v})|_{L^3} |(\varrho_t, \partial_x \varrho_t, \mathbf{v}_t, \partial_x \mathbf{v}_t)|_{L^2} |\varrho_t|_{L^6}$$

$$\leq C|(\varrho, \partial_x \varrho, \mathbf{v}, \partial_x \mathbf{v})|_{H^1} |(\varrho_t, \partial_x \varrho_t, \mathbf{v}_t, \partial_x \mathbf{v}_t)|^2_{L^2}$$

$$\leq C\delta |(\partial_x \varrho, \partial_x^2 \varrho, \partial_x \mathbf{v}, \partial_x^2 \mathbf{v}, \partial_x^3 \mathbf{v}, \partial_x \mathbf{H}, \partial_x^2 \mathbf{H})|^2_{L^2}$$

$$\leq C\delta N(0,t)^2,$$

where we can get the L^2-norm estimates of ϱ_t, $\partial_x \varrho_t$, $\mathbf{v}_t$, and $\partial_x \mathbf{v}_t$ with the aid of the equation $(19)_1$–$(19)_2$ and (24). Similarly we have that

$$\langle \partial_t S_2, \mathbf{v}_t \rangle \leq C|\langle \partial_t(\varrho \operatorname{curl}\mathbf{H} \times \mathbf{H}, \varrho \Delta \mathbf{v}, \varrho \nabla \operatorname{div}\mathbf{v}, \mathbf{v} \cdot \nabla \mathbf{v}, \varrho \cdot \nabla \varrho), \mathbf{v}_t \rangle|$$

$$\leq C|(\varrho, \varrho_t, \mathbf{v}, \partial_x \mathbf{v}, \mathbf{H}, \partial_x \mathbf{H})|_{L^3}$$

$$\times |(\partial_x \varrho, \partial_x \varrho_t, \mathbf{v}_t, \partial_x^2 \mathbf{v}, \partial_x \mathbf{v}_t, \partial_x^2 \mathbf{v}_t, \partial_x \mathbf{H}, \mathbf{H}_t, \partial_x \mathbf{H}_t)|_{L^2} |\mathbf{v}_t|_{L^6}$$

$$\leq C|(\varrho, \partial_x \varrho, \mathbf{v}, \partial_x \mathbf{v}, \mathbf{H}, \partial_x \mathbf{H})|_{H^1}$$

$$\times |(\partial_x \varrho, \partial_x^2 \varrho, \partial_x^3 \varrho, \partial_x \mathbf{v}, \partial_x^2 \mathbf{v}, \partial_x^3 \mathbf{v}, \partial_x^4 \mathbf{v}, \partial_x \mathbf{H}, \partial_x^2 \mathbf{H}, \partial_x^3 \mathbf{H})|^2_{L^2}$$

$$\leq C\delta |(\partial_x \varrho, \partial_x^2 \varrho, \partial_x^3 \varrho, \partial_x \mathbf{v}, \partial_x^2 \mathbf{v}, \partial_x^3 \mathbf{v}, \partial_x^4 \mathbf{v}, \partial_x \mathbf{H}, \partial_x^2 \mathbf{H}, \partial_x^3 \mathbf{H})|^2_{L^2},$$

and

$$\langle \partial_t S_3, \mathbf{H}_t \rangle \leq C|\langle \partial_t \operatorname{curl}(\mathbf{v} \times \mathbf{H}), \mathbf{H}_t \rangle|$$

$$\leq C|(\mathbf{v}, \partial_x \mathbf{v}, \mathbf{H}, \partial_x \mathbf{H})|_{L^3} |(\mathbf{v}_t, \partial_x \mathbf{v}_t, \mathbf{H}_t, \partial_x \mathbf{H}_t)|_{L^2} |\mathbf{H}_t|_{L^6}$$

$$\leq C|(\mathbf{v}, \mathbf{H})|_{H^2} |(\partial_x \varrho, \partial_x^2 \varrho, \partial_x \mathbf{v}, \partial_x^2 \mathbf{v}, \partial_x^3 \mathbf{v}, \partial_x \mathbf{H}, \partial_x^2 \mathbf{H}, \partial_x^3 \mathbf{H})|^2_{L^2}$$

$$\leq C\delta |(\partial_x \varrho, \partial_x^2 \varrho, \partial_x \mathbf{v}, \partial_x^2 \mathbf{v}, \partial_x^3 \mathbf{v}, \partial_x \mathbf{H}, \partial_x^2 \mathbf{H}, \partial_x^3 \mathbf{H})|^2_{L^2}.$$

Together with these inequalities, we can deduce the inequality (31). Hence the proof of Lemma 3.3 is complete. $\square$

Next we estimate the L^2-norm of the first derivatives of $\mathbf{v}$ and $\mathbf{H}$.

Lemma 3.4. Under the assumptions of Theorem 4, we have

$$|\partial_x(\mathbf{v}, \mathbf{H})|^2_{L^2} + \int_0^t |(\varrho_t, \mathbf{v}_t, \mathbf{H}_t)(s)|^2_{L^2} ds \leq CN(0,0)^2 \qquad (32)$$
$$+ C\delta N(0,t)^2,$$

and

$$|\partial_x(\mathbf{v}_t, \mathbf{H}_t)|^2_{L^2} + \int_0^t |(\varrho_{tt}, \mathbf{v}_{tt}, \mathbf{H}_{tt})(s)|^2_{L^2} ds \leq CN(0,0)^2$$
$$+ C\delta N(0,t)^2. \qquad (33)$$

Proof. By multiplying ϱ_t, $\mathbf{v}_t$ and $\mathbf{H}_t$ to the equations $(19)_1$–$(19)_3$ respectively, integrating over $\mathbb{R}^3_+$ and adding the resulting equations, we have

$$\frac{d}{dt} \int \left(\frac{\mu}{2} |\partial_x \mathbf{v}|^2 + \frac{\mu}{2} |\operatorname{div}\mathbf{v}|^2 + \frac{1}{2} |\partial_x \mathbf{H}|^2 - \gamma \varrho \operatorname{div}\mathbf{v} \right) dx$$

$$+ \int \left(\varrho_t^2 + |\mathbf{v}_t|^2 + |\mathbf{H}_t|^2 + 2\gamma \varrho_t \operatorname{div}\mathbf{v} \right) dx$$

$$= \int \langle S_1, \varrho_t \rangle + \langle S_2, \mathbf{v}_t \rangle + \langle S_3, \mathbf{H}_t \rangle dx$$

$$\leq \epsilon \left(|\varrho_t|^2_{L^2} + |\mathbf{v}_t|^2_{L^2} + |\mathbf{H}_t|^2_{L^2} \right) + C(\epsilon) \left(|S_1|^2_{L^2} + |S_2|^2_{L^2} + |S_3|^2_{L^2} \right). \qquad (34)$$

Since

$$|S_1|^2_{L^2} \leq C|(\varrho, \mathbf{v})|_{L^\infty} |\partial_x(\varrho, \mathbf{v})|^2_{L^2} \leq C\delta |\partial_x(\varrho, \mathbf{v})|^2_{L^2},$$

$$|S_2|^2_{L^2} \leq C|(\varrho, \mathbf{v}, \mathbf{H})|_{L^\infty} |(\partial_x \varrho, \partial_x \mathbf{v}, \partial_x^2 \mathbf{v}, \partial_x \mathbf{H})|^2_{L^2}$$
$$\leq C\delta |(\partial_x \varrho, \partial_x \mathbf{v}, \partial_x^2 \mathbf{v}, \partial_x \mathbf{H})|^2_{L^2},$$

and

$$|S_3|^2_{L^2} \leq C|(\mathbf{v}, \mathbf{H})|_{L^\infty} |\partial_x(\mathbf{v}, \mathbf{H})|^2_{L^2} \leq C\delta |\partial_x(\mathbf{v}, \mathbf{H})|^2_{L^2},$$

thus by integrating (34) in time and with the aid of the smallness of ϵ and Cauchy's inequality, we obtain

$$|\partial_x(\mathbf{v}, \mathbf{H})|^2_{L^2} + \int_0^t |(\varrho_t, \mathbf{v}_t, \mathbf{H}_t)(s)|^2_{L^2} ds$$

$$\leq CN(0,0)^2 + C\delta N(0,t)^2 + C \left(|\varrho|^2_{L^2} + \int_0^t |\partial_x \mathbf{v}(s)|^2_{L^2} ds \right),$$

which together with (30) yields (32). And we can also get (33) in the similar way. The proof of Lemma 3.4 is complete. $\square$

Next we estimate the L^2-Norms of the first derivatives of ϱ. We shall divide the estimates into two parts. Firstly we denote the tangential derivatives by $\partial = (\partial_1, \partial_2)$. And it is easy to see that the tangential derivatives of the solution of (19) satisfy the same boundary conditions in (19).

By taking ∂ to the equations $(19)_1$–$(19)_2$ and multiplying them with $\partial \varrho$ and $\partial \mathbf{v}$ respectively, we can deduce that

$$\frac{1}{2} \frac{d}{dt} \int |\partial \varrho|^2 + |\partial \mathbf{v}|^2 dx + \int \mu |\partial \partial_x \mathbf{v}|^2 + \mu |\partial \operatorname{div}\mathbf{v}|^2 dx$$

$$= \langle \partial S_1, \partial \varrho \rangle + \langle \partial S_2, \partial \mathbf{v} \rangle. \qquad (35)$$

Define the material derivative $\frac{d\rho}{dt} = \rho_t + \nabla \rho \cdot \mathbf{u}$. Then by the continuity equation $(19)_1$ and the formula of variable substitution (18), we have

$$\frac{D\varrho}{Dt} = -\gamma \operatorname{div}\mathbf{v} - \mu \gamma \varrho \operatorname{div}\mathbf{v}, \qquad (36)$$

where $\dfrac{D\varrho}{Dt} = \dfrac{d\rho}{dt} = \varrho_t + \mu\gamma\nabla\varrho \cdot \mathbf{v}$. By taking ∂ to the equation (36), we can obtain the following inequality:

$$\left|\partial\left(\frac{D\varrho}{Dt}\right)\right|_{L^2}^2 \le C\left(|\partial\partial_x\mathbf{v}|_{L^2}^2 + |\partial(\varrho\,\mathrm{divv})|_{L^2}^2\right),$$

then by multiplying a small enough constant α to it, together with (35), and integrating in time, we have

$$|\partial(\varrho,\mathbf{v})|_{L^2}^2 + \int_0^t \left|\partial\left(\frac{D\varrho}{Dt},\partial_x\mathbf{v}\right)\right|_{L^2}^2 ds$$

$$\le CN(0,0)^2 + C\left|\int_0^t \langle\partial S_1, \partial\varrho\rangle + \langle\partial S_2, \partial\mathbf{v}\rangle ds\right|$$

$$+ C\int_0^t |\partial(\varrho\,\mathrm{divv})|_{L^2}^2 ds$$

$$\le CN(0,0)^2 + CN(0,t)^2 + C\left|\int_0^t \langle\partial S_1, \partial\varrho\rangle + \langle\partial S_2, \partial\mathbf{v}\rangle ds\right|.$$

Similarly, we can obtain the following lemma and we omit the proof of it.

Lemma 3.5. Under the assumptions of Theorem 4, we have for k = 1, 2, 3 that

$$|\partial^k(\varrho,\mathbf{v})|_{L^2}^2 + \int_0^t \left|\partial^k\left(\frac{D\varrho}{Dt},\partial_x\mathbf{v}\right)\right|_{L^2}^2 ds$$

$$\le CN(0,0)^2 + CN(0,t)^2 + C\left|\int_0^t \langle\partial^k S_1, \partial^k\varrho\rangle \right. \tag{37}$$

$$\left. + \langle\partial^k S_2, \partial^k\mathbf{v}\rangle ds\right|.$$

Next we have to obtain the estimates for the normal derivatives of solution. We can derive the following equations from $(19)_1$–$(19)_2$.

$$\partial_3\varrho_t + \gamma\partial_3\mathrm{divv} = -\mu\gamma\partial_3(\varrho\,\mathrm{divv}) - \mu\gamma\partial_3(\nabla\varrho \cdot \mathbf{v}),$$

$$\mathbf{v}_t^3 + \gamma\partial_3\varrho - \mu\Delta\mathbf{v}^3 - \mu\partial_3\mathrm{divv} = S_2^3,$$

where $S_2 = \left(S_2^1, S_2^2, S_2^3\right)^T$. To eliminate the term $\partial_{33}\mathbf{v}^3$, we have the following equality from the above equalities

$$\frac{2\mu}{\gamma}\partial_3\varrho_t + \gamma\partial_3\varrho = -\mathbf{v}_t^3 + \mu\left(\partial_{11}\mathbf{v}^3 + \partial_{22}\mathbf{v}^3 + \partial_{13}\mathbf{v}^1 + \partial_{23}\mathbf{v}^2\right)$$

$$- 2\mu^2\partial_3(\varrho\,\mathrm{divv}) - 2\mu^2\partial_3(\nabla\varrho \cdot \mathbf{v}) + S_2^3. \tag{38}$$

And we can also derive the following equality from (36)

$$\frac{2\mu}{\gamma}\partial_3\left(\frac{D\varrho}{Dt}\right) + \gamma\partial_3\varrho = -\mathbf{v}_t^3 + \mu\left(\partial_{11}\mathbf{v}^3 + \partial_{22}\mathbf{v}^3 + \partial_{13}\mathbf{v}^1 + \partial_{23}\mathbf{v}^2\right)$$

$$- 2\mu^2\partial_3(\varrho\,\mathrm{divv}) + S_2^3. \tag{39}$$

By multiplying (38) with $\partial_3\varrho$ and integrating on $\mathbb{R}_+^3$, we have

$$\frac{\mu}{\gamma}\frac{d}{dt}|\partial_3\varrho|_{L^2}^2 + \gamma|\partial_3\varrho|_{L^2}^2$$

$$= \int \left\{-\mathbf{v}_t^3 + \mu\left(\partial_{11}\mathbf{v}^3 + \partial_{22}\mathbf{v}^3 + \partial_{13}\mathbf{v}^1 + \partial_{23}\mathbf{v}^2\right)\right.$$

$$\left. - 2\mu^2\partial_3(\varrho\,\mathrm{divv}) - 2\mu^2\partial_3(\nabla\varrho \cdot \mathbf{v}) + S_2^3\right\}\partial_3\varrho\,dx$$

$$\le \frac{\gamma}{2}|\partial_3\varrho|_{L^2}^2 + C(\gamma)|\,(\mathbf{v}_t, \partial\partial_x\mathbf{v}, \partial_3(\varrho\,\mathrm{divv}), S_2)\,|_{L^2}^2$$

$$+ C\left|\int \partial_3(\nabla\varrho \cdot \mathbf{v})\partial_3\varrho\,dx\right|, \tag{40}$$

then integrating this in time, it implies

$$|\partial_3\varrho|_{L^2}^2 + \int_0^t |\partial_3\varrho(s)|_{L^2}^2 ds$$

$$\le CN(0,0)^2 + C\int_0^t |(\mathbf{v}_t, \partial\partial_x\mathbf{v}, \partial_3(\varrho\,\mathrm{divv}), S_2)|_{L^2}^2$$

$$+ |\int \partial_3(\nabla\varrho \cdot \mathbf{v})\partial_3\varrho\,dx|ds$$

$$\le CN(0,0)^2 + C\int_0^t |(\mathbf{v}_t, \partial\partial_x\mathbf{v})|_{L^2}^2 ds + C\delta N(0,t)^2. \tag{41}$$

Similarly, by multiplying $\partial_3\left(\dfrac{D\varrho}{Dt}\right)$ to (39) and integrating it, we have

$$|\partial_3\varrho|_{L^2}^2 + \int_0^t |\partial_3\left(\frac{D\varrho}{Dt}\right)(s)|_{L^2}^2 ds$$

$$\le CN(0,0)^2 + C\int_0^t |(\mathbf{v}_t, \partial\partial_x\mathbf{v}, \partial_3(\varrho\,\mathrm{divv}), S_2)|_{L^2}^2$$

$$\le CN(0,0)^2 + C\delta N(0,t)^2 + C\int_0^t |(\mathbf{v}_t, \partial\partial_x\mathbf{v})|_{L^2}^2.$$

This together with (41) yields

$$|\partial_3\varrho|_{L^2}^2 + \int |\partial_3\varrho(s)|_{L^2}^2 + |\partial_3\left(\frac{D\varrho}{Dt}\right)(s)|_{L^2}^2 ds$$

$$\le CN(0,0)^2 + C\delta N(0,t)^2 + C\int_0^t |(\mathbf{v}_t, \partial\partial_x\mathbf{v})|_{L^2}^2 ds. \tag{42}$$

Similarly, by taking $\partial^k\partial_3^l$ to (38) and (39), multiplying with $\partial^k\partial_3^{1+l}\varrho$ and $\partial^k\partial_3^{1+l}\left(\dfrac{D\varrho}{Dt}\right)$ respectively, we will get the general form of (42) as follows:

Lemma 3.6. Under the assumptions of Theorem 4, we have for $k + l = 0, 1, 2$ that

$$|\partial^k \partial_3^{1+l} \varrho|_{L^2}^2 + \int_0^t |\partial^k \partial_3^{1+l} \varrho|_{L^2}^2 + |\partial^k \partial_3^{1+l} \left(\frac{D\varrho}{Dt}\right)(s)|_{L^2}^2 ds$$

$$\leq CN(0,0)^2 + C\delta N(0,t)^2 + C \int_0^t |\partial^k \partial_3^l (\mathbf{v}_t, \partial \partial_x \mathbf{v})|_{L^2}^2 ds. \tag{43}$$

Proof. Here we only to prove the case when $k + l = 2$. As in (42), we can obtain

$$|\partial^k \partial_3^{1+l} \varrho|_{L^2}^2 + \int_0^t |\partial^k \partial_3^{1+l} \varrho|_{L^2}^2 + |\partial^k \partial_3^{1+l} \left(\frac{D\varrho}{Dt}\right)(s)|_{L^2}^2 ds$$

$$\leq CN(0,0)^2 + C \int_0^t |\partial^k \partial_3^l (\mathbf{v}_t, \partial \partial_x \mathbf{v}, \partial_3 (\varrho \operatorname{div} \mathbf{v}), S_2)|_{L^2}^2$$

$$+ |\int \partial^k \partial_3^{1+l} (\nabla \varrho \cdot \mathbf{v}) \cdot \partial^k \partial_3^{1+l} \varrho dx| ds. \tag{44}$$

By (17), we have

$$\int_0^t |\partial_x^3 (\varrho \operatorname{div} \mathbf{v})|_{L^2}^2 ds \leq C \int_0^t |\varrho, \operatorname{div} \mathbf{v}|_{L^\infty}^2 |\partial_x^3 \varrho, \partial_x^3 \operatorname{div} \mathbf{v}|_{L^2}^2 ds$$

$$\leq C\delta N(0,t)^2.$$

Similarly,

$$\int_0^t |\partial_x^2 S_2|_{L^2}^2 ds \leq C \int_0^t |\varrho, \mathbf{v}, \mathbf{H}|_{L^\infty}^2 |\partial_x^3 \varrho, \partial_x^3 \mathbf{v}, \mathbf{H}|_{L^2}^2 ds$$

$$+ C \int_0^t |\varrho, \partial_x \varrho|_{L^\infty}^2 |\partial_x^2 \mathbf{v}|_1^2 + |\partial_x^2 \varrho|_{L^4}^2 |\partial_x^2 \mathbf{v}|_{L^4}^2 ds$$

$$\leq C\delta N(0,t)^2 + C \int_0^t |\partial_x^2 \varrho|_1^2 |\partial_x^2 \mathbf{v}|_1^2 ds$$

$$\leq C\delta N(0,t)^2.$$

Moreover, by using the identity

$$(\partial^k \partial_3^{1+l} \nabla \varrho \cdot \mathbf{v}) \cdot \partial^k \partial_3^{1+l} \varrho = \nabla \frac{|\partial^k \partial_3^{1+l} \varrho|^2}{2} \cdot \mathbf{v}$$

and integration by parts, we shall obtain

$$\int_0^t \int \partial^k \partial_3^{1+l} (\nabla \varrho \cdot \mathbf{v}) \cdot \partial^k \partial_3^{1+l} \varrho dx ds \leq C\delta N(0,t)^2.$$

Substituting these inequalities into (44), we get (43). □

Last by taking $\partial_x^l \partial^k$ to (28), and by Lemma 3.2, we have

Lemma 3.7. Under the assumptions of Theorem 4, we have for $k + l = 0, 1, 2$ that

$$\int_0^t |\partial_x^{2+l} \partial^k \mathbf{v}|_{L^2} + |\partial_x^{1+l} \partial^k \varrho|_{L^2} ds \leq C \int_0^t |\partial^k \mathbf{v}_t|_l$$

$$+ |\partial^k \left(\frac{D\varrho}{Dt}\right)|_{1+l} ds + C\delta N(0,t)^2. \tag{45}$$

Now we will finish the proof of Theorem 4 by doing the estimates for the lowest-order and highest-order derivatives. In the sequel, we divide the a priori estimates into three parts.

Part 1: estimates for the lowest derivatives of $\varrho, \mathbf{v}, \mathbf{H}$

Proposition 3.2. Under the assumptions of Theorem 4, we have

$$|(\varrho, \mathbf{v}, \mathbf{H})|_{L^2}^2 + \int_0^t |\partial_x (\varrho, \mathbf{v}, \mathbf{H})(s)|_{L^2}^2 ds \leq CN(0,0)^2 + C\delta N(0,t)^2. \tag{46}$$

Proof. Due to (30), we have only to estimate $\int_0^t |\partial_x \varrho(s)|_{L^2}^2 ds$. First by Lemma 3.5, $k = 0, 1$, we have

$$\int_0^t |\frac{D\varrho}{Dt}(s)|_{L^2}^2 + |\partial \left(\frac{D\varrho}{Dt}\right)(s)|_{L^2}^2 + |\partial \partial_x \mathbf{v}|_{L^2}^2 ds$$

$$\leq CN(0,0)^2 + C\delta N(0,t)^2 + C\delta \int_0^t |(S_1, S_2, S_3)(s)|_{L^2}^2 ds$$

$$\leq CN(0,0)^2 + C\delta N(0,t)^2. \tag{47}$$

And by (42), we have

$$\int_0^t |\partial_3 \left(\frac{D\varrho}{Dt}\right)|_{L^2}^2 ds \leq CN(0,0)^2 + C \int_0^t |(\mathbf{v}_t, \partial \partial_x \mathbf{v})|_{L^2}^2 ds$$

$$+ C\delta N(0,t)^2. \tag{48}$$

Thus (47)–(48) together with (32) implies

$$\int_0^t |\frac{D\varrho}{Dt}(s)|_1^2 ds \leq CN(0,0)^2 + C\delta N(0,t)^2. \tag{49}$$

By Lemma 3.4, $k = l = 0$, and integrating in time, we obtain

$$\int_0^t |\partial_x \varrho|_{L^2}^2 ds \leq C \int_0^t |\mathbf{v}_t|_{L^2}^2 + |\left(\frac{D\varrho}{Dt}|_1^2\right) ds + C\delta N(0,t)^2.$$

This combining with (49) and (32), yields the estimate for $\int_0^t |\partial_x \varrho(s)|_{L^2}^2 ds$. □

Part 2: estimates for the highest derivatives of ϱ, v, H

Proposition 3.3. Under the assumptions of Theorem 4, we have

$$|\partial_x^3(\varrho, \mathbf{v}, \mathbf{H})|_{L^2}^2 + \int_0^t |\left(\partial_x^3 \varrho, \partial_x^4 \mathbf{v}, \partial_x^4 \mathbf{H}\right)(s)|_{L^2}^2 ds \leq CN(0,0)^2 + C\delta N(0,t)^2.$$

(50)

Proof. We divide the proof into three steps as follows.

i) By Lemma 2.8, $k = 3$ and Lemma 2.9, $k = 2, l = 0$, we have

$$|\partial^2 \partial_x \varrho|_{L^2}^2 + \int_0^t \left|\left(\partial^2 \partial_3 \varrho, \partial^2 \partial_x \left(\frac{D\varrho}{Dt}\right), \partial^3 \partial_x \mathbf{v}\right)\right|_{L^2}^2 ds$$

$$\leq CN(0,0)^2 + C\delta N(0,t)^2 + C\left|\int_0^t \left\langle \partial^k S_1, \partial^k \varrho \right\rangle\right.$$

$$+ \left\langle \partial^k S_2, \partial^k \mathbf{v} \right\rangle ds\left| + C \int_0^t |\partial^2 \mathbf{v}_t|_{L^2}^2\right.$$

$$\leq CN(0,0)^2 + C\delta N(0,t)^2 + C \int_0^t |\partial^2 \mathbf{v}_t|_{L^2}^2 ds,$$

(51)

where we get the last inequality as in the proof of Lemma 3.6. By Lemma 3.7, $k = 2, l = 0$, we have

$$\int_0^t |\left(\partial_x^2 \partial^2 \mathbf{v}, \partial_x \partial^2 \varrho\right)|_{L^2}^2 ds \leq C \int_0^t \left|\left(\partial^2 \mathbf{v}_t, \partial^2 \partial_x \left(\frac{D\varrho}{Dt}\right)\right)\right|_{L^2}^2 ds$$

$$+ C\delta N(0,t)^2.$$

This together with (51) yields

$$|\partial^2 \partial_x \varrho|_{L^2}^2 + \int_0^t \left|\left(\partial^2 \partial_x \varrho, \partial^2 \partial_x \left(\frac{D\varrho}{Dt}\right), \partial^2 \partial_x^2 \mathbf{v}\right)\right|_{L^2}^2 ds$$

$$\leq CN(0,0)^2 + C\delta N(0,t)^2 + C \int_0^t |\partial^2 \mathbf{v}_t|_{L^2}^2 ds.$$

(52)

By Lemma 3.6, $k = l = 1$, and (52), we have

$$|\partial \partial_x^2 \varrho|_{L^2}^2 + \int_0^t \left|\left(\partial \partial_x^2 \varrho, \partial \partial_x^2 \left(\frac{D\varrho}{Dt}\right), \partial^2 \partial_x^2 \mathbf{v}\right)\right|_{L^2}^2 ds$$

$$\leq CN(0,0)^2 + C\delta N(0,t)^2 + C \int_0^t |\partial_x^2 \mathbf{v}_t|_{L^2}^2 ds.$$

(53)

Then by Lemma 3.7, $k = l = 1$, and with the help of (15), (31) and (47), we have

$$\int_0^t |\left(\partial_x^3 \partial \mathbf{v}, \partial_x^2 \partial \varrho\right)|_{L^2}^2 ds$$

$$\leq C \int_0^t \left|\left(\partial_x \mathbf{v}_t, \partial_x^2 \mathbf{v}_t, \partial\left(\frac{D\varrho}{dt}\right), \partial\partial_x\left(\frac{D\varrho}{dt}\right), \partial\partial_x^2\left(\frac{D\varrho}{Dt}\right)\right)\right|_{L^2}^2 ds$$

$$+ C\delta N(0,t)^2$$

$$\leq CN(0,0)^2 + C\delta N(0,t)^2 + C \int_0^t \left|\partial_x^2 \mathbf{v}_t, \partial\partial_x^2 \left(\frac{D\varrho}{Dt}\right)\right|_{L^2}^2 ds,$$

which together with (53) yields

$$|\partial\partial_x^2 \varrho|_{L^2}^2 + \int_0^t \left|\left(\partial\partial_x^2 \varrho, \partial\partial_x^2 \left(\frac{D\varrho}{Dt}\right), \partial\partial_x^3 \mathbf{v}\right)\right|_{L^2}^2 ds$$

$$\leq CN(0,0)^2 + C\delta N(0,t)^2 + C \int_0^t |\partial_x^2 \mathbf{v}_t|_{L^2}^2 ds.$$

(54)

By Lemma 3.6, $k = 0, l = 2$, and by (54), we have

$$|\partial_x^3 \varrho|_{L^2}^2 + \int_0^t \left|\left(\partial_x^3 \varrho, \partial_x^3 \left(\frac{D\varrho}{Dt}\right), \partial\partial_x^3 \mathbf{v}\right)\right|_{L^2}^2 ds$$

$$\leq CN(0,0)^2 + C\delta N(0,t)^2 + C \int_0^t |\partial_x^2 \mathbf{v}_t|_{L^2}^2 ds.$$

(55)

Thus by Lemma 3.7, $k = 0, l = 2$, we have

$$\int_0^t |\left(\partial_x^4 \mathbf{v}, \partial_x^3 \varrho\right)|_{L^2}^2 ds \leq CN(0,0)^2 + C\delta N(0,t)^2$$

$$+ C \int_0^t \left|\left(\partial_x^2 \mathbf{v}_t, \partial_x^3 \left(\frac{D\varrho}{Dt}\right)\right)\right|_{L^2}^2 ds,$$

together it with (55), then we have got

$$|\partial_x^3 \varrho|_{L^2}^2 + \int_0^t \left|\left(\partial_x^3 \varrho, \partial_x^3 \left(\frac{D\varrho}{Dt}\right), \partial_x^4 \mathbf{v}\right)\right|_{L^2}^2 ds \leq CN(0,0)^2$$

$$+ C\delta N(0,t)^2 + C \int_0^t |\partial_x^2 \mathbf{v}_t|_{L^2}^2 ds.$$

(56)

ii) By Lemma 3.1, $k = 3$, we have

$$|\partial_x^3(\mathbf{v}, \mathbf{H})|_{L^2}^2 \leq C \left(|(\partial_x \varrho, \mathbf{v}_t, \mathbf{H}_t)|_1^2 + |(S_2, S_3)|_1^2 + |(\mathbf{v}, \partial_x \mathbf{H})|_{L^2}^2\right)$$

$$\leq C\delta N(0,t)^2 + C|\varrho, \partial_x^3 \varrho, \mathbf{v}, \mathbf{v}_t, \partial_x \mathbf{v}_t, \partial_x \mathbf{H}, \mathbf{H}_t, \partial_x \mathbf{H}_t|_{L^2}^2,$$

(57)

which together with (30)–(33), (46) and (56) implies

$$|\partial_x^3(\mathbf{v}, \mathbf{H})|_{L^2}^2 \leq CN(0,0)^2 + C\delta N(0,t)^2 + C \int_0^t |\partial_x^2 \mathbf{v}_t|_{L^2}^2 ds.$$

(58)

iii) By Lemma 3.1, $k = 4$, we have

$$\int_0^t \left|\partial_x^4 \mathbf{H}\right|_{L^2}^2 ds \leq C \int_0^t |\mathbf{H}_t|_2^2 + |S_3|_2^2 + |\partial_x \mathbf{H}|_{L^2}^2 ds$$

$$\leq CN(0,0)^2 + C\delta N(0,t)^2 + \int_0^t \left|\partial_x^2 \mathbf{H}_t\right|_{L^2}^2 ds,$$

this together with (56) and (58), then it implies

$$\left|\partial_x^3(\varrho, \mathbf{v}, \mathbf{H})\right|_{L^2}^2 + \int_0^t \left|\left(\partial_x^3 \varrho, \partial_x^4 \mathbf{v}, \partial_x^4 \mathbf{H}\right)(s)\right|_{L^2}^2 ds$$

$$\leq CN(0,0)^2 + C\delta N(0,t)^2 + C\int_0^t \left|\partial_x^2(\mathbf{v}_t, \mathbf{H}_t)\right|_{L^2}^2 ds. \tag{59}$$

We can derive from Lemma 3.1 that

$$\left|\partial_x^2(\mathbf{v}_t, \mathbf{H}_t)\right|_{L^2}^2 \leq C(|(\partial_x \varrho_t, \mathbf{v}_t, \mathbf{v}_{tt}, \partial_x \mathbf{H}_t, \mathbf{H}_{tt}, \partial_t S_2, \partial_t S_3)|_{L^2}^2.$$

Here with the help of $(19)_1$, we have

$$\int_0^t |\partial_x \varrho_t|_{L^2}^2 ds \leq C \int_0^t |\partial_x^2 \varrho|_{L^2}^2 + |\partial_x^2 \mathbf{v}|_{L^2}^2 ds$$

$$\leq C(\varepsilon) \int_0^t |\partial_x \varrho, \partial_x \mathbf{v}|_{L^2}^2 ds + \varepsilon \int_0^t |\partial_x^3 \varrho, \partial_x^4 \mathbf{v}|_{L^2}^2 ds,$$

where the terms in the right-hand side can be absorbed by (46) and (59). Thus by (30)–(33), we obtain

$$\int_0^t \left|\partial_x^2(\mathbf{v}_t, \mathbf{H}_t)\right|_{L^2}^2 ds \leq CN(0,0)^2 + C\delta N(0,t)^2,$$

which together with (59) implies (50).   □

Part 3: Conclusion

Combining the inequalities in Proposition 3.2 and Proposition 3.3, it yields

$$\left|(\varrho, \partial_x^3 \varrho, \mathbf{v}, \partial_x^3 \mathbf{v}, \mathbf{H}, \partial_x^3 \mathbf{H})\right|_{L^2}^2 + \int_0^t \left|(\partial_x \varrho, \partial_x^3 \varrho, \partial_x \mathbf{v}, \partial_x^4 \mathbf{v}, \partial_x \mathbf{H}, \partial_x^4 \mathbf{H})\right|_{L^2}^2 ds$$

$$\leq CN(0,0)^2 + C\delta N(0,t)^2. \tag{60}$$

Thanks to (15), we can obtain that the left-hand side of (60) is equivalent to $N(0,t)^2$. Thus by the smallness of δ, it finished the proof of Theorem 4.   □

Proof of Theorem 2

In this section we shall prove the decay rates of the solution obtained in Theorem 1 to finish the proof of Theorem 2.

Some elementary decay-in-time estimates

We shall consider the convergence rates of the solution $(\varrho, \mathbf{v}, \mathbf{H})$ for the linearized problem (19). To use the L^p-L^q estimates of the linear problem for the nonlinear problem

$(19)_1$–$(19)_2$, we rewrite the solution of $(19)_1$–$(19)_2$ as

$$U(t) = E(t)U_0 + \int_0^t E(t-s)F(U(s), \mathbf{H}(s))ds, \tag{61}$$

where we have used the notations

$$U = [\varrho, \mathbf{v}]^T, U_0 = [\varrho_0, \mathbf{v}_0]^T, F = [S_1, S_2]^T, \tag{62}$$

and the fact that $E(t)$ is the solution semigroup defined by $E(t) = e^{-tA}$, $t \geq 0$, with A being a matrix-valued differential operator given by

$$A = \begin{pmatrix} 0 & \gamma \nabla^T \\ \gamma \nabla & -\mu \Delta - \mu \nabla \text{div} \end{pmatrix}.$$

The semigroup $E(t)$ has the following decay-in-time properties which can be found in (Kagei and Kobayashi 2005; Kobayashi 2002).

Lemma 3.8. There exist positive constants C and C_0 such that for any $t \geq 1$ and $l = 1, 2$, we have
i)

$$|E(t)U_0|_{L^2} \leq C \left\{ t^{-\frac{3}{4}} |U_0|_{L^1} + e^{-C_0 t} \left(|\varrho_0|_{L^2} + |\mathbf{v}_0|_{L^2} \right) \right\}, \tag{63}$$

$$\left|\partial_x^l E(t)U_0\right|_{L^2} \leq C \left\{ t^{-\frac{3}{4}-\frac{l}{2}} |U_0|_{L^1} + e^{-C_0 t}(|\varrho_0|_l + |\mathbf{v}_0|_{l-1}) \right\}, \tag{64}$$

$$\left|\partial_x^3 E(t)U_0\right|_{L^2} \leq C \left\{ t^{-\frac{15}{8}} |U_0|_{L^1} + e^{-C_0 t}(|\varrho_0|_3 + |\mathbf{v}_0|_2) \right\}, \tag{65}$$

and

$$|E(t)U_0|_{L^\infty} \leq C \left\{ t^{-\frac{3}{2}} |U_0|_{L^1} + e^{-C_0 t}(|\varrho_0|_2 + |\mathbf{v}_0|_1) \right\}. \tag{66}$$

ii)

$$\left|\partial_x^l \partial_t E(t)U_0\right|_{L^2} \leq C \left\{ t^{-\frac{5}{4}-\frac{l}{2}} |U_0|_{L^1} + e^{-C_0 t}(|\varrho_0|_l + |\mathbf{v}_0|_{l-1}) \right\}. \tag{67}$$

Lemma 3.9. For any $k \in \mathbb{Z}$, $k \geq 0$, the following inequalities hold uniformly in $0 < t \leq 1$,

$$|E(t)U_0|_{H^1 \cap L^2} \leq C|U_0|_{H^1 \cap L^2}, \tag{68}$$

$$\left|\partial_x^k \partial_x E(t)U_0\right|_{L^2} \leq Ct^{-\frac{1}{2}} |U_0|_{H^{k+1} \cap H^k}, \tag{69}$$

and

$$|E(t)U_0|_{L^\infty} \leq Ct^{-(1-\bar{\varepsilon})} |U_0|_{H^2 \cap H^1}, \tag{70}$$

for some $\bar{\varepsilon} > 0$. Here $|\cdot|_{X \cap Y} = |\cdot|_X + |\cdot|_Y$.

To treat the magnetic field, we notice that the solution to the heat equation $(19)_3$ has the following convergence estimates which one can refer to (Kagei and Kobayashi 2005; Kobayashi 2002).

Lemma 3.10. For the solution $\mathbf{H}$ to the heat equation $(19)_3$ with the initial data $\mathbf{H}(x,0) = \mathbf{H}_0$ and the boundary condition $\mathbf{H}|_{\{x_3=0\}} = 0$, there exists a constant C such that

$$|\mathbf{H}|_{L^q} \leq C(1+t)^{-\frac{3}{2}\left(1-\frac{1}{q}\right)} |\mathbf{H}_0|_{L^1 \cap L^q}$$
$$+ C \int_0^t (1+t-s)^{-\frac{3}{2}(1-\frac{1}{q})} |S_3(\cdot,s)|_{L^1 \cap L^q} ds, \tag{71}$$

and

$$|\partial_x \mathbf{H}|_{L^2} \leq C(1+t)^{-\frac{5}{4}} |\mathbf{H}_0|_{L^1 \cap H^1} + C \int_{t-1}^t (t-s)^{-\frac{1}{2}} |S_3|_{L^2} ds$$
$$+ C \int_0^{t-1} (1+t-s)^{-\frac{5}{4}} |S_3(\cdot,s)|_{L^1} ds, \tag{72}$$

for any $t \geq 1$, $2 \leq q \leq +\infty$.

Convergence rates of the lower-order derivatives

Proposition 3.4. Let $s \geq 3$. Under the assumptions of Proposition 3.1, if there exists $\delta_1' > 0$ such that the initial data $(\varrho_0, \mathbf{v}_0, \mathbf{H}_0) \in H^s \cap L^1$ and

$$|(\varrho_0, \mathbf{v}_0, \mathbf{H}_0)|_{H^s \cap L^1} \leq \delta_1',$$

then the solution $(\varrho, \mathbf{v}, \mathbf{H})$ of (19) satisfies

$$|(\varrho, \mathbf{v}, \mathbf{H})(t)|_{L^p} = O\left(t^{-\frac{3}{2}(1-\frac{1}{p})}\right), \forall p \in [2, \infty].$$

and

$$|\partial_x(\varrho, \mathbf{v}, \mathbf{H})(t)|_{L^2} = O\left(t^{-\frac{5}{4}}\right)$$

as $t \to \infty$.

Proof. We firstly estimate the L^2-estimate of the solution and its derivatives in Proposition 3.4. For simplicity, we shall introduce some notation in the sequel. Set

$$M_\sigma^s(t) = \sup_{0 \leq \tau \leq t} (1+\tau)^\sigma |\partial_x^s(\varrho, \mathbf{v}, \mathbf{H})(\tau)|_{L^2},$$

We will show that

$$M_{\frac{3}{4}}^0(t) + M_{\frac{5}{4}}^1(t) \leq C \left\{ |(\varrho_0, \mathbf{v}_0, \mathbf{H}_0)|_{L^1} + N(0,t) + M_{\frac{3}{4}}^0(t)^2 \right.$$
$$\left. + M_{\frac{5}{4}}^1(t)^2 \right\}, \tag{73}$$

where $t \geq 0$. Thus we can derive the L^2 estimates in Proposition 3.4 from (73) by using a standard method under assumption of the smallness of initial data.

Now we shall consider the L^2 estimates of $U(t)$ and $\mathbf{H}(t)$ respectively. Thanks to Proposition 3.1, we only have to show the decay rate part of Proposition 3.4 for the case $t \geq 1$. Thus we assume that $t \geq 1$ and decompose $U(t)$ as

$$U(t) = E(t)U_0 + \int_{t-1}^t E(t-s)F(U(s), H(s))ds$$
$$+ \int_0^{t-1} E(t-s)F(U(s), H(s))ds$$
$$= I_0(t) + I_1(t) + I_2(t). \tag{74}$$

By Lemma 3.8, we have

$$\left|\partial_x^l I_0(t)\right|_{L^2} \leq Ct^{-\frac{3}{4}-\frac{l}{2}} |U_0|_{L^1 \cap H^1} \leq C(1+t)^{-\frac{3}{4}-\frac{l}{2}}$$
$$\times \left\{|U_0|_{L^1} + N(0,t)\right\}, l = 0, 1. \tag{75}$$

$\square$

Lemma 3.11. Under the assumptions of Proposition 3.4, the following inequalities hold for all $t \geq 0$,

$$|S_1, \partial_x S_1, S_2, S_3|_{L^2} \leq C(1+t)^{-\frac{5}{4}} \left\{ M_{\frac{1}{4}}^1(t)^2 + N(0,t)^2 \right\}, \tag{76}$$

and

$$|S_1, S_2, S_3|_{L^1} \leq C(1+t)^{-\frac{11}{8}} \left\{ M_{\frac{3}{4}}^0(t)^2 + M_{\frac{1}{4}}^1(t)^2 + N(0,t)^2 \right\}. \tag{77}$$

Proof. Since by (16) and (17), we have

$$|S_1, S_3|_{L^2} \leq C|(\varrho, \mathbf{v}, \mathbf{H})|_{L^\infty} |\partial_x(\varrho, \mathbf{v}, \mathbf{H})|_{L^2}$$
$$\leq C(1+t)^{-\frac{5}{4}} M_{\frac{5}{4}}^1(t)N(0,t)$$
$$\leq C(1+t)^{-\frac{5}{4}} \left\{ M_{\frac{5}{4}}^1(t)^2 + N(0,t)^2 \right\},$$

$$|\partial_x S_1|_{L^2} \leq C|\partial_x(\varrho, \mathbf{v})|_{L^\infty} |\partial_x(\varrho, \mathbf{v})|_{L^2} + |\varrho, \mathbf{v}|_{L^6} |\partial_x^2(\varrho, \mathbf{v})|_{L^6}^{\frac{1}{2}} |\partial_x^2(\varrho, \mathbf{v})|_{L^2}^{\frac{1}{2}}$$
$$\leq C(1+t)^{-\frac{5}{4}} M_{\frac{5}{4}}^1(t)N(0,t) + |\partial_x(\varrho, \mathbf{v})|_{L^2} |\partial_x^2 \varrho, \partial_x^3 \varrho, \partial_x^2 \mathbf{v}, \partial_x^3 \mathbf{v}|_{L^2}$$
$$\leq C(1+t)^{-\frac{5}{4}} \left\{ M_{\frac{5}{4}}^1(t)^2 + N(0,t)^2 \right\},$$

and

$$|S_2|_{L^2} \leq C|(\varrho, \mathbf{v}, \mathbf{H})|_{L^\infty} |\partial_x(\varrho, \mathbf{v}, \mathbf{H})|_{L^2} + C |\partial_x^2 \mathbf{v}|_{L^2}^{\frac{1}{2}} |\partial_x^2 \mathbf{v}|_{L^6}^{\frac{1}{2}} |\varrho|_{L^6}$$
$$\leq C(1+t)^{-\frac{5}{4}} \left\{ M_{\frac{5}{4}}^1(t)^2 + N(0,t)^2 \right\},$$

together with these inequalities, we can easily get (76). Moreover, (77) follows from

$$|S_1, S_3|_{L^1} \leq C|(\varrho, \mathbf{v}, \mathbf{H})|_{L^2}|\partial_x(\varrho, \mathbf{v}, \mathbf{H})|_{L^2}$$
$$\leq C(1+t)^{-2}M^0_{\frac{3}{4}}(t)M^1_{\frac{5}{4}}(t) \leq C(1+t)^{-2}$$
$$\times \left\{M^0_{\frac{3}{4}}(t)^2 + M^1_{\frac{5}{4}}(t)^2\right\},$$

and

$$|S_2|_{L^1} \leq C|(\varrho, \mathbf{v}, \mathbf{H})|_{L^2}|(\partial_x\varrho, \partial_x\mathbf{v}, \partial_x\mathbf{H})|_{L^2} + C|\varrho|_{L^2}\left|\partial_x^2\mathbf{v}\right|_{L^2}$$
$$\leq C(1+t)^{-2}M^0_{\frac{3}{4}}(t)M^1_{\frac{5}{4}}(t) + C|\varrho|_{L^2}\left||\partial_x\mathbf{v}|^{\frac{1}{2}}_{L^2}\right|\partial_x^3\mathbf{v}|^{\frac{1}{2}}_{L^2}$$
$$\leq C(1+t)^{-2}\left\{M^0_{\frac{3}{4}}(t)^2 + M^1_{\frac{5}{4}}(t)^2\right\}$$
$$+ C(1+t)^{-\frac{11}{8}}M^0_{\frac{3}{4}}(t)M^1_{\frac{5}{4}}(t)^{\frac{1}{2}}N^{\frac{1}{2}}(0,t)$$
$$\leq C(1+t)^{-\frac{11}{8}}\left\{M^0_{\frac{3}{4}}(t)^2 + M^1_{\frac{5}{4}}(t)^2 + N(0,t)^2\right\}.$$

$\square$

Now we have to estimate $I_1(t)$ and $I_2(t)$. For $l = 0, 1$, and by Lemma 3.9 and Lemma 3.11, we have

$$|\partial_x^l I_1(t)|_{L^2} \leq C\int_{t-1}^t (t-\tau)^{-\frac{l}{2}}(|S_1|_1 + |S_2|_{L^2})d\tau$$
$$\leq C\int_{t-1}^t (t-\tau)^{-\frac{l}{2}}(1+\tau)^{-\frac{5}{4}}d\tau\left\{M^1_{\frac{5}{4}}(t)^2 + N(0,t)^2\right\}$$
$$\leq C(1+t)^{-\frac{5}{4}}\left\{M^1_{\frac{5}{4}}(t)^2 + N(0,t)^2\right\},$$

(78)

where the last inequality is obtained by taking $l = 1$ and $l = 2$ respectively.

For $I_2(t)$, we can derive from Lemma 3.8 and Lemma 3.9 that

$$|\partial_x^l I_2(t)|_{L^2} \leq C\int_0^{t-1}(t-\tau)^{-\frac{3}{2}-\frac{l}{2}}|F(U(\tau), \mathbf{H}(\tau))|_{L^1}$$
$$+ e^{-C_0(t-\tau)}|U(\tau)|_1 d\tau$$
$$\leq C\int_0^{t-1}(t-\tau)^{-\frac{3}{2}-\frac{l}{2}}(1+\tau)^{-\frac{11}{8}}d\tau$$
$$\times \left\{M^0_{\frac{3}{4}}(t)^2 + M^1_{\frac{5}{4}}(t)^2 + N(0,t)^2\right\}$$
$$+ C\int_0^{t-1}e^{-C_0(t-\tau)}(1+\tau)^{-\frac{5}{4}}d\tau$$
$$\times \left\{M^1_{\frac{5}{4}}(t)^2 + N(0,t)^2\right\}$$
$$\leq C(1+t)^{-\frac{3}{4}-\frac{l}{2}}\left\{M^0_{\frac{3}{4}}(t)^2 + M^1_{\frac{5}{4}}(t)^2 + N(0,t)^2\right\}.$$

(79)

Now we turn to do the estimates for $\mathbf{H}(t)$. By (71) in Lemma 3.10, we can deduce that

$$|\mathbf{H}|_{L^2} \leq C(1+t)^{-\frac{3}{4}}|\mathbf{H}_0|_{L^1\cap L^2} + \int_0^t(1+t-s)^{-\frac{3}{4}}|S_3|_{L^1\cap L^2}ds$$
$$\leq C(1+t)^{-\frac{3}{4}}\left\{|\mathbf{H}_0|_{L^1} + N(0,t)\right\}$$
$$+ C\left\{M^0_{\frac{3}{4}}(t)^2 + M^1_{\frac{5}{4}}(t)^2 + N(0,t)^2\right\}$$
$$\times \int_0^t(1+t-s)^{-\frac{3}{4}}(1+s)^{-\frac{11}{8}}ds$$
$$\leq C(1+t)^{-\frac{3}{4}}\left\{|\mathbf{H}_0|_{L^1} + N(0,t) + M^0_{\frac{3}{4}}(t)^2 + M^1_{\frac{5}{4}}(t)^2\right\},$$

and similarly by (72), we have

$$|\partial_x\mathbf{H}|_{L^2} \leq C(1+t)^{-\frac{5}{4}}|\mathbf{H}_0|_{L^1\cap H^1} + \int_{t-1}^t(t-s)^{-\frac{1}{2}}|S_3|_{L^2}ds$$
$$+ \int_0^{t-1}(1+t-s)^{-\frac{5}{4}}|S_3|_{L^1}ds$$
$$\leq C(1+t)^{-\frac{5}{4}}\left\{|\mathbf{H}_0|_{L^1} + N(0,t) + M^0_{\frac{3}{4}}(t)^2 + M^1_{\frac{5}{4}}(t)^2\right\}.$$

Combining these with (75), (78) and (79), it implies (73). Now we shall do the L^∞ estimate. Define

$$M_\infty(t) = \sup_{0\leq\tau\leq t}(1+\tau)^{\frac{3}{2}}|(\varrho, \mathbf{v}, \mathbf{H})(\tau)|_{L^\infty},$$

and

$$\mathcal{M}^s_\sigma(t) = \sum_{k=0}^s M^k_\sigma(t) = \sum_{k=0}^s \sup_{0\leq\tau\leq t}(1+\tau)^\sigma|\partial_x^k(\varrho, \mathbf{v}, \mathbf{H})(\tau)|_{L^2}.$$

By Lemma 3.8, we have

$$|I_0|_{L^\infty} \leq Ct^{-\frac{3}{2}}(|U_0|_{L^1} + |\varrho_0|_2 + |\mathbf{v}_0|_1)$$
$$\leq C(1+t)^{-\frac{3}{2}}(|U_0|_{L^1} + N(0,t)).$$

(80)

To estimate $I_1(t)$ and $I_2(t)$, we state the following estimates for S_1 and S_2.

Lemma 3.12. Under the assumptions of Proposition 3.4, there exists a sufficiently small constant $\widetilde{\varepsilon} > 0$ such that for any $0 < \varepsilon < \widetilde{\varepsilon}$, the following inequalities hold uniformly in $t \geq 0$.

i)

$$|S_1|_2, |S_2|_1 \leq C(1+t)^{-\frac{3}{2}}\left\{M_\infty(t)^2 + M^1_{\frac{5}{4}}(t)^2 + \mathcal{M}^3_{\frac{3}{4}-\varepsilon}(t)^2\right\},$$

(81)

and

$$|S_1, S_2|_{L^1} \le C t^{-(\frac{7}{4}-\varepsilon)} \left\{ M_{\frac{3}{4}}^0(t)^2 + M_{\frac{5}{4}}^1(t)^2 + \mathcal{M}_{\frac{3}{4}-\varepsilon}^3(t)^2 \right\}.$$
(82)

ii)

$$|S_1|_3, |S_2|_2 \le C(1+t)^{-\frac{3}{2}} \left\{ M_\infty(t)^2 + M_{\frac{5}{4}}^1(t)^2 + \mathcal{K}_{\frac{3}{4}-\varepsilon}^4(t)^2 \right\},$$
(83)

and

$$|\partial_t S_1|_1, |\partial_t S_2|_{L_2} \le C(1+t)^{-\frac{5}{4}} \left\{ \mathcal{K}_{\frac{3}{4}-\varepsilon}^4(t)^2 + M_\infty(t)^2 + M_{\frac{5}{4}}^1(t)^2 \right\},$$
(84)

here we define

$$\mathcal{K}_\sigma^s(t) = \mathcal{M}_\sigma^s(t) + \sum_{j=1}^{[\frac{s}{2}]} \sup_{0 \le \tau \le t} (1+\tau)^\sigma |\partial_\tau^j(\mathbf{v}, \mathbf{H})(\tau)|_{s-2j}$$

$$+ \sum_{j=1}^{[\frac{s+1}{2}]} \sup_{0 \le \tau \le t} (1+\tau)^\sigma |\partial_\tau^j \varrho(\tau)|_{s+1-2j}.$$

Proof. For $l = 0, 1, 2$, we have

$$|\partial_x^l S_1|_{L^2} \le C |\partial_x^{l+1}(\varrho \mathbf{v})|_{L^2}$$

$$\le C \left(|\varrho, \mathbf{v}|_{L^\infty} |\partial_x(\varrho, \mathbf{v})|_2 + |\partial_x(\varrho, \mathbf{v})|_{L^2}^{\frac{1}{2}} |\partial_x^2(\varrho, \mathbf{v})|_1^{\frac{3}{2}} \right)$$

$$\le C \left\{ (1+t)^{-(\frac{9}{4}-\varepsilon)} M_\infty(t) \mathcal{M}_{\frac{3}{4}-\varepsilon}^3(t) \right.$$

$$\left. + (1+t)^{-(\frac{7}{4}-\frac{3}{2}\varepsilon)} M_{\frac{5}{4}}^1(t)^{\frac{1}{2}} \mathcal{M}_{\frac{3}{4}-\varepsilon}^3(t)^{\frac{3}{2}} \right\}$$

$$\le C(1+t)^{-\frac{3}{2}} \left\{ M_\infty(t)^2 + M_{\frac{5}{4}}^1(t)^2 + \mathcal{M}_{\frac{3}{4}-\varepsilon}^3(t)^2 \right\},$$

and for $l = 0, 1$, we have

$$|\partial_x^l S_2|_{L^2} \le C \left(|\varrho, \mathbf{v}, \mathbf{H}|_{L^\infty} |\partial_x(\varrho, \mathbf{v}, \mathbf{H})|_2 \right.$$

$$\left. + |\partial_x(\varrho, \mathbf{v}, \mathbf{H})|_{L^2}^{\frac{1}{2}} |\partial_x(\varrho, \mathbf{v}, \mathbf{H})|_1^{\frac{3}{2}} \right)$$

$$\le C(1+t)^{-\frac{3}{2}} \left\{ M_\infty^2 + M_{\frac{5}{4}}^1(t)^2 + \mathcal{M}_{\frac{3}{4}-\varepsilon}^3(t)^2 \right\}.$$

Then we get (81). Similarly,

$$|S_1|_{L^1} \le C |\varrho, \mathbf{v}|_{L^2} |\partial_x(\varrho, \mathbf{v})|_{L^2} \le C(1+t)^{-2} \left\{ M_{\frac{3}{4}}^0(t)^2 + M_{\frac{5}{4}}^1(t)^2 \right\},$$

and

$$|S_2|_{L^1} \le C \left(|\varrho, \mathbf{v}, \mathbf{H}|_{L^2} |\partial_x(\varrho, \mathbf{v}, \mathbf{H})|_{L^2} + |\varrho|_{L^2} |\partial_x^2 \mathbf{v}|_{L^2} \right)$$

$$\le C \left(|\varrho, \mathbf{v}, \mathbf{H}|_{L^2} |\partial_x(\varrho, \mathbf{v}, \mathbf{H})|_{L^2} + |\varrho|_{L^2} |\partial_x \mathbf{v}|_{L^2}^{\frac{1}{2}} |\partial_x^3 \mathbf{v}|_{L^2}^{\frac{1}{2}} \right)$$

$$\le C \left\{ (1+t)^{-2} M_{\frac{3}{4}}^0(t) M_{\frac{5}{4}}^1(t) \right.$$

$$\left. + (1+t)^{-(\frac{7}{4}-\varepsilon)} M_{\frac{3}{4}}^0(t) M_{\frac{5}{4}}^1(t) \mathcal{M}_{\frac{3}{4}-\varepsilon}^3(t) \right\}$$

$$\le C(1+t)^{-(\frac{7}{4}-\varepsilon)} \left\{ M_{\frac{3}{4}}^0(t)^2 + M_{\frac{5}{4}}^1(t)^2 + \mathcal{M}_{\frac{3}{4}-\varepsilon}^3(t)^2 \right\}.$$

Thus we obtain (82). We shall get the inequalities in
ii) in the similar way. Here we only estimate $|\partial_x^3 S_1|_{L^2}$ and
$|\partial_t(\varrho \Delta \mathbf{v})|_{L^2}$ as follows.

$$|\partial_x^3 S_1|_{L^2} \le C \left(|\varrho, \mathbf{v}|_{L^\infty} |\partial_x^4(\varrho, \mathbf{v})|_{L^2} \right)$$

$$\le C \left\{ (1+t)^{-(\frac{9}{4}-\varepsilon)} M_\infty(t) \mathcal{K}_{\frac{3}{4}-\varepsilon}^4(t) \right\}$$

$$\le C(1+t)^{-\frac{3}{2}} \left\{ M_\infty^2 + \mathcal{K}_{\frac{3}{4}-\varepsilon}^4(t)^2 \right\},$$

and

$$|\partial_t(\varrho \Delta \mathbf{v})|_{L^2} \le |\varrho|_{L^\infty} |\Delta \mathbf{v}_t|_{L^2} + |\varrho_t|_1 |\partial_x^3 \mathbf{v}|_{L^2}$$

$$\le C \left\{ (1+t)^{-(\frac{9}{4}-\varepsilon)} M_\infty(t) \mathcal{K}_{\frac{3}{4}-\varepsilon}^4(t) \right.$$

$$\left. + (1+t)^{-(\frac{3}{2}-2\varepsilon)} \mathcal{K}_{\frac{3}{4}-\varepsilon}^4(t)^2 \right\}$$

$$\le C(1+t)^{-\frac{5}{4}} \left\{ M_\infty^2 + \mathcal{K}_{\frac{3}{4}-\varepsilon}^4(t)^2 \right\}.$$

$\square$

By Lemma 3.9 and (81), we have

$$|I_1|_{L^\infty} \le C \int_{t-1}^t (t-\tau)^{-(1-\bar\varepsilon)} (|S_1|_2 + |S_2|_1) d\tau$$

$$\le C \left\{ M_\infty^2 + M_{\frac{5}{4}}^1(t)^2 + \mathcal{M}_{\frac{3}{4}-\varepsilon}^3(t)^2 \right\}$$

$$\times \int_{t-1}^t (t-\tau)^{-(1-\bar\varepsilon)} (1+\tau)^{-\frac{3}{2}} d\tau$$

$$\le C(1+t)^{-\frac{3}{2}} \left\{ M_\infty^2 + M_{\frac{5}{4}}^1(t)^2 + \mathcal{M}_{\frac{3}{4}-\varepsilon}^3(t)^2 \right\}.$$
(85)

As for I_2, we can deduce from Lemma 3.8 that

$$|I_2(t)|_{L^\infty} \leq C \left\{ \int_0^{t-1} (t-\tau)^{-\frac{3}{2}} |(S_1, S_2)(\tau)|_{L^1} \right.$$

$$+ \int_0^{t-1} e^{-C_0(t-\tau)} (|S_1(\tau)|_2 + |S_2(\tau)|_1) d\tau \right\}$$

$$\leq C \left\{ M_{\frac{3}{4}}^0(t)^2 + M_{\frac{5}{4}}^1(t)^2 + \mathcal{M}_{\frac{3}{4}-\varepsilon}^3(t)^2 \right\}$$

$$\times \int_0^{t-1} (t-\tau)^{-\frac{3}{2}} (1+\tau)^{-(\frac{7}{4}-\varepsilon_1)} d\tau$$

$$+ C \left\{ M_\infty^2 + M_{\frac{5}{4}}^1(t)^2 + \mathcal{M}_{\frac{3}{4}-\varepsilon}^3(t)^2 \right\}$$

$$\times \int_0^{t-1} e^{-C_0(t-\tau)} (1+\tau)^{-\frac{3}{2}} d\tau$$

$$\leq C(1+t)^{-\frac{3}{2}} \left\{ M_\infty^2 + M_{\frac{3}{4}}^0(t)^2 + M_{\frac{5}{4}}^1(t)^2 \right.$$

$$\left. + \mathcal{M}_{\frac{3}{4}-\varepsilon}^3(t)^2 \right\}. \tag{86}$$

At last, we have to estimate $|H|_{L^\infty}$ to finish the L^∞ estimate. By Lemma 3.3, For $q = \infty$, we have

$$|H|_{L^\infty}$$

$$\leq C(1+t)^{-\frac{3}{2}} |H_0|_{L^1 \cap L^\infty} + C \int_0^t (1+t-\tau)^{-\frac{3}{2}} |S_3(\tau)|_{L^1 \cap L^\infty} d\tau$$

$$\leq C(1+t)^{-\frac{3}{2}} \left\{ |H_0|_{L^1} + N(0,t) \right\}$$

$$+ C \left\{ M_{\frac{3}{4}}^0(t)^2 + M_{\frac{5}{4}}^1(t)^2 + M_\infty(t)^2 + N(0,t)^2 \right\}$$

$$\times \int_0^t (1+t-\tau)^{-\frac{3}{2}} (1+\tau)^{-\frac{3}{2}} d\tau$$

$$\leq C(1+t)^{-\frac{3}{2}} \left\{ |H_0|_{L^1} + N(0,t) + M_\infty(t)^2 + M_{\frac{3}{4}}^0(t)^2 + M_{\frac{5}{4}}^1(t)^2 \right\}, \tag{87}$$

where we have used

$$|S_3|_{L^1} \leq C(1+t)^{-2} \left\{ M_{\frac{3}{4}}^0(t)^2 + M_{\frac{5}{4}}^1(t)^2 \right\},$$

which is in the proof of Lemma 3.12, and

$$|S_3|_{L^\infty} \leq |v, H|_{L^\infty} |v, H|_3 \leq C(1+t)^{-\frac{3}{2}} \left\{ M_\infty(t)^2 + N(0,t)^2 \right\}.$$

Thus, (80) and (85)–(87) imply

$$M_\infty \leq C \left\{ |\varrho_0, v_0, H_0|_{L^1} + N(0,t) + M_\infty(t)^2 + M_{\frac{3}{4}}^0(t)^2 \right.$$

$$\left. + M_{\frac{5}{4}}^1(t)^2 + \mathcal{M}_{\frac{3}{4}-\varepsilon}^3(t)^2 \right\}.$$

This together with (73), the estimate of $\mathcal{K}_{\frac{3}{4}-\varepsilon}^4(t)$ in Proposition 3.5 in the next subsection, it yields the L^∞-estimate in Proposition 3.4. Hence by interpolation, we

get the L^p-estimate for $2 \leq p \leq \infty$. The proof of Proposition 3.4 is complete. $\qquad \square$

Convergence rates of the higher-order derivatives

Now we shall do the estimates for the higher-order derivatives to finish the proof of Theorem 2.

Proposition 3.5. There exists a positive number $\widetilde{\varepsilon}$ such that under the assumptions of Proposition 3.4, the solution (ϱ, v, H) of (19) satisfies

$$|(\varrho, v, H)(t)|_s \leq C(|\varrho_0, v_0, H_0|_{L^1} + |\varrho_0, v_0, H_0|_s)$$

$$\times (1+t)^{-(\frac{3}{4}-\varepsilon_1)}, \quad \forall 0 < \varepsilon_1 < \widetilde{\varepsilon}$$

for all $t \geq 0$.

Set

$$\mathcal{L}_\sigma^s = \left\{ \int_0^t (1+\tau)^{2\sigma} (|\partial_x(v,H)|_{s+1} + |\partial_x \varrho|_s) d\tau \right.$$

$$\left. + \sum_{j=1}^{[\frac{s+1}{2}]} \int_0^t (1+\tau)^{2\sigma} |\partial_\tau^j(\varrho, v, H)|_{s+1-2j} d\tau \right\}^{\frac{1}{2}},$$

and

$$\mathcal{N}_\sigma^s = \left\{ \sum_{j=0}^{[\frac{s-1}{2}]} \sup_{0 \leq \tau \leq t} (1+\tau)^{2\sigma} |\partial_\tau^j S_1(\tau)|_{s-1-2j}^2 \right.$$

$$+ \sum_{j=0}^{[\frac{s}{2}]} \int_0^t (1+\tau)^{2\sigma} |\partial_\tau^j(\varrho \operatorname{div} v)(\tau)|_{s-2j}^2 d\tau$$

$$+ \sum_{j=0}^{[\frac{s-1}{2}]} \int_0^t (1+\tau)^{2\sigma} |\partial_\tau^j(v \cdot \nabla \varrho)|_{s-1-2j}^2 d\tau$$

$$+ \sum_{j=0}^{[\frac{s-2}{2}]} \sup_{0 \leq \tau \leq t} (1+\tau)^{2\sigma} |\partial_\tau^j(S_2, S_3)(\tau)|_{s-2-2j}^2$$

$$+ \sum_{j=0}^{[\frac{s-1}{2}]} \int_0^t (1+\tau)^{2\sigma} |\partial_\tau(S_2, S_3)|_{s-1-2j}^2 d\tau$$

$$+ \sum_{0 \leq k+2j \leq s} \int_0^t (1+\tau)^{2\sigma} |\langle \partial^k \partial_\tau^j S, \partial^k \partial_\tau^j V \rangle| d\tau$$

$$+ \sum_{0 \leq k+2j \leq s-1} \int_0^t (1+\tau)^{2\sigma} |\langle \partial^k \partial_\tau^j S, \partial^k \partial_\tau^{j+1} V \rangle| d\tau$$

$$+ \sum_{0 \leq k+l+2j \leq s-1} \int_0^t (1+\tau)^{2\sigma} \gamma\mu |\langle \partial^k \partial_3^{l+1} \partial_\tau^j \right.$$

$$\left. \times (v \cdot \nabla \varrho), \partial^k \partial_3^{l+1} \partial_\tau^j \varrho \rangle| d\tau \right\}^{\frac{1}{2}},$$

where $S = (S_1, S_2, S_3)^T$ and $V = (\varrho, \mathbf{v}, \mathbf{H})^T$. As in the proof of (11.6) in (Kagei and Kobayashi 2005), we have the following inequality in the similar way:

Lemma 3.13. Under the assumptions of Proposition 3.4, the following inequality holds uniformly in $t \geq 0$ and for $\sigma \geq \frac{1}{2}$,

$$
\mathcal{K}_\sigma^s(t)^2 + \mathcal{L}_\sigma^s(t)^2 \leq C|\varrho_0, \mathbf{v}_0, \mathbf{H}_0|_s^2 + C\mathcal{L}_0^s(t)^2
$$
$$
+ C\left\{\mathcal{N}_\sigma^s(t)^2 + \int_0^t (1+\tau)^{2\sigma-1}|(\varrho, \mathbf{v}, \mathbf{H})(\tau)|_{L^2}^2 d\tau\right\}.
$$
$$
(88)
$$

Proof. Here we only estimate the magnetic field. We shall show that,

$$
(1+t)^{2\sigma}|\mathbf{H}(t)|_s + \sum_{j=1}^{[\frac{s}{2}]} (1+t)^{2\sigma}|\partial_t^j \mathbf{H}(t)|_{s-2j}
$$
$$
+ \sum_{j=1}^{[\frac{s+1}{2}]} \int_0^t (1+\tau)^{2\sigma}|\partial_\tau^j \mathbf{H}(t)|_{s+1-2j} d\tau
$$
$$
\leq C\left\{|\varrho_0, \mathbf{v}_0, \mathbf{H}_0|_s^2 + \mathcal{L}_0^s(t)^2 + \mathcal{N}_\sigma^s(t)^2 \right.
$$
$$
\left. + \int_0^t (1+\tau)^{2\sigma-1}|\mathbf{H}|_{L^2}^2 d\tau\right\}.
$$
$$
(89)
$$

Then we estimate these as follows:

i) Taking ∂_t^j to $(19)_3$, multiplying $(1+t)^{2\sigma}\partial_t^j \mathbf{H}$, and integrating on $\mathbb{R}_+^3$, it implies

$$
\frac{1}{2}\frac{d}{dt}\int (1+t)^{2\sigma}|\partial_t^j \mathbf{H}(t)|^2 dx + (1+t)^{2\sigma}\int |\partial_x \partial_t^j \mathbf{H}|^2 dx
$$
$$
= \sigma(1+t)^{2\sigma-1}\int |\partial_t^j \mathbf{H}(t)|^2 dx + (1+t)^{2\sigma}\left\langle \partial_t^j S_3, \partial_t^j \mathbf{H}\right\rangle.
$$

Integrating this from 0 to t yields

$$
(1+t)^{2\sigma}|\partial_t^j \mathbf{H}(t)|_{L^2}^2 + \int_0^t (1+\tau)^{2\sigma}|\partial_x \partial_t^j \mathbf{H}|_{L^2}^2 d\tau
$$
$$
\leq C|\varrho_0, \mathbf{v}_0, \mathbf{H}_0|_s^2 + C\int_0^t \sigma(1+\tau)^{2\sigma-1}|\partial_\tau^j \mathbf{H}|_{L^2}^2 d\tau
$$
$$
+ C\int_0^t (1+\tau)^{2\sigma}\left|\left\langle \partial_\tau^j S_3, \partial_\tau^j \mathbf{H}\right\rangle\right| d\tau,
$$
$$
(90)
$$

for $2j \leq s$.

ii) Taking ∂_t^j to $(19)_3$, multiplying $(1+t)^{2\sigma}\partial_t^{j+1}\mathbf{H}$, and integrating on $\mathbb{R}_+^3$, it holds

$$
\int (1+t)^{2\sigma}|\partial_t^{j+1}\mathbf{H}(t)|^2 dx + \frac{1}{2}\frac{d}{dt}\int (1+t)^{2\sigma}|\partial_x \partial_t^j \mathbf{H}(t)|^2 dx
$$
$$
= \sigma(1+t)^{2\sigma-1}\int |\partial_x \partial_t^j \mathbf{H}(t)|^2 dx + (1+t)^{2\sigma}\left\langle \partial_t^j S_3, \partial_t^{j+1}\mathbf{H}\right\rangle.
$$

By integrating this from 0 to t, we get

$$
(1+t)^{2\sigma}|\partial_x \partial_t^j \mathbf{H}(t)|_{L^2}^2 + \int_0^t (1+\tau)^{2\sigma}|\partial_\tau^{j+1}\mathbf{H}|_{L^2}^2 d\tau
$$
$$
\leq C|\varrho_0, \mathbf{v}_0, \mathbf{H}_0|_s^2 + C\int_0^t \sigma(1+\tau)^{2\sigma-1}|\partial_x \partial_\tau^j \mathbf{H}|_{L^2}^2 d\tau
$$
$$
+ C\int_0^t (1+\tau)^{2\sigma}|\left\langle \partial_\tau^j S_3, \partial_\tau^{j+1}\mathbf{H}\right\rangle| d\tau,
$$
$$
(91)
$$

for $2j+1 \leq s$.

iii) Due to Lemma 3.7, we have

$$
(1+t)^{2\sigma}\left|\partial_x^{k+2}\partial_t^j \mathbf{H}(t)\right|_{L^2}^2 \leq C(1+t)^{2\sigma}
$$
$$
\times \left(\left|\partial_t^j \mathbf{H}\right|_k^2 + \left|\partial_t^j S_3\right|_k^2 + \left|\partial_x \partial_t^j \mathbf{H}\right|_{L^2}^2\right),
$$
$$
(92)
$$
for $k+2j+2 \leq s$

and

$$
\int_0^t (1+\tau)^{2\sigma}|\partial_x^{k+2}\partial_\tau^j \mathbf{H}|_{L^2}^2 \leq C\int_0^t (1+\tau)^{2\sigma}
$$
$$
\times \left(|\partial_\tau^j \mathbf{H}|_k^2 + |\partial_\tau^j S_3|_k^2 + |\partial_x \partial_\tau^j \mathbf{H}|_{L^2}^2\right) d\tau,
$$
$$
(93)
$$

for $k+2j+1 \leq s$.

By Cauchy inequality and $\sigma \geq \frac{1}{2}$, we have $(1+t)^{2\sigma-1} \leq C(\alpha) + \alpha(1+t)^{2\sigma}, \forall \alpha > 0$. This combining with (90)–(91), we get

$$
\sum_{k=0}^{[\frac{s}{2}]}\left\{(1+t)^{2\sigma}|\partial_t^k \mathbf{H}(t)|_{L^2}^2 + \int_0^t (1+\tau)^{2\sigma}|\partial_x \partial_\tau^k \mathbf{H}|_{L^2}^2 d\tau\right\}
$$
$$
+ \sum_{j=0}^{[\frac{s-1}{2}]}\left\{(1+t)^{2\sigma}|\partial_x \partial_t^j \mathbf{H}(t)|_{L^2}^2 + \int_0^t (1+\tau)^{2\sigma}|\partial_\tau^{j+1}\mathbf{H}|_{L^2}^2 d\tau\right\}
$$
$$
\leq C\left(|\varrho_0, \mathbf{v}_0, \mathbf{H}_0|_s^2 + \int_0^t (1+\tau)^{2\sigma-1}|\mathbf{H}|_{L^2}^2 d\tau + \sum_{k=1}^{[\frac{s}{2}]}\int_0^t |\partial_\tau^k \mathbf{H}|_{L^2}^2 d\tau \right.
$$
$$
+ \sum_{j=0}^{[\frac{s-1}{2}]}\int_0^t |\partial_x \partial_\tau^j \mathbf{H}|_{L^2}^2 d\tau + \sum_{k=0}^{[\frac{s}{2}]}\int_0^t (1+\tau)^{2\sigma}|\left\langle \partial_\tau^k S_3, \partial_\tau^k \mathbf{H}\right\rangle| d\tau
$$
$$
\left. + \sum_{j=0}^{[\frac{s-1}{2}]}\int_0^t (1+\tau)^{2\sigma}|\left\langle \partial_\tau^j S_3, \partial_\tau^{j+1}\mathbf{H}\right\rangle| d\tau\right).
$$
$$
(94)
$$

Multiplying (92) and (93) for $k = 0, 1$ with a sufficiently positive number β, and combining with (94), it yields

$$
\sum_{2j+k\leq s,\, k\leq 3}\left\{(1+t)^{2\sigma}|\partial_t^j\mathbf{H}(t)|_k^2 + \sum_{2j+k\leq s+1,\, k\leq 3}\int_0^t(1+\tau)^{2\sigma}|\partial_\tau^j\mathbf{H}|_k^2 d\tau\right\}
$$

$$
\leq C\left(|\varrho_0,\mathbf{v}_0,\mathbf{H}_0|_s^2 + \int_0^t(1+\tau)^{2\sigma-1}|\mathbf{H}|_{L^2}^2 d\tau + \sum_{k=1}^{[\frac{s}{2}]}\int_0^t|\partial_\tau^k\mathbf{H}|_{L^2}^2 d\tau\right.
$$

$$
+ \sum_{j=0}^{[\frac{s-1}{2}]}\int_0^t|\partial_x\partial_\tau^j\mathbf{H}|_{L^2}^2 d\tau + \sum_{k=0}^{[\frac{s}{2}]}\int_0^t(1+\tau)^{2\sigma}|\langle\partial_\tau^k S_3,\partial_\tau^k\mathbf{H}\rangle|d\tau
$$

$$
+ \sum_{j=0}^{[\frac{s-1}{2}]}\int_0^t(1+\tau)^{2\sigma}|\langle\partial_\tau^j S_3,\partial_\tau^{j+1}\mathbf{H}\rangle|d\tau + \sum_{2j+k\leq s,\, 2\leq k\leq 3}(1+t)^{2\sigma}|\partial_t^j S_3|_{k-2}^2
$$

$$
\left. + \sum_{2j+k+1\leq s,\, 2\leq k\leq 3}\int_0^t(1+\tau)^{2\sigma}|\partial_\tau^j S_3|_{k-2}^2 d\tau\right).
$$

$$\tag{95}$$

For $k \geq 2$, the first term in the right-hand side of (92) and (93) can be absorbed by the left-hand side of (92) and (93) for $k = k-2$. Thus together with (95), we get (89). $\quad\square$

In the proof of Subsubsection A priori estimates, we can obtain that $\mathcal{K}_0^s(t)^2 + \mathcal{L}_0^s(t)^2 \leq C|\varrho_0,\mathbf{v}_0,\mathbf{H}_0|_s^2$. Combining this with Proposition 3.4, we can derive from Lemma 3.13, for $\sigma = \frac{3}{4} - \varepsilon_1$ that

$$
\mathcal{K}_{\frac{3}{4}-\varepsilon_1}^s(t)^2 + \mathcal{L}_{\frac{3}{4}-\varepsilon_1}^s(t)^2 \leq C(\delta_1') + C\mathcal{N}_{\frac{3}{4}-\varepsilon_1}^s(t)^2, \quad (96)
$$

here $C(\delta_1') \to 0$, as $\delta_1' \to 0$. Hence it remains to estimate $\mathcal{N}_{\frac{3}{4}-\varepsilon_1}^s(t)$. We shall show that

$$
\mathcal{N}_{\frac{3}{4}-\varepsilon_1}^s(t)^2 \leq \varepsilon\mathcal{L}_{\frac{3}{4}-\varepsilon_1}^s(t)^2 + C(\varepsilon)\left\{\mathcal{K}_{\frac{3}{4}-\varepsilon_1}^s(t)^4 + \mathcal{K}_{\frac{3}{4}-\varepsilon_1}^s(t)^2\mathcal{L}_0^s(t)^2\right\}.
$$

Thus Proposition 3.5 follows. Since the other terms can be estimated in the similar way or we can see in (Kagei and Kobayashi 2005, here we only estimate $\int_0^t\left|\langle\partial^k\partial_\tau^j\left(\frac{1}{\rho}\text{curl}\mathbf{H}\times\mathbf{H}\right),\partial^k\partial_\tau^j\mathbf{v}\rangle\right|d\tau$ with $k = s - 2j$ and $s \geq 3$ as

$$
\left|\langle\partial^k\partial_\tau^j\left(\frac{1}{\rho}\text{curl}\mathbf{H}\times\mathbf{H}\right),\partial^k\partial_\tau^j\mathbf{v}\rangle\right|
$$

$$
= \left|\langle\partial^{s-1-2j}\partial_\tau^j(\frac{1}{\rho}\text{curl}\mathbf{H}\times\mathbf{H}),\partial^{s+1-2j}\partial_\tau^j\mathbf{v}\rangle\right|
$$

$$
\leq |\partial^{s-1-2j}\partial_\tau^j\left(\frac{1}{\rho}\text{curl}\mathbf{H}\times\mathbf{H}\right)|_{L^2}|\partial^{s+1-2j}\partial_\tau^j\mathbf{v}|_{L^2}
$$

$$
\leq C(\varepsilon)|\partial^{s-1-2j}\partial_\tau^j(\frac{1}{\rho}\text{curl}\mathbf{H}\times\mathbf{H})|_{L^2}^2 + \frac{\varepsilon}{2}|\partial^{s+1-2j}\partial_\tau^j\mathbf{v}|_{L^2}^2.
$$

In virtue of (17) and its general form, and with the help of the smallness assumption in Proposition 3.4, we have

i) While $j = 0$,

$$
|\partial^{s-1}\left(\frac{1}{\rho}\text{curl}\mathbf{H}\times\mathbf{H}\right)|_{L^2}^2 \leq \left|\frac{1}{\rho}\right|_{L^\infty}^2|\mathbf{H}|_{L^\infty}^2|\partial_x\mathbf{H}|_{s-1}^2
$$

$$
+ \left|\frac{1}{\rho}\right|_{L^\infty}^2|\mathbf{H}|_{s-1}^2|\partial_x\mathbf{H}|_{L^\infty}^2
$$

$$
+ \left|\partial_x\left(\frac{1}{\rho}\right)\right|_{L^\infty}^2|\mathbf{H}|_{L^\infty}^2|\partial_x\mathbf{H}|_{s-2}^2
$$

$$
+ \left|\partial_x\left(\frac{1}{\rho}\right)\right|_{L^\infty}^2|\mathbf{H}|_{s-2}^2|\partial_x\mathbf{H}|_{L^\infty}^2
$$

$$
+ \left|\partial_x\left(\frac{1}{\rho}\right)\right|_{s-2}^2|\mathbf{H}|_{L^\infty}^2|\partial_x\mathbf{H}|_{L^\infty}^2
$$

$$
\leq C|\varrho,\mathbf{H}|_{s-1}^2|\partial_x\mathbf{H}|_{s-1}^2.
$$

ii) While $1 \leq j \leq [\frac{s-1}{2}]$,

$$
\left|\partial^{s-1-2j}\partial_t^j\left(\frac{1}{\rho}\text{curl}\mathbf{H}\times\mathbf{H}\right)\right|_{L^2}^2
$$

$$
\leq \sum_{1\leq j_1+j_2+\cdots+j_k\leq j}\left|f\left(\rho;j_1,j_2,\cdots,j_{k-2}\right)\left(sgn(j_1)\partial_t^{j_1}\varrho\right)\cdots\right.
$$

$$
\times\left(sgn(j_{k-2})\partial_t^{j_{k-2}}\varrho\right)
$$

$$
\left.\times\left(\text{curl}\partial_t^{j_{k-1}}\mathbf{H}\times\partial_t^{j_k}\mathbf{H}\right)\right|_{s-1-2j}^2
$$

$$
\leq C\sum_{j=1}^{[\frac{s-1}{2}]}|\partial_t^j(\varrho,\mathbf{H})|_{s-1-2j}^2\sum_{l=1}^{[\frac{s-1}{2}]}|\partial_x\partial_t^l\mathbf{H}|_{s-1-2l}^2,
$$

where $f = \frac{c}{\rho^m}$, c is a nonzero integer and $m = m(j_1,j_2,\cdots,j_{k-2})$ is some positive integer. And the last inequality in ii) can be derived from the continuous equation $(19)_1$ and the smallness assumption of initial data. This together with the definition of $\mathcal{L}_0^s(t)$, $\mathcal{L}_{\frac{3}{4}-\varepsilon_1}^s(t)$ and $\mathcal{K}_{\frac{3}{4}-\varepsilon_1}^s(t)$, it implies

$$
\int_0^t(1+\tau)^{2(\frac{3}{4}-\varepsilon_1)t}\langle\partial^k\partial_\tau^j\left(\frac{1}{\rho}\text{curl}\mathbf{H}\times\mathbf{H}\right),\partial^k\partial_\tau^j\mathbf{v}\rangle d\tau
$$

$$
\leq \varepsilon\mathcal{L}_{\frac{3}{4}-\varepsilon_1}^s(t)^2 + C(\varepsilon)\mathcal{K}_{\frac{3}{4}-\varepsilon_1}^s(t)^2\mathcal{L}_0^s(t)^2.
$$

$\quad\square$

Now we shall finish the proof of the last part of Theorem 2.

Proposition 3.6. Let $s \geq 4$. Under the assumptions of Proposition 3.5, if there exists a positive constants δ_2 such that $|\varrho_0,\mathbf{v}_0,\mathbf{H}_0|_s + |\varrho_0,\mathbf{v}_0,\mathbf{H}_0|_{L^1} \leq \delta_2'$, then we have

$$
|\partial_x(\varrho,\mathbf{v},\mathbf{H})|_2 \leq C\delta_2'(1+t)^{-(\frac{5}{4}-\varepsilon_2)}, \forall 0 < \varepsilon_2 < \widehat{\varepsilon}
$$

for all $t \geq 0$, where $\widehat{\varepsilon} \leq \widetilde{\varepsilon}$ is some positive number.

In order to prove Proposition 3.6, we first set

$$\widetilde{M}^3_\sigma(t) = \sup_{0 \le \tau \le t} (1+\tau)^\sigma \left(\partial^2_x(\varrho, \mathbf{v}, \mathbf{H})(\tau)|_{L^2} + |\partial_\tau(\varrho, \mathbf{v}, \mathbf{H})(\tau)|_1 \right),$$

$$\widetilde{\mathcal{K}}^s_\sigma = \left\{ \sum_{k=3}^s M^k_\sigma + \sum_{j=1}^{[\frac{s}{2}]} \sup_{0 \le \tau \le t} (1+\tau)^{\frac{3}{4}-\varepsilon} |\partial^j_\tau \mathbf{v}(\tau)|_{s-2j} \right.$$
$$\left. + \sum_{j=1}^{[\frac{s+1}{2}]} \sup_{0 \le \tau \le t} (1+\tau)^{\frac{3}{4}-\varepsilon} |\partial^j_\tau \varrho(\tau)|_{s+1-2j} \right\}^{\frac{1}{2}},$$

and

$$\widetilde{\mathcal{L}}^s_\sigma = \left\{ \int_0^t (1+\tau)^{2\sigma} (|\partial_x \partial \mathbf{v}|_s + |\partial_3 \mathbf{v}|^2_{s-2} \right.$$
$$+ |\partial^2_x \varrho|_{s-2} + |\partial_x \partial_\tau (\varrho, \mathbf{v})|^2_{s-2}$$
$$\left. + \sum_{j=2}^{[\frac{s+1}{2}]} \int_0^t (1+\tau)^{2\sigma} |\partial^j_\tau(\mathbf{v}, \varrho)|_{s+1-2j}) d\tau \right\}^{\frac{1}{2}},$$

and

$$\widetilde{\mathcal{N}}^s_\sigma = \left\{ \sup_{0 \le \tau \le t} (1+\tau)^{2\sigma} |\partial^2_x S_1(\tau)|^2_{s-3} \right.$$
$$+ \sum_{j=1}^{[\frac{s-1}{2}]} \sup_{0 \le \tau \le t} (1+\tau)^{2\sigma} |\partial^j_\tau S_1(\tau)|^2_{s-1-2j}$$
$$+ \int_0^t (1+\tau)^{2\sigma} \left(|\partial_x(\varrho \mathrm{div} \mathbf{v})(\tau)|^2_{s-1} \right.$$
$$+ \sum_{j=1}^{[\frac{s}{2}]} |\partial^j_\tau (\varrho \mathrm{div} \mathbf{v})(\tau)|^2_{s-2j} \right) d\tau$$
$$+ \int_0^t (1+\tau)^{2\sigma} \left(|\partial^2_x(\mathbf{v} \cdot \nabla \varrho)(\tau)|^2_{s-3} \right.$$
$$+ \sum_{j=1}^{[\frac{s-1}{2}]} |\partial^j_\tau(\mathbf{v} \cdot \nabla \varrho)(\tau)|^2_{s-1-2j} \right) d\tau + (1+\tau)^{2\sigma} |\partial_x(S_2, S_3)(\tau)|^2_{s-3}$$
$$+ \sum_{j=1}^{[\frac{s-2}{2}]} \sup_{0 \le \tau \le t} (1+\tau)^{2\sigma} |\partial^j_\tau(S_2, S_3)(\tau)|^2_{s-2-2j}$$
$$+ \int_0^t (1+\tau)^{2\sigma} \left(|\partial_x S_2|^2_{s-2} + \sum_{j=1}^{[\frac{s-1}{2}]} |\partial^j_\tau (S_2, S_3)|^2_{s-1-2j} \right) d\tau$$
$$+ \sum_{0 \le k+2j \le s, \, (k,j) \neq (0,0)} \int_0^t (1+\tau)^{2\sigma} \left| \left\langle \partial^k \partial^j_\tau S, \partial^k \partial^j_\tau V \right\rangle \right| d\tau$$
$$+ \sum_{0 \le k+2j \le s-1, \, (k,j) \neq (0,0)} \int_0^t (1+\tau)^{2\sigma} \left| \left\langle \partial^k \partial^j_\tau S, \partial^k \partial^{j+1}_\tau V \right\rangle \right| d\tau$$
$$+ \sum_{0 \le k+l+2j \le s-1, \, (k,l,j) \neq (0,0,0)} \int_0^t (1+\tau)^{2\sigma} \gamma \mu \left| \left\langle \partial^k \partial^{l+1}_3 \partial^j_\tau \right. \right.$$
$$\left. \times (\mathbf{v} \cdot \nabla \varrho), \partial^k \partial^{l+1}_3 \partial^j_\tau \varrho \right\rangle \Big| d\tau \right\}^{\frac{1}{2}}.$$

As in the proof of Lemma 3.13 or we can see in (Kagei and Kobayashi 2005), the following inequality can be easily deduced:

Lemma 3.14. Let $s \ge 2$. Then under the assumptions of Proposition 3.5, the following inequalities hold uniformly in $t \ge 0$,

$$\widetilde{\mathcal{K}}^s_{\frac{5}{4}-\varepsilon_2}(t)^2 + \widetilde{\mathcal{L}}^s_{\frac{5}{4}-\varepsilon_2}(t)^2$$
$$\le C \left\{ \mathcal{K}^s_0(t)^2 + \mathcal{L}^s_0(t)^2 + \widetilde{\mathcal{N}}^s_{\frac{5}{4}-\varepsilon_2}(t)^2 \right.$$
$$\left. + \int_0^t (1+\tau)^{2(\frac{5}{4}-\varepsilon_2)-1} \left(|\partial_x(\varrho, \mathbf{v}, \mathbf{H})|^2_{L^2} + |\partial_\tau(\varrho, \mathbf{v}, \mathbf{H})|^2_{L^2} \right) d\tau \right\}. \tag{97}$$

And we estimate $\widetilde{\mathcal{N}}^s_{\frac{5}{4}-\varepsilon_2}(t)$ as:

Lemma 3.15. Let $s \ge 2$. Then under the assumptions of Proposition 3.5, the following inequalities hold uniformly in $t \ge 0$,

$$\widetilde{\mathcal{N}}^s_{\frac{5}{4}-\varepsilon_2}(t)^2 \le C\mathcal{K}^s_{\frac{3}{4}-\varepsilon_1}(t)^4 + \mathcal{K}^s_{\frac{3}{4}-\varepsilon_1}(t)^2 \mathcal{L}^s_{\frac{3}{4}-\varepsilon_1}(t)^2$$
$$+ C \left\{ M_\infty(t) + M^1_{\frac{5}{4}}(t) + \widetilde{M}^3_{\frac{5}{4}}(t) + \widetilde{\mathcal{K}}^s_{\frac{5}{4}-\varepsilon_2}(t) \right\}$$
$$\times \mathcal{L}^s_{\frac{3}{4}-\varepsilon_1}(t)^2. \tag{98}$$

Proof. Set $\widetilde{\mathcal{N}}^s_{\frac{5}{4}-\varepsilon_2}(t) = J_1 + J_2 + \cdots + J_{10}$. Here we only consider $s \ge 4$ since the case while $2 \le s < 4$ can be deduced more easily. Thus we will estimate J_1-J_{10} term by term.

$$J_1, J_2 \le C \sup_{0 \le \tau \le t} (1+\tau)^{2(\frac{5}{4}-\varepsilon_2)} \left(|\partial^3_x(\varrho \mathbf{v})|^2_{s-3} + \sum_{j=1}^{[\frac{s-1}{2}]} |\partial^j_\tau(\varrho \mathbf{v})|^2_{s-2j} \right)$$
$$\le C \sup_{0 \le \tau \le t} (1+\tau)^{2(\frac{5}{4}-\varepsilon_2)} \left(|\varrho, \mathbf{v}|^2_{L^\infty} |\varrho, \mathbf{v}|^2_s + \sum_{j=1}^{[\frac{s-1}{2}]} |\varrho, \mathbf{v}|^2_{L^\infty} |\partial^j_\tau(\varrho, \mathbf{v})|^2_{s-2j} \right.$$
$$\left. + \sum_{j=1}^{[\frac{s-1}{2}]} \sum_{k=1}^{j-1} |\partial^k_\tau \varrho, \partial^{j-k}_\tau \mathbf{v}|^2_{L^\infty} |\partial^k_\tau \varrho, \partial^{j-k}_\tau \mathbf{v}|^2_{s-2j} \right)$$
$$\le C \sup_{0 \le \tau \le t} (1+\tau)^{2(\frac{5}{4}-\varepsilon_2)} \left(|\varrho, \mathbf{v}|^4_s + |\varrho, \mathbf{v}|^2_s \sum_{j=1}^{[\frac{s-1}{2}]} |\partial^j_\tau(\varrho, \mathbf{v})|^2_{s-2j} \right.$$
$$\left. + \sum_{k=1}^{[\frac{s}{2}]-1} |\partial^k_\tau \varrho|^2_{s-2k-1} \sum_{l=1}^{[\frac{s-1}{2}]-1} |\partial^l_\tau \mathbf{v}|^2_{s-2l-1} \right)$$
$$\le C\mathcal{K}^s_{\frac{3}{4}-\varepsilon_1}(t)^4.$$

And this holds similarly for J_5, J_6. For the terms J_3, J_4 and J_7, we only estimate $\int_0^t (1+\tau)^{2(\frac{5}{4}-\varepsilon_2)} |\partial_x(\varrho \mathrm{div} \mathbf{v})|^2_{s-1} d\tau$

contained in J_3. The estimates of the other terms can arrive in the similar way.

$$\int_0^t (1+\tau)^{2(\frac{5}{4}-\varepsilon_2)} |\partial_x(\varrho \mathrm{div} \mathbf{v})|_{s-1}^2 d\tau$$

$$\leq C \int_0^t (1+\tau)^{2(\frac{5}{4}-\varepsilon_2)} |\varrho \mathrm{div} \mathbf{v}|_s^2 d\tau$$

$$\leq C \int_0^t (1+\tau)^{2(\frac{5}{4}-\varepsilon_2)} \left(|\varrho|_{L^\infty}^2 |\partial_x \mathbf{v}|_s^2 + |\varrho|_s^2 |\partial_x \mathbf{v}|_{L^\infty}^2 \right) d\tau$$

$$\leq C \int_0^t (1+\tau)^{2(\frac{5}{4}-\varepsilon_2)} |\varrho|_s^2 |\partial_x \mathbf{v}|_s^2 d\tau$$

$$\leq C \mathcal{K}_{\frac{3}{4}-\varepsilon_1}^s(t)^2 \mathcal{L}_{\frac{3}{4}-\varepsilon_1}^s(t)^2.$$

For J_8-J_{10}, we only estimate the term $\sum_{1\leq k\leq s-1} \int_0^t (1+\tau)^{2(\frac{5}{4}-\varepsilon_2)} |\partial_x^k S_3| |\partial_x^k \partial_\tau \mathbf{H}| d\tau$ as

$$\sum_{1\leq k\leq s-1} \int_0^t (1+\tau)^{2(\frac{5}{4}-\varepsilon_2)} |\partial_x^k S_3| |\partial_x^k \partial_\tau \mathbf{H}| d\tau$$

$$\leq C \sum_{1\leq k\leq s-1} \int_0^t (1+\tau)^{2(\frac{5}{4}-\varepsilon_2)} |\partial_x^{k+1}(\mathbf{v}\times\mathbf{H})| |\partial_x^k \partial_\tau \mathbf{H}| d\tau$$

$$\leq C \sum_{1\leq k\leq s-1} \int_0^t (1+\tau)^{2(\frac{5}{4}-\varepsilon_2)} \left(|\mathbf{v},\mathbf{H}|_{L^\infty} |\partial_x^{k+1}(\mathbf{v},\mathbf{H})|_{L^2} \right.$$

$$+ |\partial_x(\mathbf{v},\mathbf{H})|_{L^3} |\partial_x^k(\mathbf{v},\mathbf{H})|_{L^6}$$

$$\left. + \sum_{2\leq j\leq k-1} |\partial_x^j(\mathbf{v},\mathbf{H})|_{L^6} |\partial_x^{k+1-j}(\mathbf{v},\mathbf{H})|_{L^3} \right) |\partial_x^k \partial_\tau \mathbf{H}|_{L^2} d\tau$$

$$\leq C \sum_{1\leq k\leq s-1} \int_0^t (1+\tau)^{2(\frac{5}{4}-\varepsilon_2)} \left(|\mathbf{v},\mathbf{H}|_{L^\infty} + |\partial_x(\mathbf{v},\mathbf{H})|_{L^2} \right.$$

$$\left. + |\partial_x^2(\mathbf{v},\mathbf{H})|_{L^2} + \sum_{3\leq j\leq k} |\partial_x^j(\mathbf{v},\mathbf{H})|_{L^2} \right) |\partial_x^{k+1}(\mathbf{v},\mathbf{H})|_{L^2} |\partial_x^k \partial_\tau \mathbf{H}|_{L^2} d\tau$$

$$\leq C \left\{ M_\infty(t) + M_{\frac{1}{4}}^1(t) + \widetilde{M}_{\frac{3}{4}}^3(t) + \widetilde{\mathcal{K}}_{\frac{5}{4}-\varepsilon_2}^s(t) \right\} \mathcal{L}_{\frac{3}{4}-\varepsilon_1}^s(t)^2,$$

which finished the proof of Lemma 3.15. $\qquad\square$

By virtue of Lemma 3.14 and Lemma 3.15, it remains to estimate $\widetilde{M}_{\frac{5}{4}-\varepsilon_2}^3(t)$ to finish the proof of Proposition 3.6.

Lemma 3.16. Let $s \geq 4$. Then under the assumptions of Lemma 3.13, the following inequalities hold uniformly in $t \geq 0$,

$$\widetilde{M}_{\frac{3}{4}}^3(t)^2 \leq C \left\{ \mathcal{K}_{\frac{3}{4}-\varepsilon_1}^s(t)^2 + M_{\frac{3}{4}}^0(t)^2 + M_{\frac{5}{4}}^1(t)^2 + M_\infty(t)^2 \right\}. \tag{99}$$

Proof. By Lemma 3.8 $ii)$, we have

$$|\partial_x^l \partial_t I_0(t)|_{L^2} \leq C t^{-\frac{5}{4}} \left\{ |U_0|_{L^1} + |\varrho_0|_2 + |\mathbf{v}_0|_1 \right\},$$

for $t \geq 1$ and $l = 0, 1$. Now we shall estimate $\partial_x^l \partial_t I_1(t)$. Since

$$\partial_t I_1(t) = E(t-\tau)F(U,\mathbf{H})(\tau)|_{t-1}^t + \int_{t-1}^t E_t(t-\tau)F(U,\mathbf{H})(\tau)d\tau$$

$$= E(t-\tau)F(U,\mathbf{H})(\tau)|_{t-1}^t - \int_{t-1}^t E_\tau(t-\tau)F(U,\mathbf{H})(\tau)d\tau$$

$$= \int_{t-1}^t E(t-\tau)\partial_\tau F(U,\mathbf{H})(\tau)d\tau,$$

then by Lemma 3.9 and Lemma 3.12, we have

$$|\partial_x^l \partial_t I_1(t)|_{L^2} \leq C \int_{t-1}^t (t-\tau)^{-\frac{l}{2}} |\partial_\tau F(U,\mathbf{H})(\tau)|_{H^1\times L^2} d\tau$$

$$\leq C \int_{t-1}^t (t-\tau)^{-\frac{l}{2}} \{|\partial_t S_1|_1 + |\partial_t S_2|_{L^2}\} d\tau$$

$$\leq C(1+t)^{-\frac{5}{4}} \left\{ \mathcal{K}_{\frac{3}{4}-\varepsilon}^4(t)^2 + M_\infty(t)^2 + M_{\frac{1}{4}}^1(t)^2 \right\}.$$

Similarly, we can get

$$\partial_t I_2(t) = E(1)F(U,\mathbf{H})(t-1) + \int_0^{t-1} \partial_t E(t-\tau)\partial_\tau F(U,\mathbf{H})(\tau)d\tau.$$

Thus by Lemma 3.8, Lemma 3.11 and Lemma 3.12,

$$|\partial_x^l \partial_t I_2(t)|_{L^2}$$

$$\leq C|F(U,\mathbf{H})(t-1)|_1$$

$$+ C \int_0^{t-1} (t-\tau)^{-\frac{5}{4}} |F(U,\mathbf{H})(\tau)|_{L^1}$$

$$+ e^{-C_0(t-\tau)} |F(U,\mathbf{H})(\tau)|_{H^1\times L^2} d\tau$$

$$\leq C \int_{t-1}^t \tau^{-\frac{1}{2}} \left\{ |\partial_t S_1|_1 + |\partial_t S_2|_{L^2} \right\} d\tau$$

$$\leq C(1+t)^{-\frac{5}{4}} \left\{ M_\infty(t)^2 + M_{\frac{3}{4}}^0(t)^2 + M_{\frac{5}{4}}^1(t)^2 \right.$$

$$\left. + \mathcal{K}_{\frac{3}{4}-\varepsilon_1}^3(t)^2 + N(0,t)^2 \right\}.$$

It then follows that

$$|\partial_t U(t)|_1 \leq C(1+t)^{-\frac{5}{4}} \left\{ M_\infty(t)^2 + M_{\frac{3}{4}}^0(t)^2 + M_{\frac{5}{4}}^1(t)^2 \right.$$

$$\left. + \mathcal{K}_{\frac{3}{4}-\varepsilon_1}^4(t)^2 + N(0,t)^2 \right\}.$$

This together with the magnetic term which can be treated in the similar way, one has

$$|\partial_t(\varrho,\mathbf{v},\mathbf{H})|_1 \leq C(1+t)^{-\frac{5}{4}} \left\{ M_\infty(t)^2 + M_{\frac{3}{4}}^0(t)^2 + M_{\frac{5}{4}}^1(t)^2 \right.$$

$$\left. + \mathcal{K}_{\frac{3}{4}-\varepsilon_1}^4(t)^2 + N(0,t)^2 \right\}. \tag{100}$$

Next we estimate $|\partial_x^2 U(t)|_{L^2}$ and $|\partial \partial_x^2 \mathbf{v}|_{L^2}$. For $k = 1, 2$, we see from Lemma 3.8 that

$$|\partial^k \partial_x I_0(t)|_{L^2} \leq C t^{-\frac{7}{4}} \left\{ |U_0|_{L^1} + |\varrho_0|_{k+1} + |\mathbf{v}_0|_k \right\}.$$

And by Lemma 3.9 and Lemma 3.12,

$$|\partial^k \partial_x I_1(t)|_{L^2} \le C \int_{t-1}^t (t-\tau)^{-\frac{1}{2}} (|S_1|_{k+1} + |S_2|_k) d\tau$$

$$\le C \left\{ M_\infty(t)^2 + M_{\frac{1}{4}}^1(t)^2 + \mathcal{K}_{\frac{3}{4}-\varepsilon_1}^{k+2}(t)^2 \right\}$$

$$\times \int_{t-1}^t (t-\tau)^{-\frac{1}{2}} (1+\tau)^{-\frac{3}{2}} d\tau$$

$$\le C(1+t)^{-\frac{3}{2}} \left\{ M_\infty(t)^2 + M_{\frac{1}{4}}^1(t)^2 + \mathcal{K}_{\frac{3}{4}-\varepsilon_1}^{k+2}(t)^2 \right\}.$$

By Lemma 3.8 and Lemma 3.12,

$$|\partial^k \partial_x I_2(t)|_{L^2}$$

$$\le C \int_0^{t-1} (t-\tau)^{-\frac{5}{4}} |(S_1, S_2)(\tau)|_{L^1} d\tau$$

$$+ C \int_0^{t-1} e^{-C_0(t-\tau)} (|S_1(\tau)|_{k+1} + |S_2(\tau)|_k) d\tau$$

$$\le C \left\{ M_{\frac{3}{4}}^0(t)^2 + M_{\frac{1}{4}}^1(t)^2 + \mathcal{K}_{\frac{3}{4}-\varepsilon_1}^{k+2}(t)^2 \right\}$$

$$\times \int_0^{t-1} (t-\tau)^{-\frac{5}{4}} (1+\tau)^{-(\frac{7}{4}-\varepsilon_1)} d\tau$$

$$+ C \left\{ M_\infty(t)^2 + M_{\frac{1}{4}}^1(t)^2 + \mathcal{K}_{\frac{3}{4}-\varepsilon_1}^{k+2}(t)^2 \right\}$$

$$\times \int_0^{t-1} e^{-C_0(t-\tau)} (1+\tau)^{-\frac{3}{2}} d\tau$$

$$\le C(1+t)^{-\frac{5}{4}} \left\{ M_\infty(t)^2 + M_{\frac{3}{4}}^0(t)^2 + M_{\frac{1}{4}}^1(t)^2 + \mathcal{K}_{\frac{3}{4}-\varepsilon_1}^{k+2}(t)^2 \right\}.$$

Combining these with the magnetic term which has the similar estimates, it yields

$$|\partial^k \partial_x(\varrho, \mathbf{v}, \mathbf{H})|_{L^2} \le C(1+t)^{-\frac{5}{4}} \left\{ \delta_1' + M_\infty(t)^2 \right.$$

$$\left. + M_{\frac{3}{4}}^0(t)^2 + M_{\frac{1}{4}}^1(t)^2 + \mathcal{K}_{\frac{3}{4}-\varepsilon_1}^{k+2}(t)^2 \right\}$$

$$(101)$$

for $k = 1, 2$.

Next, we consider $\partial_3^2(\varrho, \mathbf{v})$. From the equation $(19)_2$, we have

$$2\mu \partial_3^2 \mathbf{v} = \mathbf{v}_t + \gamma \nabla \varrho - \mu(\partial \cdot \partial)\mathbf{v} - \mu \partial \operatorname{div} \mathbf{v}$$

$$- \mu \partial_3(\partial_1 \mathbf{v}^1 + \partial_2 \mathbf{v}^2) - S_2.$$

Thus by (100) and (101), it follows that

$$|\partial^l \partial_3^2 \mathbf{v}|_{L^2} \le C \left(|\partial^l \mathbf{v}_t|_{L^2} + |\partial^l \partial_x \varrho|_{L^2} + |\partial^{l+1} \partial_x \mathbf{v}|_{L^2} + |\partial^l S_2|_{L^2} \right)$$

$$\le C(1+t)^{-\frac{5}{4}} \left\{ M_\infty(t)^2 + M_{\frac{3}{4}}^0(t)^2 + M_{\frac{1}{4}}^1(t)^2 \right.$$

$$\left. + \mathcal{K}_{\frac{3}{4}-\varepsilon_1}^4(t)^2 + \delta_1' \right\}.$$

$$(102)$$

Similarly by $(19)_1$–$(19)_2$, we have

$$\partial_3 \varrho_t + \frac{\gamma^2}{2\mu} \partial_3 \varrho = \frac{\gamma}{2\mu} \mathbf{v}_t^3 - \frac{\gamma}{2} \partial_3(\partial_1 \mathbf{v}^1 + \partial_2 \mathbf{v}^2)$$

$$+ \frac{\gamma}{2}(\partial \cdot \partial)\mathbf{v}^3 + \partial_3 S_1 + \frac{\gamma}{2\mu} S_2.$$

Differentiating this in x_3, multiplying $\partial_3^2 \varrho$ to the result equation, and integrating on $\mathbb{R}_+^3$, we have

$$\frac{d}{dt} |\partial_3 \varrho|_{L^2} + \frac{\gamma}{\mu} |\partial_3 \varrho|_{L^2} \le C(|\partial_x \mathbf{v}_t, \partial \partial_x^2 \mathbf{v}, \partial_x^2 S_1, \partial_x S_2|_{L^2}),$$

Thus by Gronwall's inequality, and with the help of Lemma 3.12, and (87)–(89), we obtain

$$|\partial_3^2 \varrho|_{L^2}$$

$$\le e^{-\frac{\gamma}{4\mu}(t-1)} |\partial_3^2 \varrho(1)|_{L^2} + C \int_1^t e^{-\frac{\gamma}{2\mu}(t-\tau)}$$

$$\times \left(|\partial_x \mathbf{v}_t, \partial \partial_x^2 \mathbf{v}, \partial_x^2 S_1, \partial_x S_2|_{L^2} \right) d\tau$$

$$\le CN(0,t) e^{-\frac{\gamma}{4\mu}(t-1)}$$

$$+ C \left\{ M_\infty(t)^2 + M_{\frac{3}{4}}^0(t)^2 + M_{\frac{1}{4}}^1(t)^2 + \mathcal{K}_{\frac{3}{4}-\varepsilon_1}^4(t)^2 + \delta_1' \right\}$$

$$\times \int_1^t e^{-\frac{\gamma}{2\mu}(t-\tau)} (1+\tau)^{-\frac{5}{4}} d\tau$$

$$\le C(1+t)^{-\frac{5}{4}} \left\{ M_\infty(t)^2 + M_{\frac{3}{4}}^0(t)^2 + M_{\frac{1}{4}}^1(t)^2 \right.$$

$$\left. + \mathcal{K}_{\frac{3}{4}-\varepsilon_1}^4(t)^2 + \delta_1' \right\},$$

this together with (101) and (102), we get the estimate of $|\partial_x^2(\varrho, \mathbf{v})|_{L^2}$. Hence it remains to estimate $|\partial_x^2 \mathbf{H}|_{L^2}$. By Lemma 2.4 and (100), we have

$$|\partial_x^2 \mathbf{H}|_{L^2} \le C(|\mathbf{H}_t|_{L^2} + |S_3|_{L^2} + |\partial_x \mathbf{H}|_{L^2})$$

$$\le C(1+t)^{-\frac{5}{4}} \left\{ M_\infty(t)^2 + M_{\frac{3}{4}}^0(t)^2 + M_{\frac{1}{4}}^1(t)^2 \right.$$

$$\left. + \mathcal{K}_{\frac{3}{4}-\varepsilon_1}^4(t)^2 + \eta_1^2 \right\}.$$

$$(103)$$

Combining this with the above estimates, Lemma 3.16 follows. □

As in (103), we can deduce the decay rates of higher order derivatives for the magnetic field by Lemma 3.8, Lemma 3.16 and (100). Thus we state the result which is better than one in Proposition 3.6 in the following:

Proposition 3.7. Under the assumptions of Proposition 3.6, the solution $\mathbf{H}$ of $(19)_3$ satisfies that for all $t \ge 0$,

$$|\partial_x^3 \mathbf{H}(t)|_{L^2} \le C(1+t)^{-\frac{5}{4}}.$$

In view of the above established, the proof of Theorem 2 is complete.

□

Competing interests

The authors declare that they have no competing interests.

Authors' contributions

Author QC composed this paper, and others have revised it many times for publication. Both authors read and approved the final manuscript.

Acknowledgements

Q. Chen's research is supported in part by National Natural Science Foundation of China-NSAF (Nos. 11226174, 11301439). Z. Tan's research is supported in part by National Natural Science Foundation of China-NSAF (No. 11271305).

Author details

[1] School of Applied Mathematics, Xiamen University of Technology, Ligong Road, 361024 Xiamen, China. [2] School of Mathematical Sciences, Xiamen University, Siming South Road, 361005 Xiamen, China.

References

Chen Q, Tan Z (2010) Global existence and convergence rates of smooth solutions for the compressible magnetohydrodynamic equations. Nonlinear Anal Theory, Methods Appl 72:4438–4451

Chen Q, Tan Z (2012) Global existence in critical spaces for the compressible magnetohydrodynamic equations. Kinet Relat Models 5:743–787

Chen GQ, Wang DH (2002) Global solutions of nonlinear magneto-hydrodynamics with large initial data. J Differential Equations 182:344–376

Chen GQ, Wang DH (2003) Existence and continuous dependence of large solutions for the magneto-hydrodynamics equations. Z Angew Math Phys 54:608–632

Cho Y, Choe HJ, Kim H (2004) Unique solvability of the initial boundary value problems for compressible viscous fluids. J Math Pures Appl 83:243–275

Ducomet B, Feireisl E (2006) The equations of magneto-hydrodynamics: on the interaction between matter and radiation in the evolution of gaseous stars. Commun Math Phys 266:595–629

Fan JS, Yu W (2009) Strong solution to the compressible magneto-hydrodynamic equations with vacuum. Nonlinear Anal Real World Appl 10:392–409

Galdi GP, Bris CL, Lelievre T (1994) An introduction to the mathematical theory of the Navier-Stokes equations. Springer-Verlag, New York

Gerebeau JF, Bris CL, Lelievre T (2006) Mathematical methods for the magneto-hydrodynamics of liquid metals. Oxford University Press, Oxford

Kagei Y, Kobayashi T (2005) Asymptotic behavior of solutions of the compressible Navier-Stokes equations on the half space. Arch Ration Mech Anal 177:231–330

Kawashima S, Okada M (1982) Smooth global solutions for the one-dimensional equations in magneto-hydrodynamics. Proc Japan Acad Ser A Math Sci 58:384–387

Kobayashi T (2002) Some estimates of solutions for the equations of motion of compressible viscous fluid in an exterior domain in $\mathbb{R}^3$. J Differential Equations 184:587–619

Kobayashi T, Shibata Y (1999) Decay estimates of solutions for the equations of motion of compressible viscous and heat-conductive gases in an exterior domain in $\mathbb{R}^3$. Comm Math Phys 200:621–659

Matsumura A, Nishida T (1979) The initial value problem for the equations of motion of compressible viscous and heatconductive fluids. Proc Japan Acad Ser A 55:337–342

Matsumura A, Nishida T (1980) The initial value problem for the equations of motion of viscous and heat-conductive gases. J Math Kyoto Univ 20:67–104

Matsumura A, Nishida T (1983) Initial boundary value problems for the equations of motion of compressible viscous and heat-conductive fluids. Commun Math Phys 89:445–464

Ströhmer S (1990) About compressible viscous fluid flow in a bounded regione. Pacific J Math 143:359–375

Tan Z, Wang YJ (2009) Global existence and large-time behavior of weak solutions to the compressible magneto-hydrodynamic equations with coulomb force. Nonlinear Anal Theory, Methods Appl 71:5866–5884

Vol'pert AI, Hudjaev SI (1972) On the Cauchy problem for composite systems of nonlinear differential equations. Math USSR-Sb 16:517–544

3

Multiple land use change simulation with Monte Carlo approach and CA-ANN model, a case study in Shenzhen, China

Tianhong Li[1,2*] and Wenkai Li[3]

Abstract

Background: CA-ANN models which integrate Cellular Automata (CA) and Artificial Neural Networks (ANNs) for simulating land use change, usually urban and non-urban, predict the final land use type of a cell by the greatest similarity or probability after model parameters were defined in the training stage. In this study, the Monte Carlo approach was introduced into a CA-ANN model to simulate multiple land use changes with a case study in Shenzhen, China. The final land use type of a cell was jointly determined by the Monte Carlo approach and artificial neural network.

Results: The model performance were evaluated based on cell-to-cell comparison between simulated maps and actual ones by overall accuracy and kappa coefficient. The input maps of 1996, 2000 and 2004 were combined into three scenarios, the overall accuracies and kappa coefficients were all greater than 81.91% and 0.71 respectively. The land use maps of from 2004 to 2020 with 4 years interval were simulated and the results showed that build up will increase steadily while woodland will decrease. The impacts of spatial variables, neighborhood size and cell size on model performance were obtained by sensitive analysis.

Conclusions: The simulation performance were all acceptable compared with the existing studies. The model performance would increase slightly as either neighborhood size or cell size increased, and that proximities to railways and city center were the main factors driving the dynamics of land use change in the study area.

Keywords: Land use change; Cellular automata; Artificial neural network; Monte Carlo; Simulation

Background

Since the first theoretical approaches to CA-based models for simulating urban sprawl were proposed in 1980s and the first operational urban CA models were applied to real-world urban system in 1990s, lots of studies have been conducted to modify CA structure so as to address complex urban systems and Land use cover changes (LUCC) (Almeida et al. 2008; Batty 1997; Gong et al. 2009; Stevens et al. 2007; Torrens and O'Sullivan 2001; Verburg et al. 2004; Wu 1996, 1998a, b; Yang et al. 2008). The advantages of CA-based models including their simplicity, flexibility and intuitiveness and particularly their ability to incorporate the spatial and temporal dimensions of the

processes have been agreed and the studies in this issue are still proliferating.

Temporal and spatial complexities of urban land use change can be well modeled by properly defining transition rules in CA models. To obtain appropriate transition rules, besides the conventional statistical methods (De Almeida and Gleriani 2005; He et al. 2008; Wu 2002), various approaches have also been taken including fuzzy logic (Al-Ahmadi et al. 2009; Al-kheder et al. 2008; Liu and Phinn 2003; Wu 1996 and Wu 1998a), support vector machine (Yang et al. 2008) and Artificial Neural Networks (ANNs) especially Back Propagation (BP) networks (Li and Yeh 2001 and Li and Yeh 2002; Li and Liu 2006; Liu et al. 2008a; Liu et al. 2008b; Pijanowski et al. 2002; Yeh and Li 2003 and Yeh and Li 2004; Wang et al. 2011).

In the most previous studies, land use types of urban and non-urban were simulated (He et al. 2008; Li and

* Correspondence: lth@pku.edu.cn
[1]College of Environmental Sciences and Engineering, The Key Laboratory for Water and Sediment Sciences, Peking University, Beijing 100871, China
[2]Shenzhen Graduate School of Peking University, Shenzhen 518055, China
Full list of author information is available at the end of the article

Yeh 2001; Li and Liu 2006; Liu et al. 2008a and Liu et al. 2008b; Pijanowski et al. 2002; Wu 2002; Yeh and Li 2003). In recent years, city residents pay more and more attention to their living environment, therefore, in simulation, other land use types especially those with high ecological services such as forest, water body and wetland (Li et al. 2010) should not be generalized as non-urban any more. However, less than 40% among the 33 CA-based urban models reviewed by Santé et al. (2010) were capable of dealing with multiple land uses. On the other hand, land use change results from the complex interaction of many factors including policy, management, economics, culture, human behaviors and the environment. Whether a cell changes its state is not only determined by transition probability but also bears randomness. In this view, transition rules can include a stochastic component and deterministic rules (Santé et al. 2010). The most previous studies considered the deterministic rules by maximum probability. A few of them treated the stochastic component, for instance, Li and Yeh (2002) added a random variable into the probability function in ANN-CA model, Wu (2002) combined Monte Carlo approach with probability generalized by statistical method in urban and non-urban simulations.

With Shenzhen City, a typical rapid urbanization area in China or even in the whole world, as the case, this study aims to simulate multiple land use dynamics by combining Monte Carlo approach to CA-ANN model, and to choose appropriate model parameters with sensitivity analysis on predicating variables, neighborhood size and cell size.

Results and discussions
Model performance
The simulation accuracy was evaluated based on cell-by-cell comparison which was typically used (Li and Yeh 2002; Yang et al. 2008). It evaluates the similarity between the actual and simulated situations at the scale of a single cell. The results of the cell-by-cell comparison with the test cells for three scenarios were given in Table 1.

For Scenarios I, II and III, the overall accuracy is 81.91%, 83.13% and 85.33% respectively. The kappa

coefficients were calculated to quantify the actual degree of agreement (Congalton 1991). The coefficient was 0.71, 0.73 and 0.76 for Scenarios I, II and III in the same order. The similar accuracies for the two periods suggested that the land use change mechanism is relatively stable in this region and that the models trained by different senarios can be used to forecast future land use changes. Yang et al. (2008) used SVM and CA to simulate urban and non-urban land use in Shenzhen with different data sets (1993 and 2004) and reported that the overall accuracy were 87.25% and 84.90% and kappa coefficients were 0.70 and 0.68. Though their overall accuracies are a little higher than our results, but their kappa coefficients are all lower than those of the proposed model.

The actual maps and simulated maps are shown in Figures 1 and 2. Generally, the simulated maps are spatially conformity to the actual maps. The simulated maps in Figure 2 showed that Build-up and Woodland dominated the study area with their areal percentage being both around 44%, while the Water body, Cropland and Wetland occupied around 8%, 2% and 2% of the whole study area. They are generally consistent with the real land use maps. In the Figure 2 (I), the kappa coefficients of the Wetland, Water body and Woodland were all great than 0.70, suggesting these types of land use have been well simulated. In Figure 2 (II), the kappa coefficients of Woodland, Water body and Build-up were all great than 0.70. In Figure 2 (III), the coefficients of Cropland is 0.46 while those of the other four land uses were all greater than 0.70. On the other hand, the random land use type in Figure 2 (II) and Figure 2 (III) have decreased compared to Figure 2 (I).

Sensitivity analysis
With the model trained by Scenarios III which produced relatively best performance, the impacts of spatial variables, neighborhood size and cell size were discussed.

Spatial variables
The input layer of BP network contained fifteen neurons associated with the variables, while the output layer contained five neurons associated with the map of transition

Table 1 Class kappa coefficients and overall accuracy for each scenarios

Scenario*	Woodland	Cropland	Wetland	Water body	Build-up	Overall Kappa	Overall accuracy (%)
I	0.72	0.49	0.87	0.79	0.68	0.71	81.91
II	0.76	0.33	0.55	0.71	0.77	0.73	83.13
III	0.79	0.46	0.76	0.78	0.74	0.76	85.33

*Scenario I: training samples from 1996 and 2000 land use maps are input and output of the model respectively, the map of 1996 was input to the model and the map of 2000 was predicted and evaluated with the actual map of 2000 using testing samples.
Scenario II: training samples from 1996 and 2000 land use maps are input and output of the model respectively, the map of 2000 was input to the model and the map of 2004 was predicted and evaluated with the actual map of 2004 using testing samples.
Scenario III: training samples from 2000 and 2004 land use maps are input and output of the model respectively, the map of 2000 was input to the model and the map of 2004 was predicted and evaluated with the actual map of 2004 using testing samples.

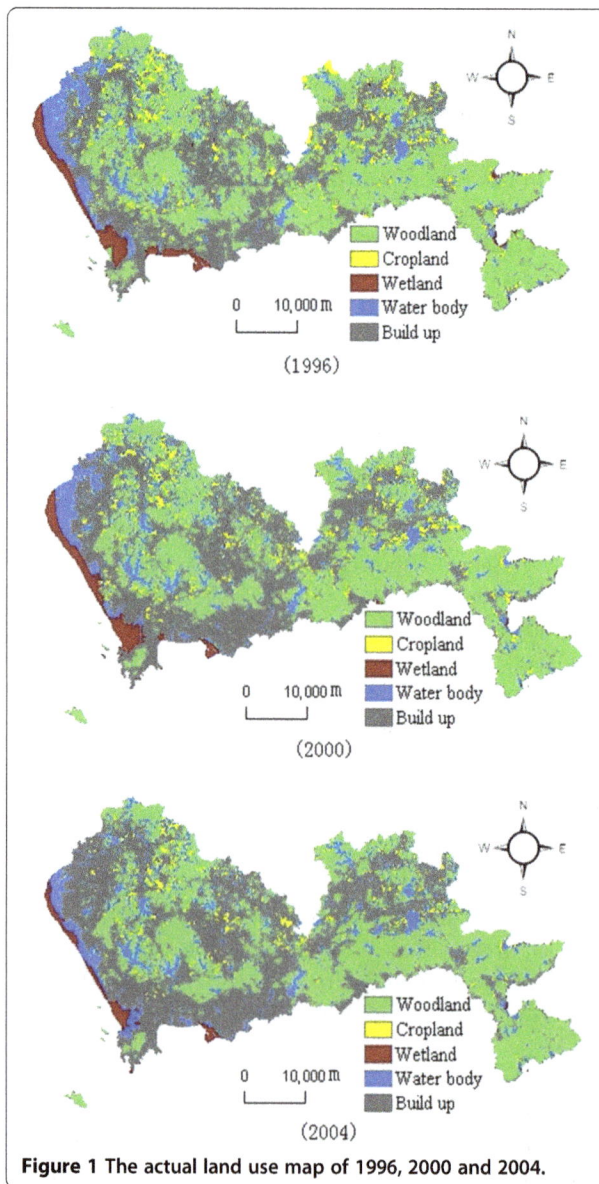

Figure 1 The actual land use map of 1996, 2000 and 2004.

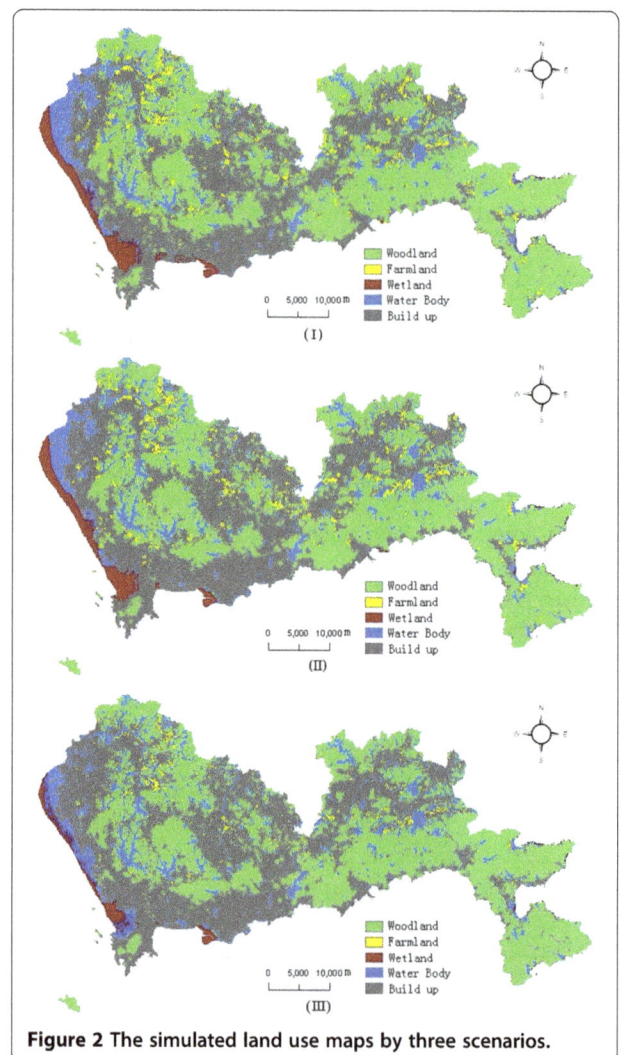

Figure 2 The simulated land use maps by three scenarios.

probabilities for the considered types of land use (Table 2). In order to investigate the effects of spatial variables on model performance, we compared the model fitness of the nine versions of the reduced-variable model with that with the full predictor variables. In Table 3, row "p" represented the model performance with complete predictor variables, while row "pi = 0" represented that with variable "pi" removed. The relative contribution of each predictor variable on the model performance was also ranked according to their differences between goodness of fit for the model with full variables and that for each of the reduced-variable models.

According to Table 3, the overall accuracy and kappa coefficients with complete predictor variables were higher than those with removed variables, indicating that

the variables selected in this study had positive effect on the model performance. The contribution of each predictor variable on the model performance was ranked as: p7 (Proximity to railways) > p13 (Proximity to city center) > p9 (Proximity to roads) > p14 (Proximity to towns) > p10 (Proximity to streets) > p8 (Proximity to expressways) > p12 (Proximity to lakes or reservoirs) > p15 (population density) > p11 (Proximity to rivers). This result showed that Proximities to railways and city center were the main factors which have driven the dynamics of land use. This result was also approved in Shi's study (Shi et al. 2000).

Neighborhood size

Neighborhood interactions represent one of the main driving factors in a large group of land use change models based on cellular automata (Verburg et al. 2004). In many studies, the size of the neighborhood was

Table 2 Input and output of the BP network

P (input)		T (output)	
p1	States of the cell (e.g. 1, 2, 3, 4, 5)	t1	Transition probability for woodland
P2	Cell number of woodland in neighborhood		
P3	Cell number of cropland in neighborhood		
P4	Cell number of wetland in neighborhood	t2	Transition probability for cropland
P5	Cell number of water body in neighborhood		
P6	Cell number of build-up in neighborhood		
P7	Proximity to railways	t3	Transition probability for wetland
P8	Proximity to expressways		
P9	Proximity to roads		
p10	Proximity to streets	t4	Transition probability for water body
p11	Proximity to rivers		
p12	Proximity to lakes or reservoirs		
p13	Proximity to city center	t5	Transition probability for build-up
p14	Proximity to towns		
p15	Population density		

chosen arbitrarily and only the direct neighborhood of a location was taken into account (e.g. Von Neumann or Moore neighborhoods), while others have argued that human activities are influenced by wider spaces, this makes an appropriate definition of the neighborhood essential (Verburg et al. 2004; White and Engelen 2000). In this study, three sizes of Moore neighborhoods, 3 × 3, 5 × 5, and 7 × 7, were used to understand the effects of neighborhood size on the model performance.

The simulation map of land use in 2004 was cell-to-cell compared to the actual map to evaluate the fitness of models with different neighborhood sizes. According to Table 4, the overall accuracy and Kappa coefficients increased slightly as the neighborhood size increased. However, this increase is too slight to make a significant impact. Considering the computing time cost, the smaller neighborhood size can be used.

Table 3 Effects of spatial variables on model performance

Spatial variables	Overall accuracy (%)	kappa	Rank
p	0.8533	0.7576	
p7 = 0	0.7896	0.6614	1
p8 = 0	0.8467	0.7471	6
p9 = 0	0.8338	0.7229	3
p10 = 0	0.8416	0.7364	5
p11 = 0	0.8512	0.7561	9
p12 = 0	0.8496	0.7513	7
p13 = 0	0.8226	0.7098	2
p14 = 0	0.836	0.7279	4
p15 = 0	0.8515	0.7541	8

Cell size

The spatial scale is also an important factor which should be taken into account in the modeling of spatial problems. Thus, the fitness of models for different cell sizes (100 m × 100 m, 250 m × 250 m, and 500 m × 500 m) was evaluated based on cell-to-cell comparison as shown in Table 5.

In Table 5, the individual kappa coefficient for each type of land use increased as the cell size increased, and so did the overall accuracy and kappa coefficients, which indicated that increasing the cell size may slightly improve the fitness of the model. However, it was not favorite for guiding land use planning if the cell size was too large. As a result, a tradeoff was made between model performance and high spatial resolution output maps in actual simulations. In the following simulation, the cell size of 100 m × 100 m was adopted because it will produce more detailed land use maps without apparently lowering predicting accuracy.

Land use simulation

With the map of land use in 2000 used as the input map and with the model parameters selected by the previous section (cell size = 100 m × 100 m, neighborhood size = 3 × 3, 9 GIS layers of predictor variables or constraints),

Table 4 Fitness of models with different neighborhood sizes

Size of neighborhood	Overall accuracy (%)	Kappa coefficients
3 × 3	85.31	0.7572
5 × 5	85.33	0.7576
7 × 7	85.34	0.7578

Table 5 Simulation accuracies with different cell sizes

Cell size	Individual Kappa coefficient					Kappa coefficient	Overall accuracy
	Woodland	Cropland	Wetland	Water body	Build-up		
100 m	0.79	0.46	0.76	0.78	0.74	0.7576	85.33
250 m	0.78	0.46	0.79	0.81	0.74	0.7590	85.39
500 m	0.82	0.54	0.84	0.82	0.75	0.7830	86.72

the tested model (Scenario II) ran for 5 cycles to simulate the land use changes from 2004 to 2020 in Shenzhen in every four years. The areal changing trends of the five land use types is shown in Figure 3.

Figure 3 show that build-up will become the dominating land use type in 2020, its area exceeding any of the other land use types. Decrease of farmland was the most fast, with the rate of 5.52% per year, followed by wetland with the decreasing rate of 2.59% per year. This result is consistent with the results of a prior study in Shenzhen (Bai 2000), in which a System Dynamics (SD) approach was used and predicted that Shenzhen would experience rapid urban growth from 1980 to 2030 and that areas of build-up would increase significantly while cropland would decline to almost zero.

Conclusions

The Monte Carlo approach was introduced to combine with transition probabilities generalized by ANN to decide the states of cells. The modified ANN-CA model proved capable of simulating multiple land use changes in Shenzhen with good performance. The variables selected in this study had positive effects on the model performance, among which the proximities to transportation and to city center were recognized as the major factors which drove the dynamics of land use. Moreover, model performance increased slightly as either neighborhood size or cell size increased. However, a compromise should be made between model performance, provision of spatial information details and computing time. The cell size of 100 m × 100 m and 3 × 3 neighborhood size was chosen for the simulation of land use changes in the study area and land use maps from 2008 to 2020 with 4 year interval were forecasted and showed that the areas of woodland, cropland, wetland and water body tend to decline while the area of built-up tend to increase through 2004 to 2020. These results can be referenced by local decision makers in pursuing ecological and human-centered urbanization.

Methods

Cellular automata in land use simulation

A cellular automata system consists of a regular grid of cells, each of which can be in one of a finite number of possible states, updated synchronously in discrete time steps according to a local, identical interaction rule. The

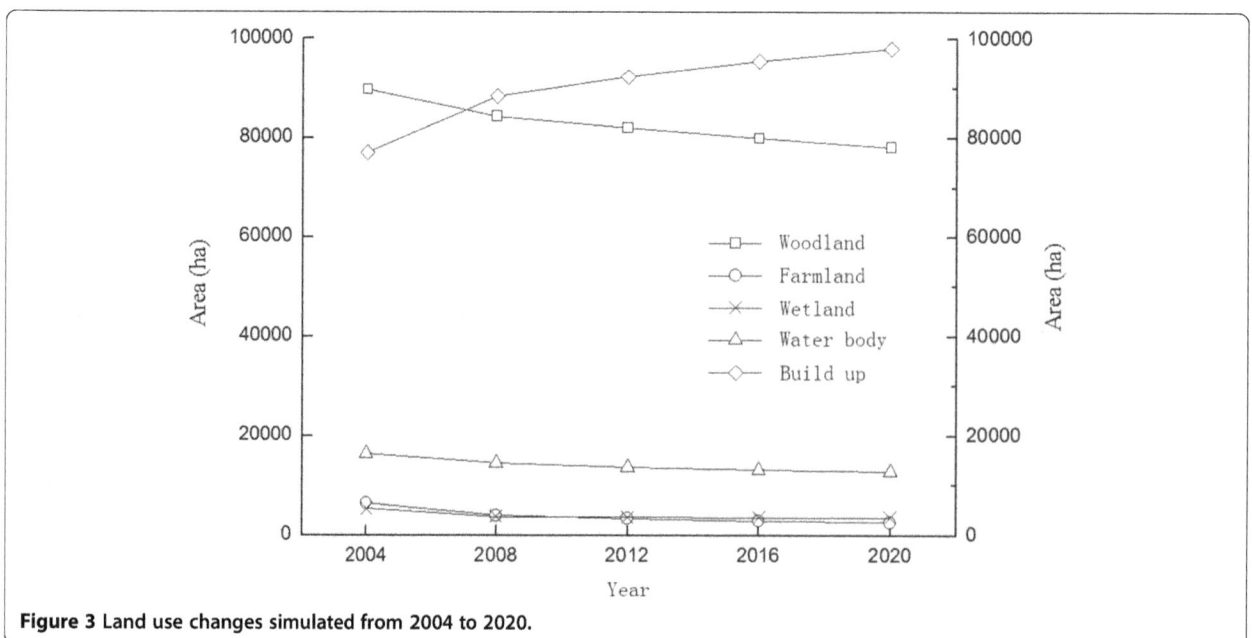

Figure 3 Land use changes simulated from 2004 to 2020.

state of a cell is determined by the previous states of a surrounding neighborhood of cells (Wolfram 1984). The principle of cellular automata can be illustrated by Eq. 1:

$$S_{t+1} = f(S_t, N) \tag{1}$$

Where S_t and S_{t+1} represents one of the finite number of possible states at time t and time $(t + 1)$ respectively, N represents the combined states of the neighborhood, and f represents the transition rules.

A basic CA model consists of five fundamental elements: lattice, cell states, neighborhood, transition rules, and time steps. Because of the complexity of urban systems, the framework of basic CA is too simplified and constrained to represent real cities. To simulate complex geographical phenomena such as urban systems, it is necessary do some modification to the basic CA model, especially to modify the elicitation process of transition rules (Santé et al. 2010; Torrens 2000).

The lattice is the space where all of the cells are located. Theoretically, the lattice can be the Euclidean space with any dimension and any geometric shape though the lattice of rectangular grids is widely used.

The cell is the minimum unit in the CA system and it is regularly located on the Euclidean space, with discrete and finite states. To simulate the dynamics of multiple land use types, the state "s" can be defined as s = {1, 2, 3,..., n}, representing n types of land use.

There are three common types of neighborhoods in conventional CA models: Von Neumann, Moore, and Circular neighborhoods (Kocabas and Dragicevic 2006; Liu and Phinn 2003).

The transition rules can be defined as the theoretical functions which determine the state of each cell according to the previous states of its own and its neighbors. Since changes of land use may result from the complex interaction of many factors, including policy, management, economics, culture, human behavior, and the environment (Pijanowski et al. 2002), the classic CA transition rules should be modified. However, it is not possible to recognize and incorporate all of the factors which drive land use changes. In practice, therefore, variables were usually chosen with the help of literature review and with considering the data availability.

ANN-based CA model

When ANNs were integrated to CA models, their predictive abilities can be improved (De Almeida and Gleriani 2005; Li and Yeh 2001; Pijanowski et al. 2002). In this study, the widely used CA-ANN model is modified in three aspects, namely, the multiple land use maps can be input and predicted, a threshold parameter is introduced to make the model controllable and the Monte Carlo approach is introduced to determine the final state of the

cell. The learning aspect was based on a BP network involving training and testing with different datasets. Figure 4 illustrates the framework of the proposed model.

Monte Carlo

Since the process of land use change in the real world is complex and stochastic, the deterministic transition rules in conventional CA models were modified to probability functions (Eq. 2), and the converted states of each cell in the next step were obtained using the Monte Carlo approach (Eq. 3).

$$S_i^{t+1} = \left\{ p_1, p_2, ..., p_j, ..., p_n \right\} \tag{2}$$

$$S'^{t+1}_i = MonteCarlo\left(S_i^{t+1}\right) \tag{3}$$

p_j is the probabilities that a certain cell would transition to jth land use type. A random number θ between 0 and 1 was obtained in the Monte Carlo approach.

For the state of a cell can be determined by the following rules, given $p_i > up$, i $\epsilon\{1,2,3,..., n\}$.

$$S'^{t+1}_i = MonteCarlo\left(S_i^{t+1}\right) = 1 \, \text{if} \, 0 {\leq} \theta < p_1 \tag{4}$$

$$S'^{t+1}_i = MonteCarlo\left(S_i^{t+1}\right) = j \, \text{if} \, \sum\nolimits_{k=1}^{j-1} p_k {\leq} \theta < \sum\nolimits_{k=1}^{j} p_k \tag{5}$$

When combining Monte Carlo approach with transition potentials generalized by ANN, this study introduced a parameter "up" to make the model controllable. This parameter controls the number of cells on which Monte Carlo approach will be used. For all the land use types, if the transition probability generalized by BP network to one land use type of a cell is smaller than "up", then the state of the cell would remain the same in the iterative cycle. Otherwise, the state of the cell was then determined by Monte Carlo approach. Thus, if "up" is set zero, all of the cells will be treated according to the rules of combination of Monte Carlo approach and transition probabilities. If "up" is set one, none of the cells will be converted. This process was iterated and land use map was predicted. Since only those cells with transition probabilities larger than "up" would be treated by the Monte Carlo approach and transition probabilities, as the value of "up" increases, the effect of Monte Carlo approach will wane, the output land use map will have less insular land use patches mainly near the boundary between land use types. Theoretically, if a cell has the same possibilities of changing into (or remaining unchanged) other land use types, the probability value should be the reciprocal of the number of land use types. To compromise, a value around the reciprocal of the number of land use types is recommended for "up".

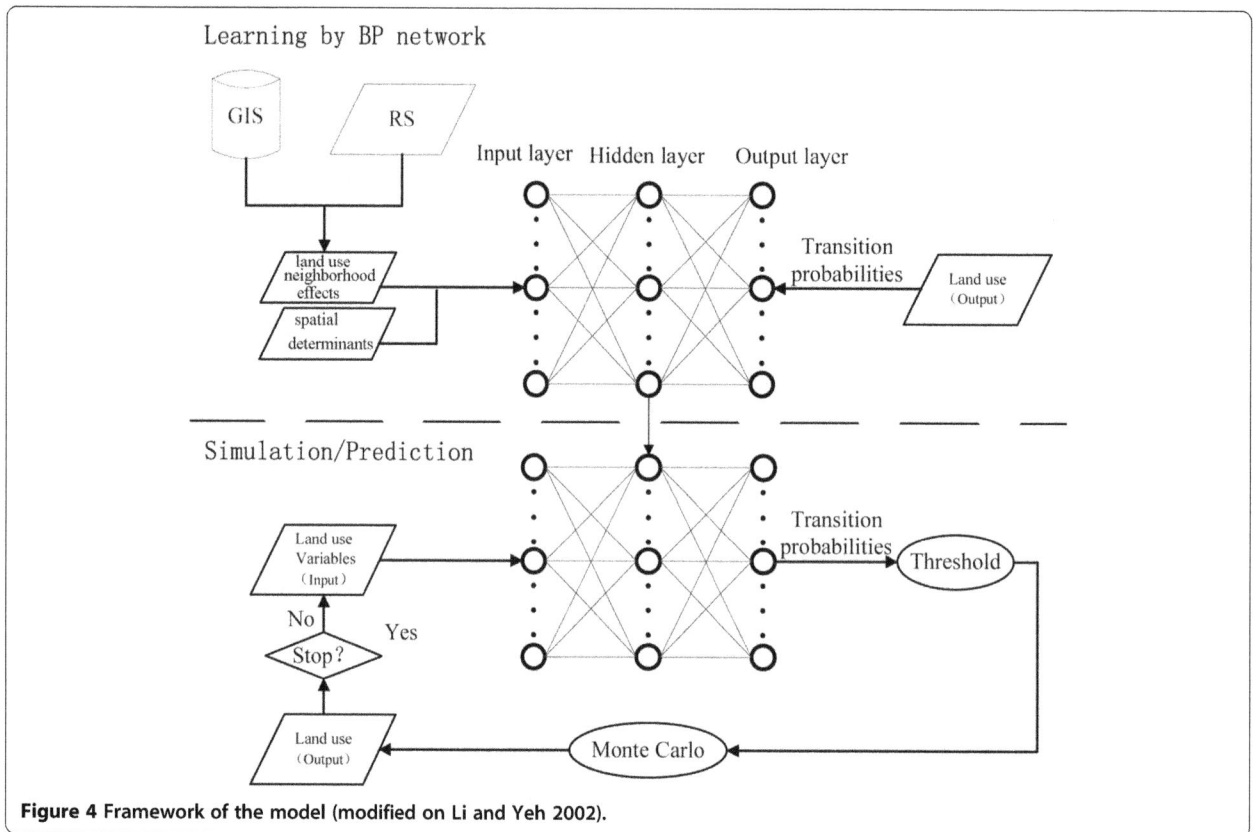

Figure 4 Framework of the model (modified on Li and Yeh 2002).

The exact value of "*up*" can be set by trial simulations for a certain study area.

CA will evolve in a sequence of discrete time steps. The real time of each step can be calibrated by historical data. Suppose that the periods of input and output data for CA training be T_1 and T_2, respectively. The interval "*N*" between two consecutive time steps can be calculated: $N = T_2 - T_1$. Thus, the sequent time of each step can be calculated: $T_3 = T_2 + N$, $T_4 = T_3 + N$, $T_5 = T_4 + N$....After the structure, parameter, and transition rules are defined, the CA model will evolve dynamically.

The proposed model was implemented through the programming language of Visual Basic (VB) 6.0 and the technology of Component of Object Model (COM) on the platform MATLAB 6.5.

The study area

Shenzhen (22°27′ N ~22°52′ N and 113°46′ E ~ 114°37′ E) is a coastal city in the Pearl River Delta Region, southern China and locates directly north of Hong Kong (Figure 5). The total terrestrial area of Shenzhen is 0.19 million ha, and the coastline of the city totals 229.96 kilometers. Shenzhen has a mild, subtropical maritime climate with plenty of rain and sunshine. Since it was established in 1979 as the China's first Special Economic Region, Shenzhen has experienced rapid urbanization and economic development, and its land use structure has experienced, and continues to experience dramatic changes.

Data sets and preprocessing

The spatial data selected for this study consisted of land use datasets (on a scale of 1:50,000) in the year of 1996, 2000 and 2004, transportation network maps, river system and lake distribution maps, and administrative borderline maps (including the city center, towns, and residential blocks). These datasets were officially acquired mainly by aerial photograph interpretation aided by ground surveys and historical land use maps. In the data preparation, the same data interpretation methods and the same classification categories were adopted. The data were then edited, calibrated, and converted into raster data in ArcGIS for the subsequent spatial analysis.

The original land use categories included two levels of Chinese National Standard Categories. In this study, the datasets were reclassified into five categories: woodland, cropland, wetland, water body, and build-up. Therefore, the values of raster grids were coded as 1, 2, 3, 4, and 5, representing the five land use types in the order mentioned above.

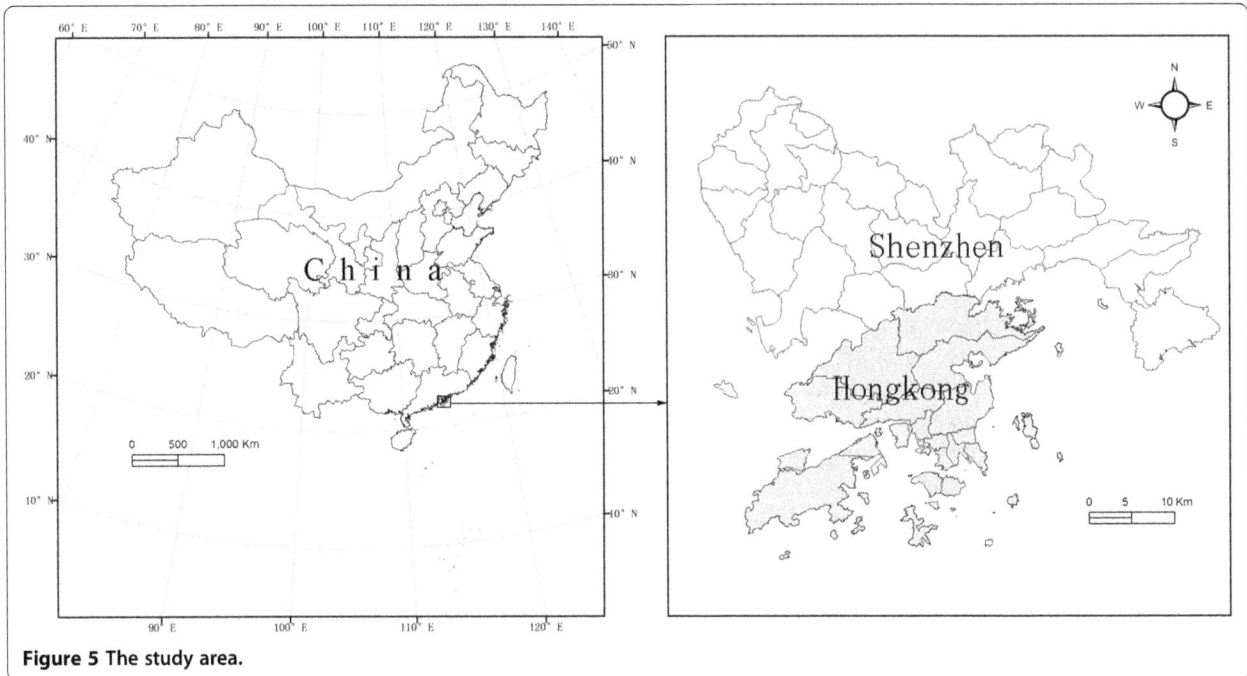

Figure 5 The study area.

Moore neighborhood, which is the most frequent used, was adopted in this case study.

To build the transaction rules, variables were chosen with the help of literature review (Kocabas and Dragicevic 2007; Santé et al. 2010; Singh 2003; Shi et al. 2000; Yang et al. 2008) with considering the data availability. They are $Tr0$ (proximity to railways), $Tr1$ (proximity to expressways), $Tr2$ (proximity to roads), $Tr3$ (proximity to streets), C (proximity to city centers), T (proximity to towns), and P (population density). Because major rivers and lakes are ecologically sensitive areas in the study area, they should be protected and any developing activities should not be restricted in these areas. To this end, two proximities, R (proximity to rivers) and L (proximity to lakes), were selected as constraints of the model. The modified CA rules can be described by

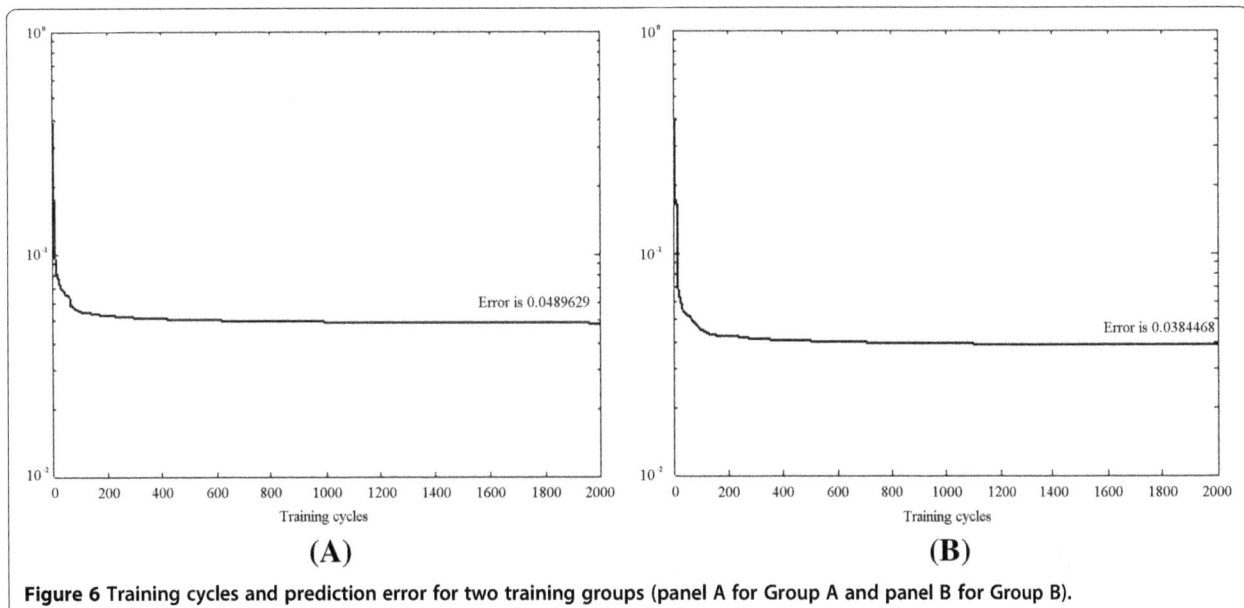

Figure 6 Training cycles and prediction error for two training groups (panel A for Group A and panel B for Group B).

Eq. 6. As these spatial variables range extensively, they tend to be normalized to the same data range (e.g. integer values from 1 to 9).

$$S_i^{t+1} = f\left(S_i^t, N_i^t, Tr0_i^t, Tr1_i^t, Tr2_i^t, Tr3_i^t, R_i^t, L_i^t, C_i^t, T_i^t, P_i^t\right) \tag{6}$$

i refers to cell "i", t refers to time steps, S refers to cell states, and N refers to the combined states of the neighborhood, i.e. the pixel numbers of each land use category in the whole neighborhood.

For the eight proximity variables, the minimum Euclidean distance to each feature was calculated in ArcGIS and eight raster maps were produced. Since it was highly difficult to acquire the population for each cell, it was assumed that population locations complied with residential blocks. Thus, the variable of population density was obtained indirectly by calculation of residential block density based on the Kernel density. As these proximity variables range extensively and have different dimensions, these proximity and density variables were reclassified into 9 classes by Natural Breaks in ArcGIS and denoted by nine integers from 1 to 9.

Since only those cells with transition probabilities larger than up would change their states in the next step through the Monte Carlo approach, the bigger value of up would produce fewer fragments while reduce the effect of Monte Carlo approach. Compromise is necessary, as above discussed. Because 5 land use types are simulated in this study, the value of up can be around 0.2, after several trial simulations, up was set to 0.15.

Model training

The CA transition potentials were learned through a BP network with training data. In this study, the proposed BP network had four layers: one input layer, two hidden layers, and one output layer. The neurons of input layer and output layer were define in Table 2. Through trials and experiments, the number of neurons in the first and second hidden layers was 15 and 8 respectively. Since the output of the neural network represented the transition probabilities for land use types, the Log-Sigmoid transfer function was adopted in order to guarantee of the output falling into the close range of [0, 1].

It is more appropriate to normalize and convert the input data into the close range of [0,1] using Eq. 7 since this scaling procedure makes the input values compatible with the Sigmoid activation function that produces a value between 0 and 1 (Gong 1996).

$$P_i' = \frac{P_i - P_{\min}}{P_{\max} - P_{\min}} \tag{7}$$

The land use maps of 1996, 2000 and 2004 were combined into two groups, namely, from 1996 to 2000 (Group A) and from 2000 to 2004 (Group B). The neighborhood size was set as 3×3, and the cell size was set to $100\ \text{m} \times 100\ \text{m}$. There are 194,989 cells in each map, of which 140,852 were chosen by stratified random sampling for training and the rest 54,137 cells were used for model test.

The algorithm of scaled conjugate gradient (SCG) (Moller 1993) was used to conduct the training. After about 2,000 running cycles, the ANN was stabilized and the error was reduced to 0.049 and 0.038 for Group A, B respectively (Figure 6).

Competing interests
The authors declare that they have no competing interests.

Authors' contributions
TL have made substantial contributions to conception and design, acquisition of data and data preprocessing, and manuscript drafting. WL have made substantial contributions to method implementation with programming, case study, data processing and manuscript drafting. Both of them have read and approved the final manuscript.

Authors' information
Tianhong Li got his Ph.D at Peking University in 1998, and now, he is an associated professor at Peking University. His research interests include water environment and ecological rehabilitation of Basins, and LUCC and its eco-environmental effects. He has published more than 70 papers in peer reviewed journals.
Wenkai Li got his M.S. at Peking University in 2008 and got his Ph.D at University of California Merced, USA in 2013. His major research interests include remote sensing and its applications to environmental studies. He has published more than 10 peer reviewed papers. Now, he is an associated professor at Sun Yat-Sen University.

Acknowledgements
Financial support was provided by the National Natural Scientific Fund under grant No.50979003. We would like to thank the anonymous reviewers also.

Author details
[1]College of Environmental Sciences and Engineering, The Key Laboratory for Water and Sediment Sciences, Peking University, Beijing 100871, China. [2]Shenzhen Graduate School of Peking University, Shenzhen 518055, China. [3]Geography and Planning School, Sun Yat-sen University, Guangzhou 510275, China.

References
Al-Ahmadi K, See L, Heppenstall A, Hogg J (2009) Calibration of a fuzzy cellular automata model of urban dynamics in Saudi Arabia. Ecol Complex 6(2):80–101, doi:10.1016/j.ecocom.2008.09.004
Al-kheder S, Wang J, Shan J (2008) Fuzzy inference guided cellular automata urban-growth modelling using multi-temporal satellite images. Int J Geogr Inf Sci 22(11–12):1271–93, doi:10.1080/13658810701617292
Almeida CM, Gleriani JM, Castejon EF, Soares BS (2008) Using neural networks and cellular automata for modeling intra-urban land-use dynamics. Int J Geogr Inf Sci 22(9):943–63, doi:10.1080/13658810701731168
Bai WQ (2000) Analysis on land use dynamics of Shenzhen. J Nat Resour 15 (2):112–6 (in Chinese)
Batty M (1997) Cellular automata and urban form: a primer. J Am Plan Assoc 63:266–74, doi:10.1080/01944369708975918
Congalton RG (1991) A review of assessing the accuracy of classification of remotely sensed data. Remote Sens Environ 37:35–46, doi:10.1016/0034-4257(91)90048-B
De Almeida C M, Gleriani J M (2005) Cellular automata and neural networks as a modeling framework for the simulation of urban land use change. Anais XII Simpósio Brasileiro de Sensoriamento Remoto, Goiânia, Brasil, 16–21 abril, INPE: 3697–3705.

Gong P (1996) Integrated analysis of spatial data from multiple sources: using evidential reasoning and artificial neural network techniques for geological mapping. Photogramm Eng Remote Sens 62:513–23

Gong JZ, Liu YS, Xia BC, Zhao GW (2009) Urban ecological security assessment and forecasting based on a cellular automata model: a case study of Guangzhou, china. Ecol Model 220:3612–20, doi:10.1016/j.ecolmodel.2009.10.018

He C, Okada N, Zhang Q, Shi P, Li J (2008) Modelling dynamic urban expansion processes incorporating a potential model with cellular automata. Landsc Urban Plan 86:79–91, doi:10.1016/j.landurbplan.2007.12.010

Kocabas V, Dragicevic S (2006) Assessing cellular automata model behavior using sensitivity analysis approach. Comput Environ Urban Syst 30(6):921–53, doi:10.1016/j.compenvurbsys.2006.01.001

Kocabas V, Dragicevic S (2007) Enhancing a GIS cellular automata model of land use change: Bayesian networks, influence diagrams and causality. Trans GIS 11(5):681–702, doi:10.1111/j.1467-9671.2007.01066.x

Li X, Liu X (2006) An extended cellular automaton using case-based reasoning for simulating urban development in a large complex region. Int J Geogr Inf Sci 20(10):1109–36, doi:10.1080/13658810600816870

Li X, Yeh AG (2001) Calibration of cellular automata by using neural networks for the simulation of complex urban systems. Environ Plan A 33:1445–62, doi:10.1068/a33210

Li X, Yeh AG (2002) Neural-network-based cellular automata for simulating multiple land use changes using GIS. Int J Geogr Inf Sci 16:323–43, doi:10.1080/13658810210137004

Li TH, Li WK, Qian ZH (2010) Variations in ecosystem service value in response to land use changes in Shenzhen. Ecol Econ 69(7):1427–35, doi:10.1016/j.ecolecon.2008.05.018

Liu Y, Phinn SR (2003) Modeling urban development with cellular automata incorporating fuzzy-set approaches. Comput Environ Urban Syst 27:637–58, doi:10.1016/S0198-9715(02)00069-8

Liu X, Li X, Liu L, He J, Ai B (2008a) A bottom-up approach to discover transition rules of cellular automata using ant intelligence. Int J Geogr Inf Sci 22(11–12):1247–69, doi:10.1080/13658810701757510

Liu X, Li X, Shi X, Wu S, Liu T (2008b) Simulating complex urban development using kernel-based non-linear cellular automata. Ecol Model 211:169–81, doi:10.1016/j.ecolmodel.2007.08.024

Moller MFA (1993) A scaled conjugate gradient algorithm for fast supervised learning. Neural Netw 6:525–33

Pijanowski BC, Brown BG, Shellito BA, Manik GA (2002) Using neural networks and GIS to forecast land use changes: a land transformation model. Comput Environ Urban Syst 26:553–75, doi:10.1016/S0198-9715(01)00015-1

Santé I, García AM, Miranda D, Crecente R (2010) Cellular automata models for the simulation of real-world urban processes: a review and analysis. Landsc Urban Plan 96:108–22, doi:10.1016/j.landurbplan.2010.03.001

Shi PJ, Chen J, Pan Y (2000) Land use change mechanism in Shenzhen city. Acta Geograph Sin 55(2):151–60 (in Chinese)

Singh A K (2003) Modeling land use land cover changes using cellular automata in a geo-spatial environment. International Institute for Geo-Information Science and Earth Observation. Vol. Master of Science, Enscheda, the Netherlands.

Stevens D, Dragicevic S, Rothley K (2007) iCity: A GIS-CA modeling tool for urban planning and decision making. Environ Model Softw 22:761–73, doi:10.1016/j.envsoft.2006.02.004

Torrens P M (2000). How cellular models of urban systems work. (1. Theory). Centre for advanced spatial analysis working Paper Series 28 (ISSN: 1467-1298@ Copyright CASA, UCL.).

Torrens PM, O'Sullivan D (2001) Editorial: cellular automata and urban simulation: where do we go from here? Environm Plan B 28:163–8, doi:10.1068/b2802ed

Verburg PH, de Nijs TCM, von Eck JR, Visser H, Jong K (2004) A method to analyze neighborhood characteristics of land use patterns. Comput Environ Urban Syst 28:667–90, doi:10.1016/j.compenvurbsys.2003.07.001

Wang H, Li XB, Long HL, Qiao YW, Li Y (2011) Development and application of a simulation model for changes in land-use patterns under drought scenarios. Comput Geosci 37:831–43, doi:10.1016/j.cageo.2010.11.014

White R, Engelen G (2000) High-resolution integrated modeling of the spatial dynamics of urban and regional systems. Comput Environ Urban Syst 24:383–400, doi:10.1016/S0198-9715(00)00012-0

Wolfram S (1984) Cellular automata as models of complexity. Nature 311:419–24, doi:10.1038/311419a0

Wu F (1996) A linguistic cellular automata simulation approach for sustainable land development in a fast growing region. Comput Environ Urban Syst 20:367–87, doi:10.1016/S0198-9715(97)00003-3

Wu F (1998a) Simulating urban encroachment on rural land with fuzzy-logic-controlled cellular automata in a geographical information system. J Environ Manag 53:293–308, doi:10.1006/jema.1998.0195

Wu F (1998b) SimLand: a prototype to simulate land conversion through the integrated GIS and CA with AHP-derived transition rules. Int J Geogr Inf Sci 12(1):63–82, doi:10.1080/136588198242012

Wu FL (2002) Calibration of stochastic cellular automata: the application to rural–urban land conversions. Int J Geogr Inf Sci 16(8):795–818, doi:10.1080/13658810210157769

Yang QS, Li X, Shi X (2008) Cellular automata for simulating land use changes based on support vector machines. Comput Geosci 34(6):592–602, doi:10.1016/j.cageo.2007.08.003

Yeh AG, Li X (2003) Simulation of development alternatives using neural networks, cellular automata, and GIS for urban planning. Photogramm Eng Remote Sens 69:1043–52

Yeh AG, Li X (2004) Integration of neural networks and cellular automata for urban planning. Geo Spatial Inf Sci Q 7:6–13

4

Suitability analysis for *Jatropha curcas* production in Ethiopia - a spatial modeling approach

Habitamu Taddese

Abstract

Background: Jatropha is an oil-bearing plant growing in tropical and subtropical regions of the world within 30°N and 35°S latitudes. It is considered as a potential solution to the prevailing shortage of fossil fuel and environmental challenges. However, in most parts of Africa including Ethiopia, traditional land allocation systems for biodiesel investment do not involve integration of multiple variables. This research tries to introduce the advantages of Geographical Information Systems (GIS) and Spatial Analytical Hierarchy Process (SAHP) to identify suitable areas for jatropha production in Ethiopia. Combination of these methods enables integration of different environmental data in a multi-criteria analysis. The study will provide basic information in the biodiesel investment, which has been susceptible to failure due to poor land allocation. The methods used in this study will also be available for similar endeavors in the future.

Results: In general, although individual factor evaluations provided varying amounts of suitability, results of weighted overlay analysis for biophysical suitability evaluation using spatial modeling methods identified 15.07% (166,082 km^2), 76.57% (844,040 km^2) and 8.36% (92114 km^2) of the land as highly suitable, moderately suitable and not suitable for jatropha production, respectively.

Conclusion: The methods used in this study provided considerably reliable estimate of suitable sites for jatropha production in Ethiopia. In this study, the main limiting factors of jatropha production identified were elevation, climate (temperature and rainfall extremities) and water logging conditions. Suitable sites do not compete with existing land use systems ensuring that biodiesel production will not risk food security programs.

Keywords: SAHP; GIS; Land suitability; Jatropha curcas L; Spatial modeling

Background

Located in the horn of Africa, Ethiopia is the second populous country in the continent with an estimated population of about 89.2 million in 2013 (PRB 2013). Its energy demand is increasing tremendously and cost of petroleum import exceeded export earnings by 2008 (NBE 2010).

Recently, declining trends in the global energy supply and consequences of climate change have created huge global concern. Due to this huge concern, many countries are making efforts in developing clean energy options (Van der Putten 2010; Achten et al. 2008; Wu et al. 2009).

In Africa, for instance, high fossil fuel prices and national security concerns have sparked interest in bio-based fuel development in different parts of the

continent (Koikai 2008; Pillay and Da Silva 2009; Nyebenge et al. 2009). Ethiopia has designed a biodiesel development strategy to promote biodiesel investment. The strategy will help the country evade its reliance on import of fossil fuels for its energy consumption and reduce impacts of climate change (CRGE 2011; Nyebenge et al. 2009; Makkar and Becker 2009). Jatropha, palm tree and castor bean were identified in the strategy as promising biodiesel bearing plants. This paper was inspired by the multiple products and services obtained from jatropha to ameliorate land degradation, negative energy balance, fertility loss and poor health condition of the rural community (Brittaine and Lutaladio 2010; Grass 2009; Heller 1996).

Jatropha curcas L. is an oil-bearing plant growing in the tropical and subtropical regions of the world within the limits of 30°N and 35°S latitudes (Jongschaap et al. 2007). The plant belongs to the Euphorbiaceae family

Correspondence: habtu1976@gmail.com
Hawassa University, Wondo Genet College of Forestry and Natural Resources, P. O. Box 128, Shashemene, Ethiopia

and sets fruits that contain seeds with an oil content of more than 30%.

Some survey reports, estimates (using conventional methods) and author's personal observation revealed presence of jatropha around home gardens and farmlands in different regions of the country. Plantations were also established by different actors; however, the methods used so far for site identification were ineffective (Nyebenge et al. 2009; Wendimu 2013; MELCA 2008). Consequently, failure accounts of investment projects have been reported (Wendimu 2013). The conventional techniques of identifying marginal land for jatropha investment lack scientific foundation (Wendimu 2013) and decisions were dependent on old data (Birega et al. 2010).

This study employs Spatial Analytic Hierarchy Process (SAHP) and Geographic Information System (GIS) to generate valuable information in land allocation for jatropha production.

SAHP is a derivative of Analytic Hierarchy Process (AHP), which is used to resolve highly complex decision making problems involving multiple factors (Saaty 1977, Saaty and Vargas 1991). Its spatial equivalent, SAHP, is now becoming an emerging tool for multi-criteria analysis in which positional relationship between features is relevant (Ghamgosar et al. 2011, Emami and Zarkesh 2011). SAHP was used by several researchers for land use site selection due to its paramount advantages. Some of the special features of SAHP were explained by Emami and Zarkesh (2011) as the ability to review both quantitative and qualitative criteria simultaneously, the possibility of simplifying complex issues into a form of hierarchy, pair-wise comparisons and weighing criteria, simple calculations and possibility of ranking the final options. It also works well with various factor weighting and quantifies experts' opinions (Zarkesh et al. 2011).

A combination of SAHP and GIS has been used in determining suitable areas for rangeland management (Jafari and Zaredar 2011), ecotourism (Zarkesh et al. 2011), municipal solid waste landfill (Javaheri 2006; Paul 2012), and forestry (Store and Kangas 2001; Babaie-Kafaky et al. 2009). This implies these methods can be customized to specific features of a particular field.

Spatial modeling technique is a useful method of overlaying multiple datasets in a GIS to assess suitability (Duc 2006). Generally, there are two approaches to model ecological suitability; namely, correlative approach and mechanistic approach (Figure 1). Correlative habitat models identify distribution of a species with environmental data like soil, temperature and topography. Examples of correlative habitat suitability models include BRT (Boosted Regression Tree), MaxEnt (Maximum Entropy), and CART (Classification and Regression Tree) models, which rely on occurrence data. These models are appropriate to identify habitat requirements of a species

(Valavanis et al. 2008) and relate that with a larger landscape dataset.

Mechanistic habitat suitability model, on the other hand, is used to determine the mechanistic link between an organism's environment and its fitness with the environmental conditions (Kearney 2006). It generates information about conditions in which the species can ideally persist based on observations made in laboratory studies or documented realities.

This study uses a mechanistic suitability modeling approach since it deals with a plant whose environmental requirements are well documented. The approach is adopted from FAO (1976).

Findings of this study will have paramount significance in supporting decision making in the biodiesel energy development sector. Local communities, universities, investors, researchers, community-based organization (CBO's) and non-governmental organizations (NGO's) will benefit from the research results. The research idea was originally derived from information requests from different actors since previous estimates were neither accessible nor dependable. Availability of this publication will answer those questions of partners and serve as an alternative source of information. The methods used in this study will contribute to replacing the existing ineffective ones; and will be useful for future efforts of research and development initiatives.

Therefore, specific objectives of this investigation were to identify factors and select criteria of growth and yield requirements of jatropha, classify and weigh environmental variables into different levels of suitability, and produce a suitability map for jatropha production in Ethiopia.

Results and discussion
Assessment of environmental requirements for growth and yield of jatropha

The factors presented in Figure 2 were identified as important criteria influencing jatropha production. They influence the amount and quality of products derived from jatropha, most importantly oil. Land suitability studies basically make a matching between land use requirements and existing land characteristics (FAO 1976). So, this portion of the study focused on assessing land use requirements of jatropha for achieving optimal growth and yield (Table 1).

Single factor suitability evaluation for *Jatropha curcas* production

Several studies have examined the correlation between jatropha production and environmental conditions; and there is a consensus that climate, terrain and soil properties are key factors determining growth and yield of jatropha (Jingura et al. 2011; Wu et al. 2009; Jongschaap et al. 2010). To grow well and give high yields, the plant needs enough water, appropriate temperature and

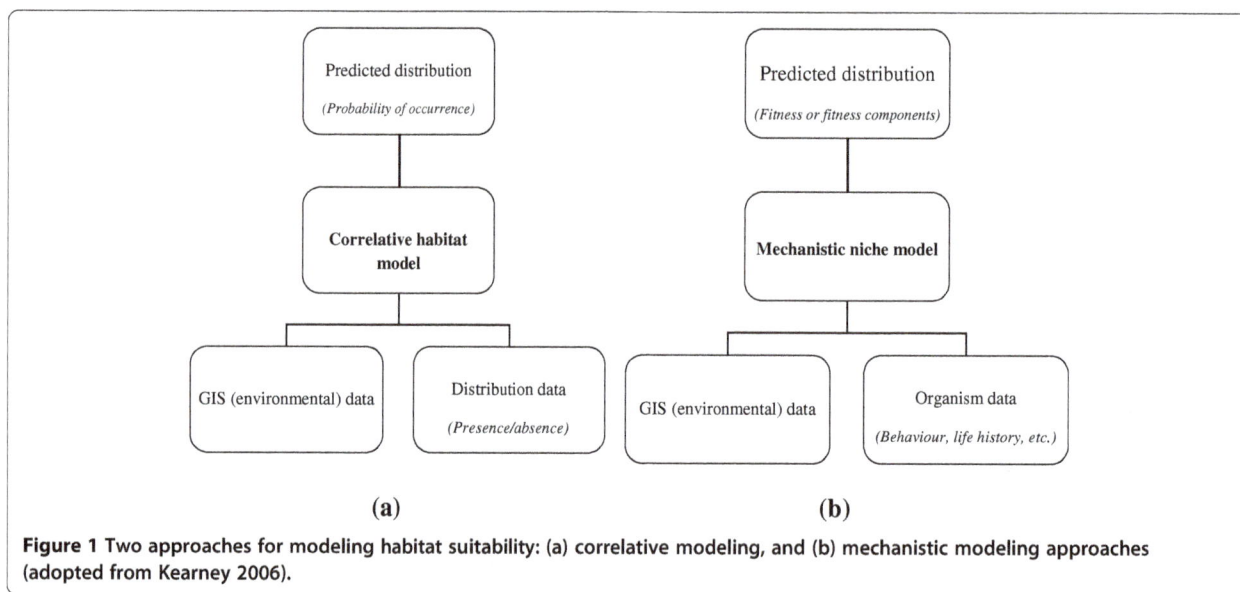

Figure 1 Two approaches for modeling habitat suitability: (a) correlative modeling, and (b) mechanistic modeling approaches (adopted from Kearney 2006).

altitude; and the soil type has to be right. Ouwens et al. (2007) indicated that ecological conditions do not only affect yield but also determine length and degree of injury by pests and diseases. So, selection of appropriate ecological conditions for proper growth and yield of jatropha was a very essential aspect of this study.

Literature review has been conducted to identify environmental requirements of *Jatropha curcas* based on experiences in tropical and subtropical regions. The information obtained from literature review has been summarized to define the different classes of suitability for each criterion or factor.

In Table 1, total range of each factor represents the environmental conditions in which the plant can survive; but production might not be possible across some portions of the range. Literature shows three major suitability ranges for assessing compatibility of an environmental condition for normal growth and yield of jatropha. The plant bears optimum production as long as it grows in the suitable (S_1) condition of each factor. Moderately suitable (S_2) condition represents friendly situations to support good production of jatropha as far as other factors are not beyond threshold ranges; otherwise, seed setting

and production of fruits will be impaired. Ranges of values of environmental variables that fall under the not suitable (N_1) category are difficult conditions where jatropha cannot survive unless improvement is made; or if it survives, seed or fruit production may not be attainable. Brief descriptions of factors influencing jatropha production are indicated below.

Rainfall

The minimum amount of annual rainfall that jatropha needs to produce fruits is 600 mm (Ouwens et al. 2007; Grass 2009). At this moisture condition, the plant will give poor yields. However, the optimal annual rainfall is between 1000 and 1500 mm (Grass 2009). If it rains more than 1500 mm, jatropha will have problems with fungal attack, root rot and other diseases (Franken 2010). Thus, rainfall data was classified into three of the suitability classes (S_1, S_2 and N_1).

Temperature

It was attested that if the annual mean temperature is less than 17°C, the area is not suitable for jatropha production (Heller 1996; Gour 2006). Low temperature

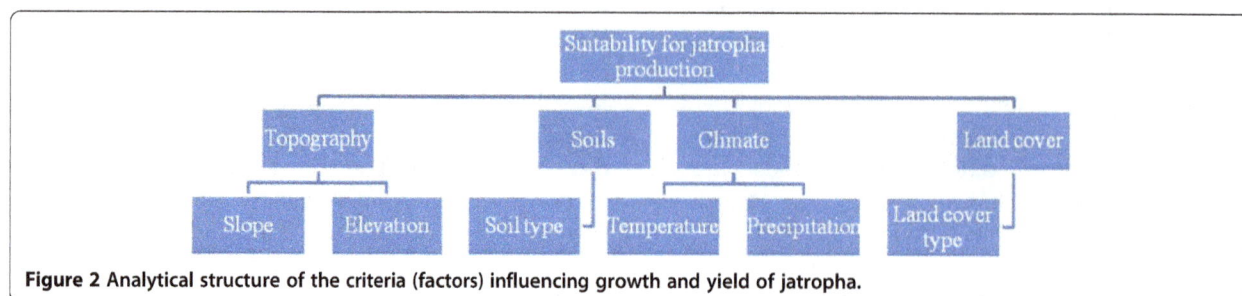

Figure 2 Analytical structure of the criteria (factors) influencing growth and yield of jatropha.

Table 1 Environmental attributes required for jatropha production categorized under three classes of suitability

Factor or criterion	Unit	Total range	Suitability classes[a] of environmental attributes			Source of information (reference)
			Suitable (S_1)	Moderately suitable (S_2)	Not suitable (N_1)	
1. Rainfall	mm/year	250 - 3000	1000 – 1500	600 – 1000	< 600 or >1500	Grass 2009; Heller 1996
2. Temperature	Degree Celsius	17 – 28	20 – 28	17 – 20	<=17 or >28	Wu et al. 2009; Achten et al. 2008; Gour 2006
3. Elevation	Meters from sea level	0 - 2150	0 – 1500	1500 – 2150	< 0 or >2150	Gour 2006; ICRAF 2009; Achten et al. 2008; Wiesenhütter 2003
4. Soils	Soil type	Any soil type without (or with little) clay content	Well drained sand and loam soils	Small proportion of clay or little water logging potential	Heavy clayey soils, which have water logging effect	Ouwens et al. 2007; Brittaine and Lutaladio 2010; Achten et al. 2008
5. Slope	Degree	0 - 30	<=15	15 – 30	>30	Achten et al. 2008; Wu et al. 2009
6. Land cover (use)	Cover type	Land cover other than waterlogged, conservation areas, settlements and water bodies	Well drained marginal lands, open grasslands, wooded grasslands	Disturbed forests and bush/shrub lands, salt and flats	Water logged, conservation sites, settlements, cultivated areas, etc.	Brittaine and Lutaladio 2010

Data source: Existing literature and experts' views.
[a]Suitability classes indicate environmental conditions in which large scale jatropha production is evaluated.

affects metabolic activities that influence germination, growth and development of most tropical plants including *Jatropha curcas* (Divakara et al. 2009; Garg et al. 2011; Liang et al. 2007). Jatropha establishment requires mean temperature between 17°C and 28°C and seedlings will be injured if temperature is lower than the optimal range (Achten et al. 2008; Ye et al. 2009). In contrast, it was observed that very high temperature depresses yields (Gour 2006; Makkar and Becker 1997). So, temperature data was classified into three classes of suitability (S_1, S_2 and N_1).

Elevation

Effect of elevation on yield of jatropha is manifested in the damage imposed by frost since frost is a direct consequence of elevation. It was explained by several researchers that jatropha is unable to withstand frost (Heller 1996; Grass 2009). Experiences in different countries indicate that the optimum elevation for growth and productivity of jatropha ranges from sea level to 1500 meters above sea level (Muok and Kallback 2008; Brittaine and Lutaladio 2010) because at this elevation, risk of frost is minimal. Altitudes from 1500 to 2150 meters above sea level are moderately suitable. All areas above 2150 meters are not suitable for jatropha production because of frost (ICRAF 2009). So, elevation data was classified into the three levels of suitability (S_1, S_2 and N_1).

Soils

A soil with good infiltration rates and without water logging tendencies is suitable for jatropha cultivation, while soils with bad infiltration rates and a high tendency for water logging are not suitable. In heavy clayey soils, root formation of jatropha is hindered (Heller 1996; Brittaine and Lutaladio 2010; Ouwens et al. 2007; Biswas et al. 2006; Singh et al. 2006; Achten et al. 2008). The best soils for jatropha are well-drained aerated sands and loams (Gour 2006; Heller 1996). It was revealed that jatropha is tolerant to saline soil condition (Sahoo et al. 2009; Gao et al. 2008). Jatropha is also known for its ability to survive in very poor dry soils in conditions considered marginal for agriculture and can even root into rock crevices though productivity may be limited (Makkar and Becker 1997). Therefore, the soil data of the study area was classified into suitable, moderately suitable and not suitable categories.

Slope

Slope is an important indicator of land suitability since it affects drainage, irrigation and soil erosion (Wu et al. 2009). Steep slopes reduce infiltration efficiency of rainfall because it facilitates runoff. Slopes up to 15° are ideal for optimum growth and yield of jatropha; whereas slopes between 15° and 30°, exhibit linear decrease in

suitability. Slopes greater than 30° are not suitable for jatropha production. Therefore, slope data was classified into three levels of suitability, which is in accordance with similar studies (Achten et al. 2008; Wu et al. 2009; Grass 2009).

Land cover (use)

The national bio-fuel development and utilization strategy clearly indicates that land allocated for bio-diesel development must not jeopardize farmers' food production needs (MME 2008). Likewise, settlers should not be evacuated; and reserve areas must not be affected by such projects. Therefore, land under cultivation, urban settlements and conservation areas are regarded as not suitable for jatropha production although they are biophysically conducive sites to the plant (Muok and Kallback 2008; Grass 2009). Furthermore, jatropha dislikes permanent wetness (Ouwens et al. 2007); thus, permanent water bodies and wetlands were masked out from land cover (use) data. Woodlands and bush land areas may be converted to jatropha investment given priority sites are exhaustively utilized and thus are considered moderately suitable for jatropha investment (Wu et al. 2009). Grass lands and marginal land are suitable for jatropha investment.

Criteria weights

In this research, weights of the selected criteria were derived using SAHP method. A pair-wise comparison matrix of the SAHP is presented in Table 2.

The numbers in the above table indicate preference (intensity of importance) of the factors being compared based on experts' opinions. These numbers were obtained during expert consultations for comparing the different factor combinations that affect jatropha production. The weight and CR column values were calculated from the intensity of importance values based on a series of procedures (Saaty 2008; Triantaphyllou and Mann 1995). The importance weight is unit-less measure of relative preference of the factors.

The weights indicate that rainfall, elevation and temperature have respectively greater importance values contributing more to the overall multivariate analysis. On the other hand, slope, land cover and soil type have importance values less than 10% each.

This shows that most of the influence to the resultant suitability comes from characteristics of rainfall, elevation and temperature of the area. These are factors affecting growth performance of the plant and thus are basic to influence analysis of suitability. Furthermore, it is evident from literature that even if there may be variation in performance, jatropha generally grows on most soils except those experiencing water logging conditions. Slope is not a limiting factor for growth and yield of jatropha. Its effect

Table 2 Weight and consistency ratio (CR) of pair-wise comparison matrix of factors that affect jatropha production

Criteria	Precipitation	Elevation	Temperature	Soils	Land cover	Slope	Weight	CR
Precipitation	1.00	2.00	4.00	6.00	8.00	9.00	0.43	0.045
Elevation	0.50	1.00	2.00	4.00	6.00	8.00	0.26	0.036
Temperature	0.25	0.50	1.00	2.00	4.00	6.00	0.15	0.024
Soils	0.17	0.25	0.50	1.00	2.00	4.00	0.08	0.012
Land cover	0.13	0.17	0.25	0.50	1.00	2.00	0.05	0.007
Slope	0.11	0.13	0.17	0.25	0.50	1.00	0.03	0.005
Column total	2.15	4.04	7.92	13.75	21.50	30.00	1.00	

Source: Expert interview data and author's calculation.

is reflected on soil moisture and fertility status of the land. Both soil moisture and fertility do not significantly influence performance of jatropha (ICRAF 2009). Land cover is a factor considered to ensure conformity of the results with existing national development strategies; thus, it is not a determinant factor to influence agronomic suitability.

Multi-factor analysis of suitability for jatropha production in Ethiopia

Suitability maps of individual factors of climate, soils, land cover and topography are shown in Figure 3. Based on single factor evaluation, landscape characteristics like slope and elevation were found extensively suitable. However, less than half of the land area of the country was found suitable in terms of rainfall, soils and land cover characteristics.

The results indicate that topographically most part of the country is suitable for jatropha production. It was revealed from this investigation that 97% of the slopes and 65.8% of the elevation of Ethiopia are suitable for jatropha. It was also apparent that high altitude areas and areas below sea level, which account for 34.2% of the landmass of Ethiopia, are not appropriate for jatropha production.

On the other hand, it appears that climatic variables are limiting factors controlling growth and yield of the target plant. Effect of rainfall on jatropha production was considerable. This effect is manifested in its influence on germination, growth, seed production and its likely impacts on attracting diseases and pests.

In terms of area, a significant amount of the country is actually suitable for jatropha production. However, more than three-fourths of the land is potentially (moderately) suitable with some limitations that may require socio-economic and environmental management mechanisms to make use of them for investment in this sector (Figure 4; Table 3).

Figure 5 shows interplay of the different factors that affect performance of jatropha. It was found that large coverage of the suitable areas of temperature and terrain factors do not notably contribute to the final suitability index. For example, although 97% of the slope is suitable for jatropha, its percentage influence was minimal (only 3%) in providing more suitable sites in the resultant suitability index. However, rainfall (only 18.8% of which is suitable) has the maximum percentage influence (43%) to the resultant suitability map (Table 2 and Figure 5).

The result of the multivariate analysis indicates that 15.07% of the land area of Ethiopia is agronomically suitable for jatropha production. This area refers to biophysically suitable sites that are compatible with the current development strategies of the country (CRGE 2011). Suitable areas for jatropha production were predominantly attributed to characteristics of rainfall, elevation and temperature with importance weights of 43%, 26% and 15% respectively (Table 2).

However, the above suitability evaluation results are just based on natural conditions. Whether or not these lands could be used to plant jatropha is still subject to social and economic evaluation. These results should be regarded as theoretically potential land that could be used for jatropha plantation. But further evaluation of social and economic factors is still important. In addition, it is obvious that suitability is subject to temporal dynamics of environmental variables. Therefore, effect of climate variability and changes in other environmental variables need to be evaluated to plan for future investment opportunities.

Conclusions

Careful selection of all the possible variables that affect growth and yield of jatropha is a basic step to make sure that the result will be consumed by decision makers. This research is intended to support investment decision making in the energy sector. Investment decisions must depend on reliable evidence since environmental, social and economic crises arising from an intervention may be devastative or irreversible. Therefore, this study identified and selected potential factors that determine growth

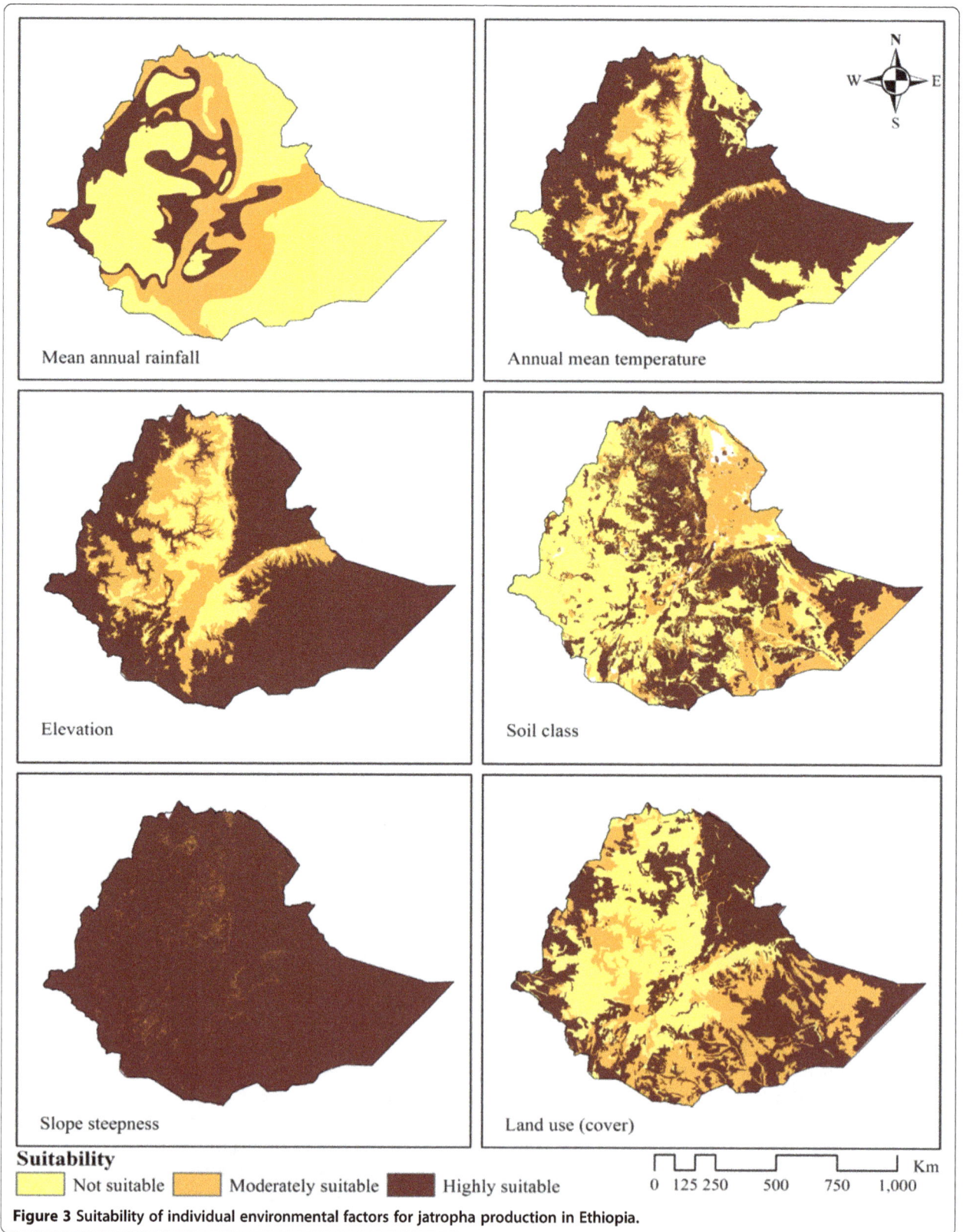

Figure 3 Suitability of individual environmental factors for jatropha production in Ethiopia.

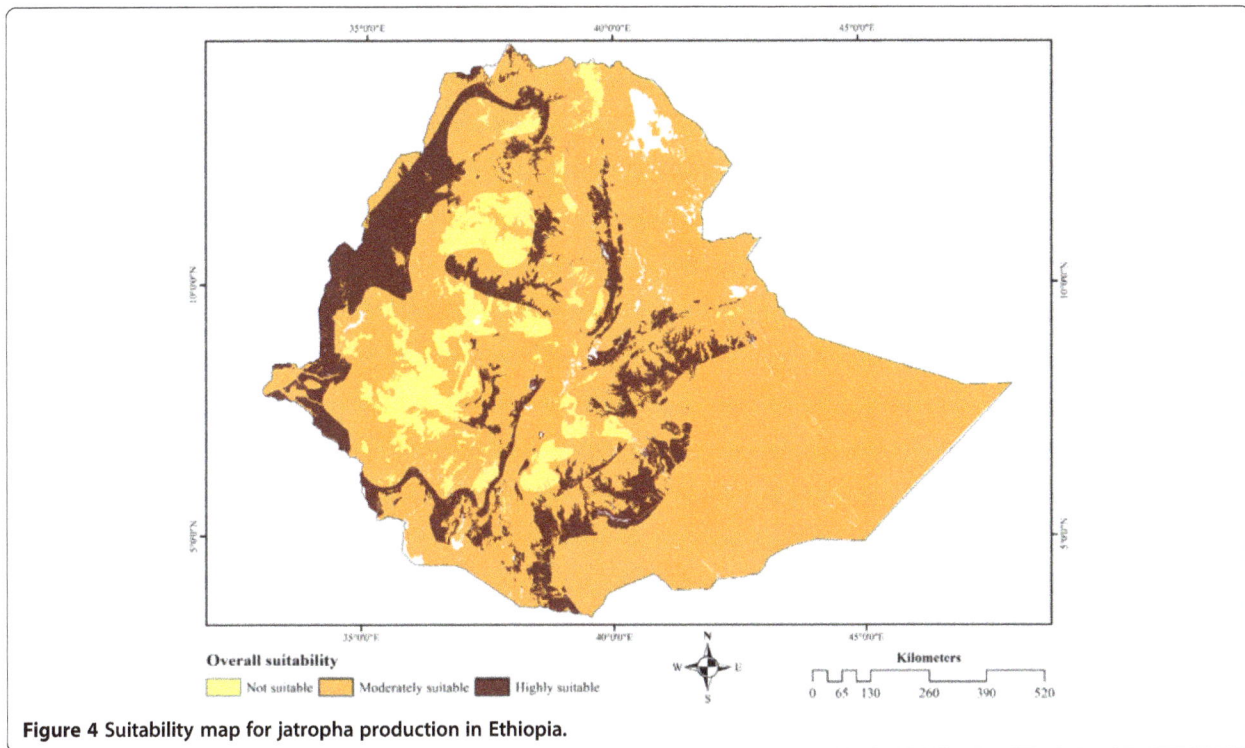

Figure 4 Suitability map for jatropha production in Ethiopia.

and yield performance of jatropha. Climate variables (temperature and rainfall), soil, land cover/use, slope and elevation were identified as the major factors.

Multi-criteria analysis techniques were used to integrate the different environmental data in a spatial modeling environment. SAHP and GIS were used for mechanistic suitability modeling of jatropha production sites in Ethiopia. This research has introduced these approaches to solve drawbacks of existing conventional techniques like remote sensing and expert opinions or judgments for assessing suitable areas. For studies like this one, mechanistic modeling is preferred over correlative modeling since it explicitly incorporates potential range-limiting processes. For instance, a mechanistic modeling can provide information on proximate constraints limiting distribution and abundance of a species.

Classification of the datasets of the identified factors into three levels of suitability enables measuring each factor in terms of fixed suitability classes. This further

enabled combination of all variables in a weighted overlay analysis. The classes used in this study were "suitable", "moderately suitable" and "not suitable".

A "suitable" area in the map shows that the area has favorable biophysical and climatic conditions for successful production of jatropha and is explained in terms of suitability with respect to all the factors considered. A "moderately suitable" area indicates a second priority for jatropha production, which must be allotted for this purpose only after detailed scrutiny of all the factors and decisions on the feasibility of this investment over other opportunities. On the other hand, "not suitable" areas represent those sites that are not appropriate for jatropha growing. They are limited by frost, water logging, inadequate rainfall, scorching temperature, terrain steepness and/ or occupied land cover/use types.

The study has shown that there is ample opportunity for jatropha investment in the country. However, considerable attention should be given to proper technologies for establishment, management and processing of jatropha products to get optimum benefit from the sector. Establishment of jatropha plantations should depend on the identified geographic locations to avoid conflicting use interests on a piece of land. Land preparation and agronomic practices must be supported with appropriate technologies. Furthermore, processing industries should be established on appropriate sites based on future trends of plantation establishment.

Table 3 Proportion of land suitability classes for jatropha production

No.	Level of suitability	Area (square kilometer)	Proportion (%)
1	Not suitable	92114.31	8.36
2	Moderately suitable	844039.72	76.57
3	Highly suitable	166082.41	15.07
	Total	**1,102,236.43**	**100.00**

Source: Own generated.

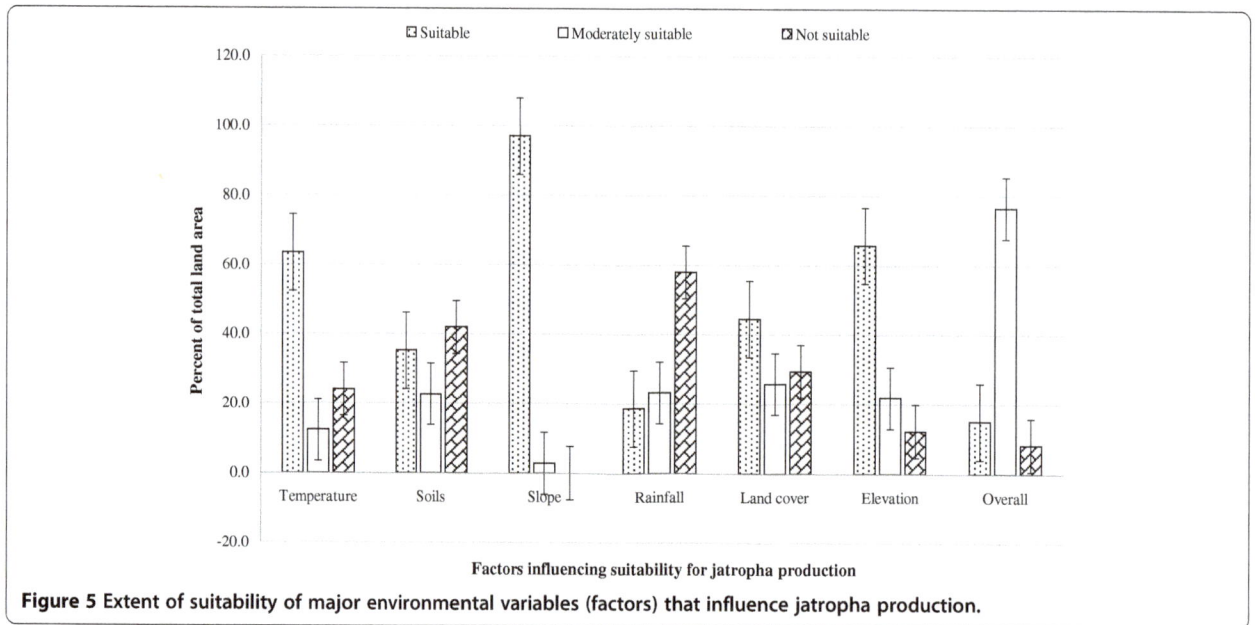

Figure 5 Extent of suitability of major environmental variables (factors) that influence jatropha production.

Methods

Site description

Ethiopia is geographically located within the tropics between 3 degrees and 15 degrees of north latitude and between 33 degrees and 48 degrees of east longitude. It has common borders with Kenya, Sudan, South Sudan Republic, Somalia, Eritrea and Djibouti (Figure 6).

There is great variation in altitude ranging from about 116 meters below sea level to 4620 meters above sea level (IBC 2007; EPA 1998). The country has an undulating topography providing ample opportunity to satisfy bio-based development interests.

The mean annual temperature of the country is 22.2°C. The lowest temperature ranges from 4°C to 15°C in the

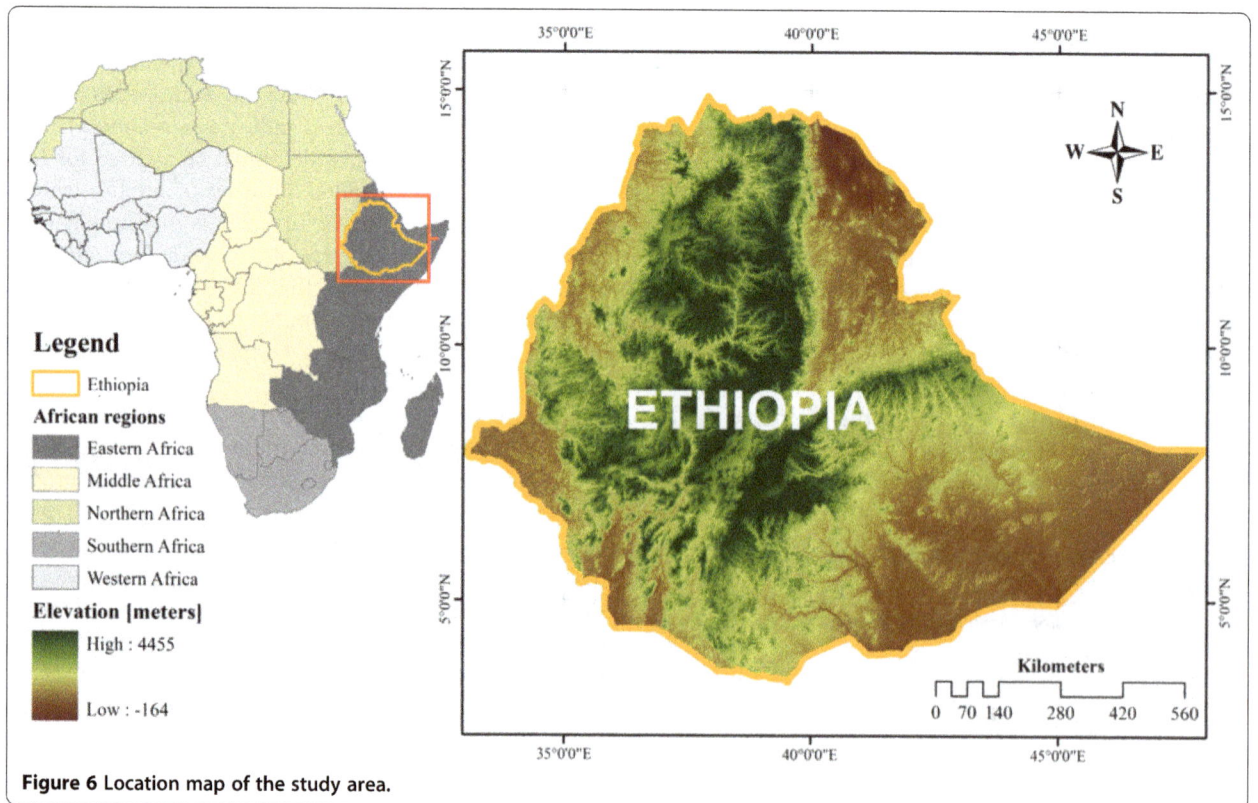

Figure 6 Location map of the study area.

Table 4 Definition of weighing scale for pair-wise comparison

Intensity of importance (numerical value)	Definition (verbal judgment of preference)
1	Equal preference
2	Somewhat moderate preference
4	Moderate preference
6	Strong preference
8	Very strong preference
9	Extreme preference
Reciprocals of the above	If criterion i has one of the above non-zero numbers assigned to it when compared with criterion j, then j has the reciprocal value when compared with i.

Source: Modified by the author from Saaty (1977).

highlands, and the highest mean temperature is 31°C in the lowlands at the Denakil Depression (Awulachew et al. 2007). The country receives mean annual rainfall of 812.4 mm, with a minimum of 91 mm and a maximum of 2,122 mm.

Relief variability and the resulting climatic characteristics make the country home to a wide range of plant, animal and microbial diversity. Consequently, the country is regarded as a centre of endemism (IBC 2007; Vivero et al. 2010).

Jatropha is one of the plant species that is traditionally used by the Ethiopian population for a number of domestic purposes. Even though there is no in-depth study of identifying optimal sites for large scale production of jatropha and other oil bearing plants in general, some sources indicate presence of oil bearing plants including jatropha in many parts of the country. For example, castor bean is located elsewhere in many parts of the country. Another oil-bearing plant, pongamia, was introduced by Indians in Beninshangul Gumuz regional state of Ethiopia. Moreover, existing plantations and wildings of jatropha in different areas signify that the country has huge potential for large scale bio-diesel production.

Selecting criteria for suitability assessment
The criteria of suitability assessment were selected through an intensive literature review on site requirements of jatropha for optimum growth and yield. Besides review of international experience from literature about the subject matter, expert consultation was a helpful tool used in the rating of factors using pair-wise

comparisons. Availability of data was also a key element considered during selection of factors for this study.

Standardization of the criteria
To compare the criteria, values of each dataset need to be transformed to the same unit of measurement scale. The different input maps (like rainfall, soil type, temperature, etc.) have various units of measurement. Each dataset was converted into raster data format. Pixels of the derived raster data represent values of the different criteria. These pixel values, though having the same unit of measurement scale, were classified into suitability classes for jatropha production. After classification, all raster data of each factor had values of 3, 2 and 1 representing "suitable", "moderately suitable" and "not suitable areas", respectively.

Weighing of the criteria
For determining the relative importance of each criterion in the resultant overlay analysis, a pair-wise comparison matrix using a modified form of Saaty's nine-point weighing scale was applied (Table 4).

For preventing bias during criteria weighing, consistency ratio was used as a tool to ensure coherent comparisons. Consistency ratio is a general measure of the comparative judgments' goodness in building up decision matrices within the AHP. It was calculated as the ratio of consistency index (CI) and random consistency index (RI). The RI is the random index representing consistency of a randomly generated pair-wise comparison matrix. Consistency ratio is a decision tool to evaluate whether an AHP is acceptable for decision making or not (Saaty 1999). It was computed from expert preference values using equations (1) and (2).

$$CI = \frac{(\lambda_{max} - n)}{(n-1)} \quad (1)$$

$$CR = \frac{CI}{RI} \quad (2)$$

Where;
n = number of items being compared, and
λ_{max} = the largest Eigen value
CR = consistency ratio
CI = consistency index
RI = is the consistency index of a randomly generated pair-wise comparison matrix.

Values of RI depend on the number of elements being compared (see Table 5).

Table 5 Random consistency index values in a pair-wise comparison matrix

Number of items being compared (n)	1	2	3	4	5	6	7	8
RI	0	0	0.52	0.9	1.12	1.24	1.32	1.41

Source: Adopted from Saaty and Vargas (1991).

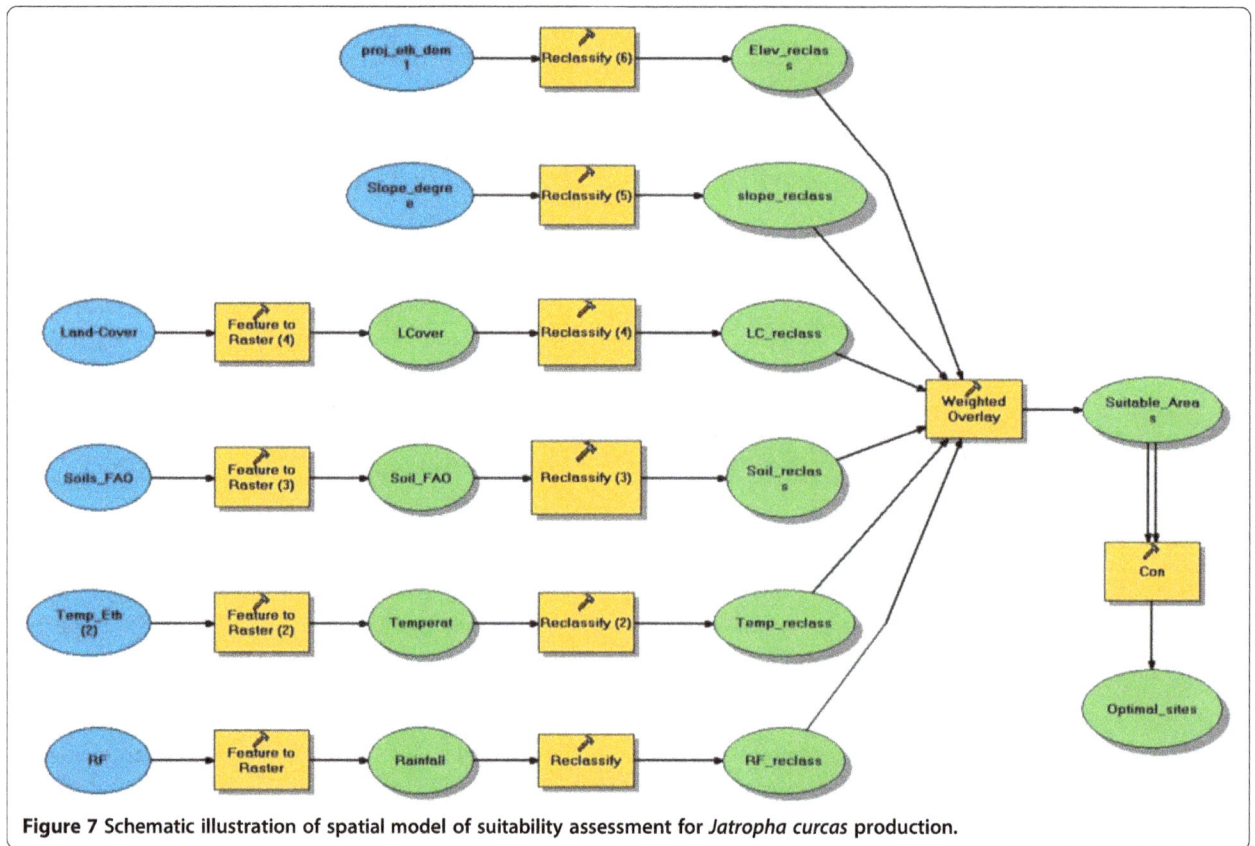

Figure 7 Schematic illustration of spatial model of suitability assessment for *Jatropha curcas* production.

Therefore, CR = CI/1.24, in this case.

Values of consistency ratio exceeding 0.10 are indicative of inconsistent judgments; whereas values of 0.10 or less indicate reasonable level of consistency in the pair-wise comparison.

Spatial modeling

Spatial model was built in ArcGIS (version 10.0); where data format conversion, reclassification and the final weighted overlay analysis were performed (see Figure 7). The various factors (i.e. precipitation, elevation, temperature, soils, land use/cover and slope) were combined to a suitability map of three levels of suitability. In the overall weighted overlay analysis, each criterion was weighed by its importance value, which reflects influence of the criteria in the overall suitability (S).

$$S = \sum_{i=1}^{n}(W_i \times C_i) \qquad (3)$$

Where W_i represents weight of each criterion (C_i).

Model feasibility

Different evaluation techniques were applied to make sure that the methods applied in this study were feasible. The first effort was ensuring the factors identified are relevant environmental variables that influence growth and yield of jatropha. This was verified through intensive literature review. Consistency ratio was the other mechanism to ensure whether the factor ranking process was reliable. Moreover, the model was verified with field realities by comparing the model results with location coordinates of actual occurrence data that were collected using global positioning system (GPS). Although the field data collection was not exhaustive, most of the suitable areas identified in this study complied with the field observation data.

The model is feasible for the current environmental conditions of the study area. However, if one or more of the environmental variables considered in this study changes, the model result will also be different. For instance, climate change may influence the patterns of suitability. Hence, this study assumes that the resultant suitable sites are identified based on the current environmental settings of the study area. This suitability index is sensitive to any change in the variables.

Abbreviations
AHP: Analytical Hierarchy Process; BRT: Boosted Regression Tree; CART: Classification and Regression Tree; CBO: Community based organization; CI: Consistency index; CR: Consistency ratio; CRGE: Climate Resilient Green Economy; DEM: Digital Elevation Model; EPA: Environmental Protection Authority; FAO: Food and Agriculture Organization (of the Unites Nations); GIS: Geographic Information Systems; GPS: Global Positioning

Systems; IBC: Institute of Biodiversity Conservation; MaxEnt: Maximum Entropy; MELCA: Movement for Ecological Learning and Community Action; MME: Ministry of Mines and Energy; NBE: National Bank of Ethiopia; NGO: Non-governmental organization; PRB: Population Reference Bureau; RI: Random Consistency Index; SAHP: Spatial Analytic Hierarchy Process.

Competing interests
The author declares that there are no competing interests associated with this research work.

Author's contribution
HT has carried out the research design, data collection, analysis, synthesis, field checking and write up of the research paper. The author organized, read and approved the final manuscript.

Author's information
Habitamu Taddese is a lecturer and researcher in Hawassa University, Wondo Genet College of Forestry and Natural Resources, Ethiopia. Habitamu has interest in spatial modeling exercises.

Acknowledgements
I am grateful of the efforts of the late Bayush Dessalegn, my mother, for the fruitful contributions she made to enable such an accomplishment. Let God rest her in peace. The researcher conveys special gratitude to the Ethiopian Meteorological Agency for providing data. I want to thank the Ministry of Agriculture of Ethiopia for additional data provision. I also appreciate the contribution of Mr. Kebede Wolka, Mr. Be'eminet Mengesha, Mr. Yidnekachew Habte and Mr. Solomon Abate who encouraged me to engage in this study. Special thanks to Tatum Branaman for helping in language editing. I also acknowledge the anonymous reviewers and the editor of this journal for their detailed and constructive comments to rectifying the overall structure and enriching the contents of the paper.

References
Achten WMJ, Verchot L, Franken YJ, Mathijs E, Singh VP, Aerts R, Muys B (2008) Jatropha bio-diesel production and use. Biomass Bioenergy 32(12):1063–1084

Awulachew SB, Yilma AD, Loulseged M, Loiskandl W, Ayana M, Alamirew T (2007) Water Resources and Irrigation Development in Ethiopia. International Water Management Institute, Colombo, Sri Lanka, Working Paper 123

Babaie-Kafaky S, Mataji A, Ahmadi SN (2009) Ecological capability assessment for multiple-use in forest areas using GIS- based multiple criteria decision making approach. Am J Environ Sci 5(6):714–721

Birega G, Maina A, Anderson T (2010) Biofuels - a Failure for Africa. African Biodiversity Network, Ethiopian Society for Consumer Protection, the Gaia Foundation

Biswas S, Kaushik N, Srikanth G (2006) Biodiesel: Technology and Business Opportunities - an Insight. In: Singh B, Swaminathan R, Ponraj V (eds) Proceedings of the Biodiesel Conference Toward Energy Independence - Focus of Jatropha. Rashtrapati Bhawan, New Delhi, India, pp 303–330

Brittaine R, Lutaladio NB (2010) Jatropha: A Smallholder Bioenergy Crop - The Potential for Pro-Poor Development. Integrated Crop Management. ISBN 1020-4555

CRGE (2011) Ethiopia's Climate-Resilient Green Economy. Federal Democratic Republic of Ethiopia, Addis Ababa, Ethiopia

Divakara BN, Upadhyaya HD, Wani SP, Laxmipathi-Gowda CL (2009) Biology and genetic improvement of Jatropha curcas L. A review. Appl Energy 87:732–742

Duc TT (2006) Using GIS and AHP Techniques for Land-use Suitability Analysis. International Symposium on Geo-informatics for Spatial Infrastructure Development in Earth and Allied Sciences, Polytechnic University of Hochiminh city, Vietnam

Emami B, Zarkesh MMK (2011) Application of spatial analytical hierarchy process in land suitability: case study on urban development of Tabriz Province, Iran. J Food Agric Environ 9(2):561–567

EPA (1998) Environmental Protection Authority. National Action Program to Combat Desertification. Federal Democratic Republic of Ethiopia, Addis Ababa

FAO (1976) Food and Agricultural Organization. A framework for land Evaluation. FAO, Rome

Franken YJ (2010) Plantation Establishment and Management. In: The Jatropha Handbook - From Cultivation to Application (9 - 29). FACT Foundation, Eindhoven. ISBN 978-90-815219-1-8

Gao S, Ouyang C, Wang S, Xu Y, Tang L, Chen F (2008) Effects of salt stress on growth, antioxidant enzyme and phenylalanine ammonialyase activities in Jatropha curcas L. seedlings. Plant Soil Environ 54(9):374–381

Garg KK, Karlberg L, Wani SP, Berndes G (2011) Jatropha production on wastelands in India: opportunities and trade-offs for soil and water management at the watershed scale. Bioenerg Water 5(4):410–430

Ghamgosar M, Haghyghy M, Mehrdoust F, Arshad N (2011) Multicriteria decision making based on analytical hierarchy process (AHP) in GIS for tourism. Middle-East J Sci Res 10(4):501–507

Gour VK (2006) Production Practices Including Post-Harvest Management of Jatropha Curcas. In: Singh B, Swaminathan R, Ponraj V (eds) Proceedings of the Biodiesel Conference Toward Energy Independence - Focus of Jatropha. Rashtrapati Bhawan, New Delhi, India, pp 223–351

Grass M (2009) Jatropha curcas L. Visions and realities. J Agric Rural Dev Trop Subtrop 110(1):29–38

Heller J (1996) Physic nut, Jatropha curcas. In: Promoting the Conservation and Use of Underutilized and Neglected Crops. International Plant Genetic Resources Institute (IPGRI), Rome, Italy

IBC (2007) Institute of Biodiversity Conservation. Ethiopia: Second Country Report on the State of PGRFA (Plant Genetic Resources for Food and Agriculture) to FAO, Addis Ababa, Ethiopia

ICRAF (2009) Jatropha Reality Check - A Field Assessment of the Agronomic and Economic Viability of Jatropha and Other Oilseed Crops in Kenya. Deutsche Gesellschaft für Technische Zusammenarbeit (GTZ) GmbH, Washington, DC and Nairobi

Jafari S, Zaredar N (2011) Land suitability analysis using multi-attribute decision making approach. Intern J Environ Sci Dev 1(5):441–445

Javaheri H (2006) Site selection of municipal solid waste landfill using analytical hierarchy process method in geographical information system technology environment in Giroft. Iran J Environ Health Sci Eng 3(3):177–184

Jingura RM, Matengaifa R, Musademba D, Musiyiwa K (2011) Characterization of land types and agro-ecological conditions for production of Jatropha as a feedstock for bio-fuels in Zimbabwe. Biomass Bio-Energy 35:2080–2086

Jongschaap REE, Corré WJ, Bindraban PS, Brandenburg WA (2007) Claims and Facts on Jatropha Curcas L. Global Jatropha Curcas Evaluation. Breeding and Propagation Programme. Plant Research International, Wageningen University & Research Centre, Wageningen, the Netherlands

Jongschaap REE, Montes OLR, De Ruijter FJ, Van ENL (2010) Highlights of the Jatropha Curcas Evaluation Program (JEP): Crop Management and the Fate of Press-Cake and Other by-Products With its Effects on Environmental Sustainability. Plant Research International Wageningen UR, Wageningen, the Netherlands, Groningen

Kearney M (2006) Habitat, environment and niche: what are we modeling? OIKOS 115(1):186–191

Koikai JS (2008) Utilizing GIS-Based Suitability Modeling to Assess the Physical Potential of Bioethanol Processing Plants in Western Kenya. Saint Mary's University of Minnesota University Central Services Press, Winona, MN, http://www2.smumn.edu/gis/GradProjects/KoikaiJ.pdf

Liang Y, Chen H, Tang M, Yang P, Shen S (2007) Responses of Jatropha curcas seedlings to cold stress: photosynthesis-related proteins and chlorophyll fluorescence characteristics. Physiol Plant 131:508–517

Makkar, H. P. S. & Becker, K. 1997. Potential of Jatropha curcas seed meal as a protein supplement to livestock feed, constraints to its utilization and possible strategies to overcome constraints. In: Biofuels and Industrial Products from Jatropha curcas (eds. Gubitz GM, Mittelbach M, Trabi M), Managua, Nicaragua, Graz, Austria, pp. 190–205

Makkar HPS, Becker K (2009) Jatropha curcas, a promising crop for the generation of biodiesel and value-added co-products. Review Article Eur J Lipid Sci Technol 111(8):773–787

MELCA (2008) Rapid assessment of biofuels development status in Ethiopia and proceedings of the National Workshop on Environmental Impact Assessment and Biofuels (eds. Anderson T, Million B), MELCA Mahiber, Addis Ababa, Ethiopia

MME (2008) Ministry of Mines and Energy. Ethiopian Bio-fuels Development and Utilization Strategy, Addis Ababa, Ethiopia

Muok B, Kallback L (2008) Feasibility Studies of Jatropha curcas as a Bio-fuel Feedstock in Kenya. http://kerea.org/wp-content/uploads/2012/12/Feasibility-Study-of-Jatropha-Curcas-as-a-Biofuel-Feedstock-in-Kenya.pdf

NBE (2010) National Bank of Ethiopia. Annual Report of 2008/2009. NBE, Addis Ababa, Ethiopia

Nyebenge M, Wanjara J, Owuor J, Theuri W (2009) Geographical Information Systems (GIS) Production and Spatial Analysis of Suitability Maps for Biofuel Feedstock for Ethiopia, Rwanda, Tanzania, Uganda, and Kenya. Final Report. ICRAF-GTZ Project. World Agro-forestry Centre (ICRAF), Nairobi, Kenya, http://www2.gtz.de/dokumente/bib-2011/giz2011-0042en-maps-biofuel.pdf

Ouwens KD, Francis G, Franken YJ, Rijssenbeek W, Riedacker A, Foidl N, Jongschaap R, Bindraban P (2007) Position Paper on *Jatropha Curcas* - State of the Art, Small and Large Scale Project Development. FACT Foundation, Wageningen

Paul S (2012) Location allocation for urban waste disposal site using multi-criteria analysis: A study on Nabadwip Municipality, West Bengal, India. Intern J Geomat Geosci 3(1):74–88

Pillay D, Da Silva E (2009) Sustainable development & bio-economic prosperity in Africa: Bio-fuels & the South African gateway. Afr J Biotechnol 8(1):2397–2408

PRB (2013) Population Reference Bureau. World Population Datasheet, Washington DC, USA

Saaty TL (1977) A scaling method for priorities in hierarchical structures. J Math Psychol 15:231–281

Saaty TL (1999) Basic theory of the analytic hierarchy process: how to make a decision. Rev R Acad Cienc Exact Fis Nat (Esp) 93(4):395–423

Saaty TL (2008) Decision making with the analytic hierarchy process. Int J Serv Sci 1(1):83–98

Saaty TL, Vargas LG (1991) Prediction, Projection and Forecasting. Kluwer Academic Publishers, Boston

Sahoo NK, Kumar A, Sharma S, Naik SN (2009) Interaction of *Jatropha curcas* plantation with ecosystem. Proceeding of International Conference on Energy and Environment. ISSN: 2070-3740.

Singh L, Bargali SS, Swamy SL (2006) Production Practices and Post-Harvest Management. In: Singh B, Swaminathan R, Ponraj V (eds) Proceedings of the biodiesel conference toward energy independence - Focus of Jatropha. Rashtrapati Bhawan, New Delhi, India, pp 252–267

Store R, Kangas J (2001) Integrating spatial multi-criteria evaluation and expert knowledge GIS-based habitat suitability modeling. Landsc Urban Plan 55:79–93

Triantaphyllou E, Mann SH (1995) Using the analytic hierarchy process for decision making in engineering applications: some challenges. Intern J of Industrial Eng 2(1):35–44

Valavanis VD, Pierce GJ, Zuur AF, Palialexis A, Saveliev A, Katara I, Wan J (2008) Modeling of essential fish habitat based on remote sensing, spatial analysis and GIS. Hydrobiologia 612:5–20

Van der Putten E (2010) General Data on Jatropha. In: FACT Foundation - The Jatropha Handbook - From Cultivation to Application (1–7). FACT Foundation, Eindhoven

Vivero JL, Ensermu K, Sebsebe D (2010) The red List of Endemic Trees & Shrubs of Ethiopia and Eritrea. Fauna & Flora International, Cambridge, UK

Wendimu MA (2013) Jatropha Potential on Marginal Land in Ethiopia: Reality or Myth? University of Copenhagen, Department of Food and Resource Economics (IFRO), Denmark

Wiesenhütter J (2003) Use of the Physic nut *(Jatropha Curcas L.)* to Combat Desertification and Reduce Poverty: Possibilities and Limitations of Technical Solutions in a Particular Socio-Economic Environment, the Case of Cape Verde. GTZ, Convention Project to Combat Desertification

Wu WG, Huang JK, Deng XZ (2009) Potential land for plantation of Jatropha curcas as feedstock for biodiesel in China. Sci China Series Earth Sci, doi: 10.1007/s11430-009-0204-y

Ye M, Li C, Francis G, Makkar HPS (2009) Current situation and prospects of Jatropha curcas as a multipurpose tree in China. Agroforest Syst 76:487–497, doi:10.1007/s10457-009-9226-x

Zarkesh MMK, Almasi N, Taghizadeh F (2011) Ecotourism land capability evaluation using spatial multi-criteria evaluation. Res J Appl Sci Eng Technol 3(7):693–700

5

Evaluation of parameter uncertainties in nonlinear regression using Microsoft Excel Spreadsheet

Wei Hu[1], Jing Xie[1], Henry Wai Chau[2] and Bing Cheng Si[1*]

Abstract

Background: Nonlinear relationships are common in the environmental discipline. Spreadsheet packages such as Microsoft Excel come with an add-on for nonlinear regression, but parameter uncertainty estimates are not yet available. The purpose of this paper is to use Monte Carlo and bootstrap methods to estimate nonlinear parameter uncertainties with a Microsoft Excel spreadsheet. As an example, uncertainties of two parameters (a and n) for a soil water retention curve are estimated.

Results: The fitted parameters generally do not follow a normal distribution. Except for the upper limit of a using the bootstrap method, the lower and upper limits of a and n obtained by these two methods are slightly greater than those obtained using the SigmaPlot software which linearlizes the nonlinear model.

Conclusions: Since the linearization method is based on the assumption of normal distribution of parameter values, the Monte Carlo and bootstrap methods may be preferred to the linearization method.

Keywords: Bootstrap; Monte Carlo; Soil water retention; Parameter uncertainty; Excel

Background

Nonlinear relationships are common in natural and environmental sciences (Wraith and Or 1998; Luo et al. 2003; Cwiertny and Roberts 2005). As a result, there are many software packages (such as SAS and MathCAD) that implement nonlinear parameter estimation. However, spreadsheet techniques are easier to learn than other specialized mathematical programs for nonlinear parameter estimation, because no programming skills are needed in spreadsheets to develop their own parameter estimation routines (Wraith and Or 1998). In addition, spreadsheets have the merits of wide accessibility and powerful computation in terms of fitting nonlinear models. For these reasons, spreadsheets such as Microsoft Excel are widely suggested to make nonlinear parameter estimation (Harris 1998; Smith et al. 1998; Wraith and Or 1998; Brown 2001; Berger 2007).

Parameter uncertainty refers to lack of knowledge regarding the exact true value of a quantity (Tong et al. 2012). Different observations are usually obtained when experiments are repeated, resulting in different values of parameters. It is usually expressed as an interval of

parameter values at a certain confidence level, say, 95%. It is also expressed as the standard error of the mean by assuming normal distribution of parameter values. Parameter uncertainty can be used to judge the degree of reliability of the parameter estimates, which is important to making decisions for environmental management. For these reasons, estimation of parameter uncertainties is significant for nonlinear parameter estimates. However, relatively less work has focused on the nonlinear parameter uncertainty estimates using spreadsheet packages.

Parameter uncertainty can be obtained exactly by assuming normal distribution of a parameter in linear regression, but not in nonlinear regression. Nonlinear regression programs usually give the parameter uncertainty by calculating the standard error of the mean, and assuming linear relationship between variables in the vicinity of the estimated parameter values and normal distribution of parameter values. Furthermore, this method usually involves evaluating a Hessian matrix (a square matrix of second-order partial derivatives of a scalar-valued function to describe the local curvature of a function of many variables) or an inequality, which makes it more complicated and time demanding (Brown 2001). More general methods such as Monte Carlo and bootstrap simulation can be used to estimate the parameter uncertainties. Both methods have their

---

* Correspondence: bing.si@usask.ca
[1]Department of Soil Science, University of Saskatchewan, Saskatoon, SK S7N 5A8, Canada
Full list of author information is available at the end of the article

own advantages: while the Monte Carlo method is based on a theoretical probability distribution of a variable, the bootstrap method has no assumption on the probability distribution of a variable and thus has no limits on sampling size. Among numerous related applications are testing fire ignition selectivity of different landscape characteristics using the Monte Carlo simulation (Conedera et al. 2011) and estimating uncertainty of greenhouse gas emissions using the bootstrap simulation (Tong et al. 2012). However, parameter uncertainties estimation in spreadsheets using the Monte Carlo and bootstrap methods has been rarely discussed.

Both nonlinear parameter values and their associated uncertainties are important for decision making and thus should be implemented in spreadsheet program like Excel. Microsoft Excel spreadsheets have other advantages including their general facility for data input and management, ease in implementing calculations, and often advanced graphics and reporting capabilities (Wraith and Or 1998). These advantages are likely to make the use of spreadsheets to quantify parameter uncertainties more desirable.

The objective of this paper is to apply the Monte Carlo and bootstrap simulations to obtain parameter uncertainties with a Microsoft Excel spreadsheet. In addition, the influences of number of simulation on uncertainty estimates are also discussed. For this, we use as an example, a common soil physical property - soil water

retention curve, which has been widely used in soil, hydrological, and environmental communities.

Results and discussion
Nonlinear regression parameters estimation
Here are the steps to estimate parameters α and n in Excel using nonlinear regression.

1. List the applied suction pressure as the independent variable in column A and measured soil water content (θ) as the dependent variable in column B (Figure 1).

2. Temporarily set the value of α as 0.1 and n as 1 in cells B19 and B20, respectively (Figure 1). It is important to set an appropriate initial value because an obviously unreasonable initial value will lead to an unanticipated value. Please refer to related document for initial value estimation (e.g., Delboy 1994). List the measured θ_r and θ_s in cells B21 and B22, respectively. Then the predicted θ value can be calculated with the van Genuchten soil water retention curve model (Eq. 5) using suction pressure and all parameter values. For example, the predicted θ in cell E2 ($\hat{\theta}_{E2}$) is calculated by the following formula:

$$\hat{\theta}_{E2} = \$B\$21 + (\$B\$22 - \$B\$21) * (1 + (\$B\$19 \\ * \$A2)^\wedge\$B\$20)^\wedge(-1 + 1/\$B\$20) \quad (1)$$

As Figure 1 shows, all the predicted θ values are 0.395 given the initial values.

	A	B	C	D	E
	Soil matrix potential (-cm)	Measured θ (cm³ cm⁻³)			Predicted θ (cm³ cm⁻³)
1					
2	0	0.37545			0.39500
3	1	0.38246			0.39500
4	2	0.38483			0.39500
5	3	0.38529			0.39500
6	4	0.38301			0.39500
7	5	0.37299			0.39500
8	6	0.37135			0.39500
9	7	0.36688			0.39500
10	15	0.24659			0.39500
11	25	0.14189			0.39500
12	50	0.09960			0.39500
13	100	0.07646			0.39500
14	300	0.03281			0.39500
15	500	0.03189			0.39500
16	1000	0.01832			0.39500
17	15000	0.01082			0.39500
18					
19	α	0.10000			
20	n	1.00000			
21	θ_r	0.01100			
22	θ_s	0.39500			
23	SSE	0.83005			

Figure 1 Data input and initial value set for α and n in a spreadsheet.

3. Calculate the sum of squared residuals (*SSE*) using Excel function SUMXMY2 in cell B23 by entering "SUMXMY2(B2 : B17, E2 : E17)". We obtain 0.83005 for *SSE* for the given initial parameter values (Figure 1).

4. The model obtains the maximum likelihood when the *SSE* is minimized, which is the principle of least-square fitting method. The *Solver* tool in Excel can be used to minimize the *SSE* values. The *Solver* tool can be found under the Data menu in Excel. If not found there, it has to be added from File menu through the path *File- > Options- > Add-Ins- > Solver Add-in*. As Figure 2 shows, the "Set Objective" box is the value to be optimized, which is the *SSE* value in cell B23. Click "Min" to minimize the objective *SSE* by changing the values of α and n as shown in the "By Changing Variable Cells".

5. The *Solver* will then find the minimum *SSE* (in cell B23) and corresponding α (in cell B19) and n (in cell B20) values (Figure 3). The measured θ values are in agreement with the predicted θ values (Figure 4), indicating a good nonlinear curve fitting.

Using Monte Carlo method to estimate parameter uncertainty

Stepwise application of the Monte Carlo method in estimating parameter uncertainties with 200 simulations is demonstrated below:

1. Resample θ using the Monte Carlo method in different columns. Take cell L2 for example, the simulated θ (θ_{L2}) is calculated by the following formula:

$$\theta_{L2} = \$E2 + NORM.INV(RAND(), 0, SQRT(SSE/df)) \tag{2}$$

where $\$E2$ refers to the corresponding predicted θ. *SSE* is the value calculated above, which is the value in cell B23 in Figure 5. The degree of freedom (*df*) equals 14. The θ values in the other rows in column L are simulated in a similar way. The simulated values for a new dependent variable θ are demonstrated in cells L2-L17 (Figure 5). The same Monte Carlo simulations are performed from column M to column HC. Therefore, a total of 200 sets of simulated θ are obtained (Figure 5).

2. Use the same procedure as introduced before to conduct nonlinear regression for each new data set of θ. Note that the new data set of θ will change during optimization, which will result in errors in fitting. Therefore, we copy the simulated θ data to a new sheet by right-clicking "Paste Special" and selecting "Values" in the dialogue of "Paste Special". For better display, predicted θ array, all the parameter values, and corresponding *SSE* value are presented in the same column for each simulation (Figure 5). Optimization of parameters α and n is made independently for each simulation using the *Solver* tool. The initial values are set as the fitted values

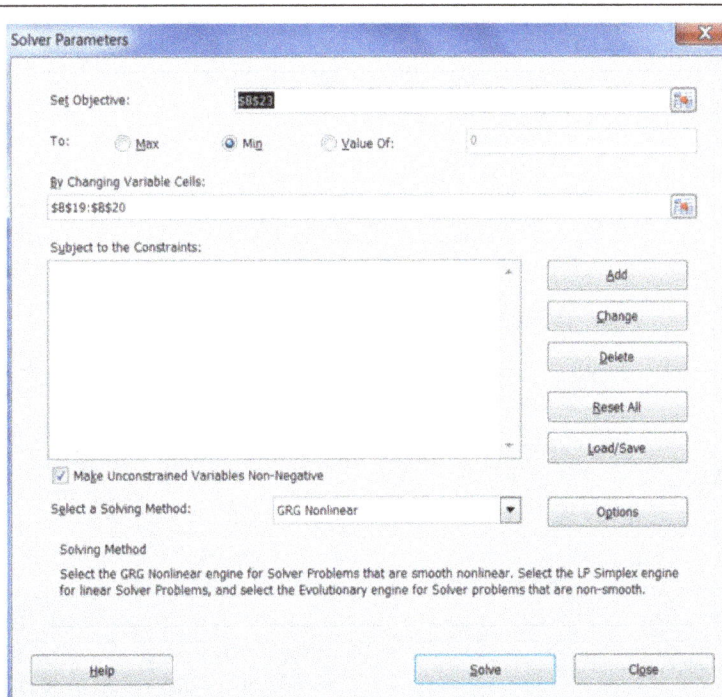

Figure 2 Solver working screen in Excel 2010.

	A	B	C	D	E
1	Soil matrix potential (-cm)	Measured θ (cm^3 cm^{-3})			Predicted θ (cm^3 cm^{-3})
2	0	0.37545			0.39500
3	1	0.38246			0.39401
4	2	0.38483			0.39080
5	3	0.38529			0.38536
6	4	0.38301			0.37788
7	5	0.37299			0.36865
8	6	0.37135			0.35804
9	7	0.36688			0.34642
10	15	0.24659			0.25062
11	25	0.14189			0.17179
12	50	0.09960			0.09245
13	100	0.07646			0.04981
14	300	0.03281			0.02266
15	500	0.03189			0.01765
16	1000	0.01832			0.01410
17	15000	0.01082			0.01116
18					
19	α	0.07988			
20	n	2.09920			
21	θ_r	0.01100			
22	θ_s	0.39500			
23	SSE	0.00319			

Figure 3 A spreadsheet for estimating nonlinear regression coefficients a and n.

obtained before for all simulations to reduce the time required during the optimization. Therefore, initial values of 0.07988 and 2.09920 are set for α and n, respectively (Figure 5). Because the maximum number of variables *Solver* can solve is 200, we can minimize the *SSE* values for 100 simulations at one time by minimizing the sum of *SSE* values of 100 simulations. For example, the parameters α and n for the first 100 simulations can be optimized by minimizing cell E19 by entering "=SUM(L18:

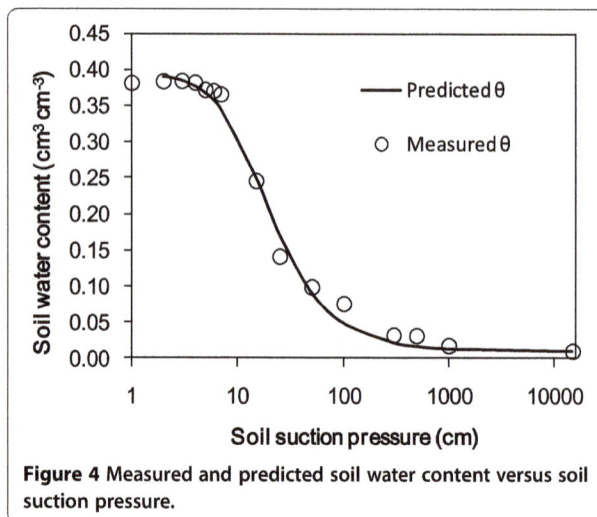

Figure 4 Measured and predicted soil water content versus soil suction pressure.

DG18)" (Figure 5). Similarly, the parameters for the second 100 simulations can be optimized in cell E20 by entering "=SUM(DH18:HC18)". Therefore, we obtain 200 values for both parameters (α and n) as shown in cells from L19 to HC20 (Figure 6). The frequency distribution of α and n are shown in Figure 7. Visually, both of them follow a normal distribution. However, the Shapiro-Wilk test shows that the parameter α does not conform to a normal distribution, whereas parameter n does. This indicates that the fitted parameters may not necessarily be normally distributed even if the dependent variable is normally distributed, due to the nonlinear relations between them.

3. Calculate the 95% confidence interval of α or n values with 200 simulations. We copy all the fitted α or n values, then paste them to a new sheet by right-clicking "Paste Special" and selecting "Transpose" in the dialogue of "Paste Special" to list all the fitted α or n values in one column. Select the transposed data, and rank them in an ascending order using *Sort* tool in *Data* tab. Find the value of α and n corresponding to the 2.5 percentile and 97.5 percentile, which are the lower limit and upper limit, respectively, at a 95% confidence. The 95% confidence interval are (0.0680, 0.0939) and (1.9185, 2.3689) for α and n, respectively (Table 1). The difference between upper limit and lower limit is 0.0259 and 0.4504 for α and n, respectively.

	A	B	C	D	E	K	L	M	N	O	P
1	Soil matrix potential (-cm)	Measured θ (cm³ cm⁻³)			Predicted θ (cm³ cm⁻³)	Monte Carlo					
2	0	0.37545			0.39500		0.40260	0.39739	0.40217	0.40302	0.38397
3	1	0.38246			0.39401		0.41745	0.37636	0.39762	0.38644	0.38687
4	2	0.38483			0.39080		0.39070	0.40722	0.37544	0.37140	0.39730
5	3	0.38529			0.38536		0.39195	0.37067	0.37169	0.39430	0.40272
6	4	0.38301			0.37788		0.40189	0.39625	0.38612	0.36903	0.37417
7	5	0.37299			0.36865		0.38383	0.37797	0.38544	0.37240	0.35964
8	6	0.37135			0.35804		0.35263	0.36020	0.37474	0.38739	0.36548
9	7	0.36688			0.34642		0.34043	0.33859	0.34283	0.34326	0.37098
10	15	0.24659			0.25062		0.26945	0.24960	0.25453	0.24270	0.23907
11	25	0.14189			0.17179		0.15018	0.17221	0.17870	0.17457	0.15425
12	50	0.09960			0.09245		0.10468	0.08979	0.10991	0.11010	0.10472
13	100	0.07646			0.04981		0.06072	0.07586	0.03827	0.06461	0.04422
14	300	0.03281			0.02266		0.02130	-0.00292	0.05046	0.02671	0.01236
15	500	0.03189			0.01765		0.04485	0.00853	0.03129	0.02363	-0.01379
16	1000	0.01832			0.01410		0.00228	0.00753	0.01668	0.01281	0.03531
17	15000	0.01082			0.01116		0.00160	0.02626	0.01119	0.00555	-0.00526
18						SSE	0.00359	0.00299	0.00259	0.00223	0.00371
19	α	0.07988	1-100	SUM_SSE(L:DG)	0.34049	α	0.07988	0.07988	0.07988	0.07988	0.07988
20	n	2.09920	101-200	SUM_SSE(DH:HC)	0.34874	n	2.09920	2.09920	2.09920	2.09920	2.09920
21	θr	0.01100				θr	0.01100	0.01100	0.01100	0.01100	0.01100
22	θs	0.39500				θs	0.39500	0.39500	0.39500	0.39500	0.39500
23	SSE	0.00319					1	2	3	4	5
24											
25				0			0.39500	0.39500	0.39500	0.39500	0.39500
26				1			0.39401	0.39401	0.39401	0.39401	0.39401
27				2			0.39079	0.39079	0.39079	0.39079	0.39079
28				3			0.38534	0.38534	0.38534	0.38534	0.38534
29				4			0.37785	0.37785	0.37785	0.37785	0.37785
30				5			0.36861	0.36861	0.36861	0.36861	0.36861
31				6			0.35799	0.35799	0.35799	0.35799	0.35799
32				7			0.34636	0.34636	0.34636	0.34636	0.34636
33				15			0.25059	0.25059	0.25059	0.25059	0.25059
34				25			0.17181	0.17181	0.17181	0.17181	0.17181
35				50			0.09250	0.09250	0.09250	0.09250	0.09250
36				100			0.04986	0.04986	0.04986	0.04986	0.04986
37				300			0.02268	0.02268	0.02268	0.02268	0.02268
38				500			0.01767	0.01767	0.01767	0.01767	0.01767
39				1000			0.01411	0.01411	0.01411	0.01411	0.01411
40				15000			0.01116	0.01116	0.01116	0.01116	0.01116

Figure 5 Resampling dependent variable θ using Monte Carlo method and initial values set (only the first 5 simulations are shown).

Using bootstrap method to estimate parameter uncertainty

Stepwise application of the bootstrap method in estimating parameter uncertainties with 200 simulations is demonstrated as follows:

1. Resample θ using the bootstrap method in different columns (Figure 8). Take cell L2 for example, the simulated θ ($θ_{L2}$) can be calculated by the following formula:

$$θ_{L2} = \$E2 + INDEX(\$C : \$C, INT(RAND() * 16 + 2)) \tag{3}$$

where the function INDEX is used to randomly select a residue value from row 2 to row 17 in column C (the residual is calculated by subtracting predicted θ from the original θ). The θ values at other rows in column L and in other columns (column M to column HC) are simulated in a similar way. Here, 2 in the right hand side of Eq. (3) means that data start at second row.

2. Similar to the Monte Carlo method, parameters α and n for all simulations are fitted by minimizing the sum of every 100 SSE values using the Solver tool (Figures 8 and 9). Therefore, we can also obtain 200 values for both parameters (α and n) as shown in cells from L19 to HC20 (Figure 9). The frequency distribution of α and n are shown in Figure 10. They also visually follow a normal distribution. However, the Shapiro-Wilk test shows that the parameter α does not conform to normal distribution, whereas parameter n does.

3. Similar to the Monte Carlo method, the 95% confidence intervals for these two parameters are calculated. They are (0.0680, 0.0925) and (1.9172, 2.3356), for α and n respectively (Table 1). The corresponding difference between upper limit and lower limit is 0.0245 and 0.4183 for α and n, respectively.

Influences of number of simulation on parameter uncertainty analysis

Datasets of fitted values with different numbers of simulations are obtained using a similar method as demonstrated

	A	B	C	D	E	K	L	M	N	O	P
	Soil matrix potential (-cm)	Measured θ (cm³ cm⁻³)			Predicted θ (cm³ cm⁻³)		Monte Carlo				
2	0	0.37545			0.39500		0.40260	0.39739	0.40217	0.40302	0.38397
3	1	0.38246			0.39401		0.41745	0.37636	0.39762	0.38644	0.38687
4	2	0.38483			0.39070		0.39080	0.40722	0.37544	0.37140	0.39730
5	3	0.38529			0.38536		0.39195	0.37067	0.37169	0.39430	0.40272
6	4	0.38301			0.37788		0.40189	0.39625	0.38612	0.36903	0.37417
7	5	0.37299			0.36865		0.38383	0.37797	0.38544	0.37240	0.35964
8	6	0.37135			0.35804		0.35263	0.36020	0.37474	0.38739	0.36548
9	7	0.36688			0.34642		0.34043	0.33859	0.34283	0.34326	0.37098
10	15	0.24659			0.25062		0.26945	0.24960	0.25453	0.24270	0.23907
11	25	0.14189			0.17179		0.15018	0.17221	0.17870	0.17457	0.15425
12	50	0.09960			0.09245		0.10468	0.08979	0.10991	0.11010	0.10472
13	100	0.07646			0.04981		0.06072	0.07586	0.03827	0.06461	0.04422
14	300	0.03281			0.02266		0.02130	-0.00292	0.05046	0.02671	0.01236
15	500	0.03189			0.01765		0.04485	0.00853	0.03129	0.02363	-0.01379
16	1000	0.01832			0.01410		0.00228	0.00753	0.01668	0.01281	0.03531
17	15000	0.01082			0.01116		0.00160	0.02626	0.01119	0.00555	-0.00526
18						SSE	0.00351	0.00298	0.00225	0.00195	0.00345
19	α	0.07988	1-100	SUM_SSE(L:DG)	0.29465	α	0.07698	0.07884	0.07677	0.08020	0.07722
20	n	2.09920	101-200	SUM_SSE(DH:HC)	0.30399	n	2.10742	2.10373	2.04938	2.01214	2.21869
21	θr	0.01100				θr	0.01100	0.01100	0.01100	0.01100	0.01100
22	θs	0.39500				θs	0.39500	0.39500	0.39500	0.39500	0.39500
23	SSE	0.00319					1	2	3	4	5
24											
25				0			0.39500	0.39500	0.39500	0.39500	0.39500
26				1			0.39410	0.39404	0.39398	0.39380	0.39428
27				2			0.39115	0.39093	0.39084	0.39023	0.39170
28				3			0.38611	0.38564	0.38565	0.38446	0.38702
29				4			0.37915	0.37834	0.37860	0.37677	0.38027
30				5			0.37050	0.36931	0.37000	0.36753	0.37164
31				6			0.36048	0.35890	0.36015	0.35710	0.36139
32				7			0.34944	0.34748	0.34939	0.34585	0.34990
33				15			0.25588	0.25237	0.25972	0.25565	0.24919
34				25			0.17630	0.17321	0.18292	0.18098	0.16458
35				50			0.09479	0.09312	0.10171	0.10239	0.08306
36				100			0.05078	0.05005	0.05588	0.05733	0.04262
37				300			0.02286	0.02269	0.02527	0.02634	0.01933
38				500			0.01774	0.01765	0.01935	0.02015	0.01547
39				1000			0.01413	0.01410	0.01504	0.01554	0.01292
40				15000			0.01116	0.01116	0.01124	0.01129	0.01107

Figure 6 Nonlinear regression fitting for resampled θ using Monte Carlo method (only the first 5 simulations are shown).

before (data not shown). According to Shapiro-Wilk test, both parameters do not follow a normal distribution with different numbers of simulations except for a few cases that the number of simulations ≤400. This may indicate that the assumption of normality in the linearization method does not hold true.

The lower limit, upper limit, and their difference change slightly with the number of simulations. However, they are almost constant beyond a certain number of simulations (Figure 11). Here, we determine the number of simulations required for both methods according to the change in difference between the upper limit and lower limit. If

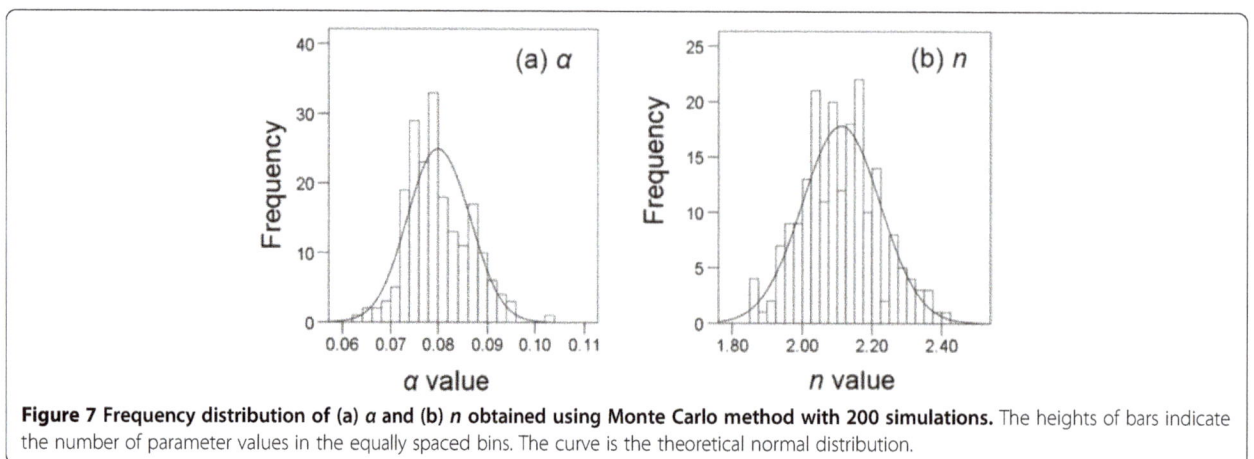

Figure 7 Frequency distribution of (a) α and (b) n obtained using Monte Carlo method with 200 simulations. The heights of bars indicate the number of parameter values in the equally spaced bins. The curve is the theoretical normal distribution.

Table 1 Comparison of parameter uncertainties calculated by different methods

Parameter		Monte Carlo	Bootstrap	Linearization
α	Lower limit	0.0680	0.0680	0.0670
	Upper limit	0.0939	0.0925	0.0928
	Upper limit-Lower limit	0.0259	0.0245	0.0258
n	Lower limit	1.9185	1.9172	1.8758
	Upper limit	2.3689	2.3356	2.3218
	Upper limit-Lower limit	0.4504	0.4184	0.4460

the relative difference (RD%) of the difference between the upper limit and lower limit under a certain number of simulation is less than 5% compared with that under 2000 simulations, the number of simulations tested is taken to be the required number of simulations. The RD% can be calculated as

$$RD\% = \left| \frac{V_m - V_{2000}}{V_{2000}} \right| * 100\% \qquad (4)$$

where V_m and V_{2000} are the differences between the upper limit and lower limit under m simulations and 2000 simulations, respectively.

For the Monte Carlo method, the RD% is less than 5% for α and n when the numbers of simulation are ≥100 and 200, respectively. For the bootstrap method, the RD% is less than 5% for α and n when the numbers of simulation are ≥500 and 400, respectively. Therefore, simulation number of 200 and 500 are needed to produce reliable data at the 95% confidence interval of parameters for the Monte Carlo and bootstrap methods, respectively. In this sense, the Monte Carlo method may be better than the bootstrap method. However, the optimal number of simulation may also differ with specific situations. For example, Efron and Tibshirani (1993) stated that a minimum of approximately 1000 bootstrap re-samples was sufficient to obtain accurate confidence

	A	B	C	D	E	K	L	M	N	O	P
	Soil matrix potential (~cm)	Measured θ (cm³ cm⁻³)	Residue (cm³ cm⁻³)		Predicted θ (cm³ cm⁻³)		Bootstrap				
1											
2	0	0.37545	-0.01955		0.39500		0.38346	0.36509	0.38903	0.38903	0.40515
3	1	0.38246	-0.01154		0.39401		0.40825	0.39914	0.40732	0.37446	0.39393
4	2	0.38483	-0.00597		0.39080		0.39514	0.38483	0.40411	0.39593	0.39501
5	3	0.38529	-0.00007		0.38536		0.40583	0.37940	0.37382	0.40583	0.40583
6	4	0.38301	0.00513		0.37788		0.38803	0.38803	0.34797	0.38301	0.40453
7	5	0.37299	0.00434		0.36865		0.37880	0.37378	0.37287	0.36858	0.38290
8	6	0.37135	0.01331		0.35804		0.37228	0.37135	0.36519	0.37228	0.38468
9	7	0.36688	0.02046		0.34642		0.35063	0.36688	0.35063	0.35357	0.35973
10	15	0.24659	-0.00403		0.25062		0.23908	0.25575	0.25575	0.26393	0.23908
11	25	0.14189	-0.02991		0.17179		0.18510	0.17692	0.16776	0.17172	0.16025
12	50	0.09960	0.00716		0.09245		0.08842	0.11291	0.09678	0.08648	0.10260
13	100	0.07646	0.02665		0.04981		0.04973	0.03026	0.05996	0.03826	0.04947
14	300	0.03281	0.01015		0.02266		0.03597	0.02232	0.01111	0.01111	0.01863
15	500	0.03189	0.01425		0.01765		0.02278	0.03189	0.02186	0.03096	0.01757
16	1000	0.01832	0.00421		0.01410		0.02741	0.02741	0.04075	0.00256	0.02126
17	15000	0.01082	-0.00034		0.01116		0.01108	0.01549	0.01108	0.01549	0.01108
18						SSE	0.00191	0.00297	0.00253	0.00196	0.00278
19	α	0.07988	1-100	SUM_SSE (L:DG)	0.30515	α	0.07988	0.07988	0.07988	0.07988	0.07988
20	n	2.09920	101-200	SUM_SSE (DH:HC)	0.32341	n	2.09920	2.09920	2.09920	2.09920	2.09920
21	θ_r	0.01100				θ_r	0.01100	0.01100	0.01100	0.01100	0.01100
22	θ_s	0.39500				θ_s	0.39500	0.39500	0.39500	0.39500	0.39500
23	SSE	0.00319					1	2	3	4	5
24											
25				0			0.39500	0.39500	0.39500	0.39500	0.39500
26				1			0.39401	0.39401	0.39401	0.39401	0.39401
27				2			0.39079	0.39079	0.39079	0.39079	0.39079
28				3			0.38534	0.38534	0.38534	0.38534	0.38534
29				4			0.37785	0.37785	0.37785	0.37785	0.37785
30				5			0.36861	0.36861	0.36861	0.36861	0.36861
31				6			0.35799	0.35799	0.35799	0.35799	0.35799
32				7			0.34636	0.34636	0.34636	0.34636	0.34636
33				15			0.25059	0.25059	0.25059	0.25059	0.25059
34				25			0.17181	0.17181	0.17181	0.17181	0.17181
35				50			0.09250	0.09250	0.09250	0.09250	0.09250
36				100			0.04986	0.04986	0.04986	0.04986	0.04986
37				300			0.02268	0.02268	0.02268	0.02268	0.02268
38				500			0.01767	0.01767	0.01767	0.01767	0.01767
39				1000			0.01411	0.01411	0.01411	0.01411	0.01411
40				15000			0.01116	0.01116	0.01116	0.01116	0.01116

Figure 8 Resampling dependent variable θ using bootstrap method and initial values set (only the first 5 simulations are shown).

	A	B	C	D	E	K	L	M	N	O	P
	Soil matrix potential (-cm)	Measured θ (cm³ cm⁻³)	Residue (cm³ cm⁻³)		Predicted θ (cm³ cm⁻³)		Bootstrap				
2	0	0.37545	-0.01955		0.39500		0.38346	0.36509	0.38903	0.38903	0.40515
3	1	0.38246	-0.01154		0.39401		0.40825	0.39914	0.40732	0.37446	0.39393
4	2	0.38483	-0.00597		0.39080		0.39514	0.38483	0.40411	0.39593	0.39501
5	3	0.38529	-0.00007		0.38536		0.40583	0.37940	0.37382	0.40583	0.40583
6	4	0.38301	0.00513		0.37788		0.38803	0.38803	0.34797	0.38301	0.40453
7	5	0.37299	0.00434		0.36865		0.37880	0.37378	0.37287	0.36858	0.38290
8	6	0.37135	0.01331		0.35804		0.37228	0.37135	0.36519	0.37228	0.38468
9	7	0.36688	0.02046		0.34642		0.35063	0.36688	0.35063	0.35357	0.35973
10	15	0.24659	-0.00403		0.25062		0.23908	0.25575	0.25575	0.26393	0.23908
11	25	0.14189	-0.02991		0.17179		0.18510	0.17692	0.16776	0.17172	0.16025
12	50	0.09960	0.00716		0.09245		0.08842	0.11291	0.09678	0.08648	0.10260
13	100	0.07646	0.02665		0.04981		0.04973	0.03026	0.05996	0.03826	0.04947
14	300	0.03281	0.01015		0.02266		0.03597	0.02232	0.01111	0.01111	0.01863
15	500	0.03189	0.01425		0.01765		0.02278	0.03189	0.02186	0.03096	0.01757
16	1000	0.01832	0.00421		0.01410		0.02741	0.02741	0.04075	0.00256	0.02126
17	15000	0.01082	-0.00034		0.01116		0.01108	0.01549	0.01108	0.01549	0.01108
18						SSE	0.00177	0.00246	0.00246	0.00129	0.00224
19	α	0.07988	1-100	SUM_SSE (L:DG)	0.26909	α	0.07524	0.07137	0.08117	0.07007	0.07234
20	n	2.09920	101-200	SUM_SSE (DM:HC)	0.28362	n	2.14942	2.17036	2.04523	2.29828	2.27779
21	θ_r	0.01100				θ_r	0.01100	0.01100	0.01100	0.01100	0.01100
22	θ_s	0.39500				θ_s	0.39500	0.39500	0.39500	0.39500	0.39500
23	SSE	0.00319					1	2	3	4	5
24											
25					0		0.39500	0.39500	0.39500	0.39500	0.39500
26					1		0.39421	0.39433	0.39385	0.39452	0.39446
27					2		0.39154	0.39201	0.39032	0.39265	0.39239
28					3		0.38688	0.38789	0.38452	0.38911	0.38852
29					4		0.38031	0.38202	0.37671	0.38381	0.38278
30					5		0.37203	0.37455	0.36724	0.37680	0.37526
31					6		0.36232	0.36568	0.35651	0.36823	0.36614
32					7		0.35151	0.35569	0.34489	0.35832	0.35569
33					15		0.25716	0.26454	0.25163	0.26211	0.25745
34					25		0.17536	0.18137	0.17552	0.17263	0.16924
35					50		0.09226	0.09483	0.09733	0.08412	0.08314
36					100		0.04849	0.04921	0.05373	0.04147	0.04145
37					300		0.02167	0.02163	0.02464	0.01836	0.01852
38					500		0.01694	0.01685	0.01900	0.01479	0.01492
39					1000		0.01368	0.01360	0.01488	0.01254	0.01262
40					15000		0.01112	0.01111	0.01123	0.01105	0.01105

Figure 9 Nonlinear regression fitting for resampled θ using bootstrap method (only the first 5 simulations are shown).

interval estimates. In order to obtain reliable confidence interval estimates, we suggest increasing the simulation times by 100 at each step, and the final results can be obtained when the values stabilize within consecutive steps.

Comparison with parameter uncertainty approximated by linear model

The values α and n are estimated to be 0.0799 and 2.0988, respectively, by SigmaPlot 10.0 (Figure 12), which are exactly the same as the estimates made by nonlinear

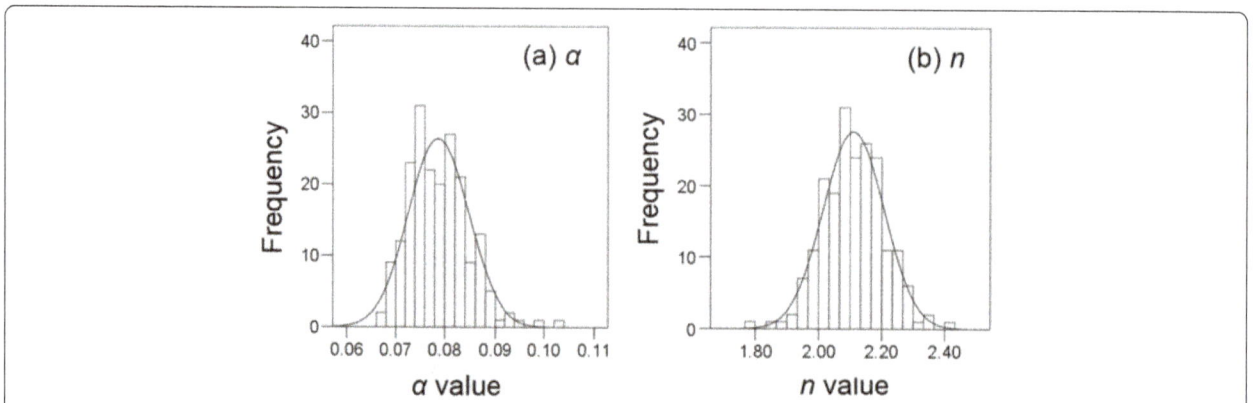

Figure 10 Frequency distribution of (a) α and (b) n obtained using bootstrap method with 200 simulations. The heights of bars indicate the number of parameter values in the equally spaced bins. The curve is the theoretical normal distribution.

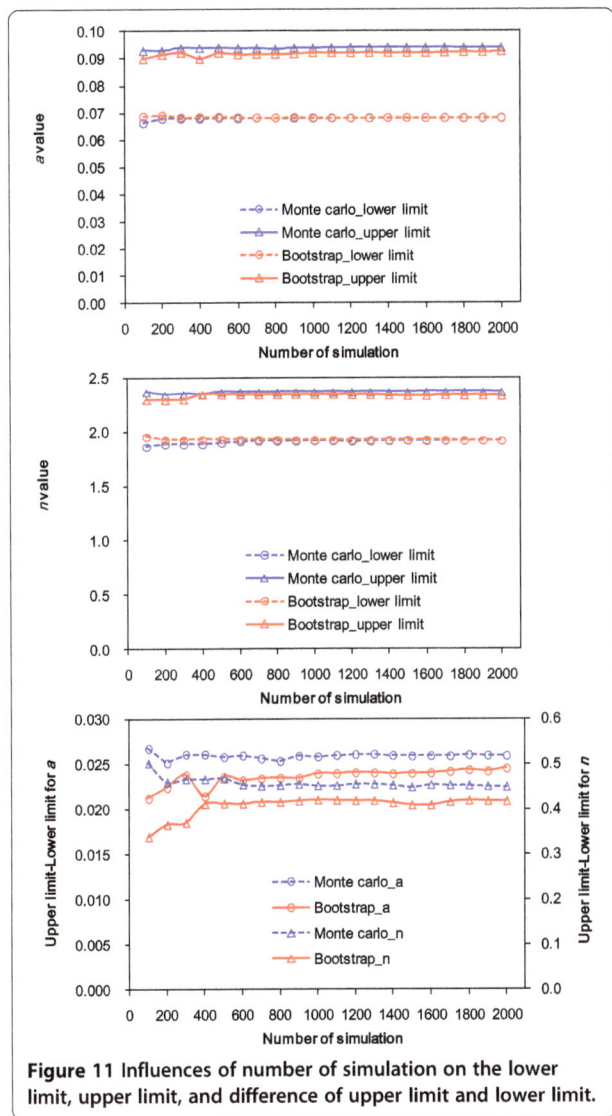

Figure 11 Influences of number of simulation on the lower limit, upper limit, and difference of upper limit and lower limit.

method produces a slightly greater uncertainty than the bootstrap method. The slight difference is due to the differences in re-sampling residues. While the Monte Carlo simulation generates residues based on a theoretical normal distribution, the bootstrap method randomly takes the residues with replacement and no assumption is made about the underlying distributions. They are also comparable to those approximated by the linear model obtained from the SigmaPlot software. However, by comparing the results of these three methods, the lower limit and upper limit of α and n obtained by the Monte Carlo and bootstrap methods are slightly greater than those obtained based on a linear assumption except for upper limit of α by the bootstrap method. Because the linearization method is based on the assumption of normal distribution of parameters and linearity at the vicinity of the estimated parameter value, and it is more complicated in terms of calculation, the Monte Carlo and bootstrap methods may be preferred to the linearization method to calculate the parameter uncertainties in spreadsheets. Furthermore, the Monte Carlo method may be preferred to the bootstrap method considering the less number of simulations required for the Monte Carlo method. However, if the number of measurements is too small to determine the probability distribution for Monte Carlo method, the bootstrap method may be superior.

Conclusions

This paper shows step-by-step how to use a Microsoft Excel spreadsheet to fit nonlinear parameters and to estimate their uncertainties using the Monte Carlo and bootstrap methods. Both Monte Carlo and bootstrap methods can be applied in Excel spreadsheets to resample a large number of measurements for dependent variable from which different values of parameters can be obtained. Our results clearly show that the Monte Carlo and bootstrap methods can be used to estimate the parameter uncertainties using spreadsheet methods. The main limitation is that one execution of standard Microsoft Excel Solver has a limit of 200 simultaneous optimizations. This limit can be overcome by multiple independent executions of Solver. Due to the wide accessibility of Microsoft Excel software and ease of use for these two methods, employing the

regression in Excel. Based on the linear model, the associated standard errors are estimated to be 0.0066 and 0.1138, respectively (Figure 12). Then the 95% confidence intervals of α and n are (0.0670, 0.0928) and (1.8758, 2.3218), respectively. As Table 1 shows, the difference between upper limit and lower limit by the Monte Carlo and bootstrap methods are comparable, although the Monte Carlo

R	Rsqr	Adj Rsqr	Standard Error of Estimate			
0.9960	0.9919	0.9913	0.0151			

	Coefficient	Std. Error	t	P	VIF
a	0.0799	0.0066	12.0888	<0.0001	2.0984
n	2.0988	0.1138	18.4450	<0.0001	2.0984

Figure 12 Estimates of parameters (α and n) and associated standard errors using SigmaPlot 10.0.

Monte Carlo and bootstrap methods in spreadsheets is strongly recommended to estimate nonlinear regression parameter uncertainties.

In this paper, we demonstrated the methodology with the van Genuchten water retention curve. The method can be applied to any mathematical functions or models that can be evaluated by Excel. Therefore, the methodology presented in this paper has wide applicability. Further, with little modification, the Monte Carlo method or bootstrap method can be used in Microsoft Excel to estimate the uncertainty of hydrologic or environmental predictions with single or multiple input parameters under different degrees of uncertainty.

Methods

Soil water retention curve

Soil water content is a function of soil matric potential ψ under equilibrium conditions, and this relationship $\theta(\psi)$ can be described by different types of water retention curves. The soil water retention curve is a basic soil property and is critical for predicting water related environmental processes (Fredlund et al. 1994). Among various soil water retention curve models, the van Genuchten (1980) model is the most widely used one (Han et al. 2010). It is highly nonlinear and can be expressed as:

$$\theta(\psi) = \theta_r + (\theta_s - \theta_r)(1 + (\alpha|\psi|)^n)^{-1+\frac{1}{n}} \tag{5}$$

where $\theta(\psi)$ is the soil water content [$cm^3\ cm^{-3}$] at soil water potential ψ (–cm of water), θ_r is the residual water content [$cm^3 cm^{-3}$], θ_s is the saturated water content [$cm^3 cm^{-3}$], α is related to the inverse of the air entry suction [cm^{-1}], and n is a measure of the pore-size distribution (dimensionless). We measured $\theta(\psi)$ at 16 soil water potentials for a sandy soil using Tempe pressure cells (at soil matrix potentials ranging from 0 to –500 cm) and pressure plates (at soil water potentials of –1000 and –15000 cm). θ_s is measured using oven drying method after saturation, and θ_r is estimated as water content of soil approaching air-dry conditions (Wang et al. 2002). θ_s and θ_r are 0.395 and 0.011, respectively. Note that the soil water content (0.375) at zero matrix potential is lower than θ_s due to the soil water movement under gravity. This paper will focus on the estimation of parameters α and n and their associated 95% confidence intervals.

Parameter uncertainty estimation by linearization of nonlinear model

We express the van Genuchten model (Eq. 5) as:

$$\theta_i = f(\beta, \psi_i) + \varepsilon_i \tag{6}$$

where θ_i is the ith observation for the dependent variable $\theta(\psi)$ ($i = 1, 2, ...16$), ψ_i is the ith observation for the predictor $|\psi|$. β is a vector of parameters which includes

parameters α and n. ε_i is a random error, which is assumed to be independent of the errors of other observations and normally distributed with a mean of zero and variance of σ^2.

The sum of squared residuals (SSE) for nonlinear regression can be written as:

$$SSE(\beta) = \sum (\theta_i - f(\beta, \psi_i))^2 \tag{7}$$

The model has the maximum likelihood when the SSE is minimized. Namely, when the partial derivative

$$\frac{\partial SSE(\beta)}{\partial \beta} = -2 \sum (\theta_i - f(\beta, \psi_i)) \frac{\partial f(\beta, \psi_i)}{\partial \beta} \tag{8}$$

is zero, parameters β are optimized. Once the optimum values of β are obtained, the parameter uncertainties can be estimated by linearizing the nonlinear model function at the optimum point using the first-order Taylor series expansion method (Fox and Weisberg 2010).

Let

$$F_{ij} = \frac{\partial f(\hat{\beta}, \psi_i)}{\partial \beta_j} \tag{9}$$

where $\hat{\beta}$ is the optimized value, j refers to the jth of parameters ($j = 1, 2$, and $\beta_1 = \alpha$, $\beta_2 = n$).

Assume matrix $F = [F_{ij}]$. In our case,

$$F = \begin{bmatrix} \dfrac{\partial f(\hat{\beta}, \psi_1)}{\partial \alpha} & \dfrac{\partial f(\hat{\beta}, \psi_1)}{\partial n} \\ \dfrac{\partial f(\hat{\beta}, \psi_2)}{\partial \alpha} & \dfrac{\partial f(\hat{\beta}, \psi_2)}{\partial n} \\ \vdots & \vdots \\ \dfrac{\partial f(\hat{\beta}, \psi_{16})}{\partial \alpha} & \dfrac{\partial f(\hat{\beta}, \psi_{16})}{\partial n} \end{bmatrix} \tag{10}$$

where $\frac{\partial f(\hat{\beta}, \psi_i)}{\partial \alpha}$ and $\frac{\partial f(\hat{\beta}, \psi_i)}{\partial n}$ can be calculated by the following formulae:

$$\frac{\partial f(\hat{\beta}, \psi_i)}{\partial \alpha} = (f((\hat{\alpha} + \Delta\hat{\alpha}), \hat{n}, \psi_i) - f((\hat{\alpha} - \Delta\hat{\alpha}), \hat{n}, \psi_i))/(2\Delta\hat{\alpha}) \tag{11}$$

$$\frac{\partial f(\hat{\beta}, \psi_i)}{\partial n} = (f(\hat{\alpha}, (\hat{n} + \Delta\hat{n}), \psi_i) - f(\hat{\alpha}, (\hat{n} - \Delta\hat{n}), \psi_i))/(2\Delta\hat{n}) \tag{12}$$

where $\Delta = 0.015$, $\hat{\alpha}$ and $\hat{n}$ are optimized value of α and n, respectively.

The estimated asymptotic covariance matrix (V) of the estimated parameters can be obtained by (Fox and Weisberg 2010):

$$V = \begin{bmatrix} \delta_{g\alpha}^2 & \delta_{gn}^2 \\ \delta_{n\alpha}^2 & \delta_{nn}^2 \end{bmatrix} = \sigma^2 \left(F'F \right)^{-1} \tag{13}$$

where $(F'F)^{-1}$ is the inverse of $F'F$, and F' is a transpose of F.

The σ^2 can be approximated by dividing the SSE by the degree of freedom, df, as in the form (Brown 2001):

$$\sigma^2 = \frac{SSE}{df} \tag{14}$$

where df is calculated as the number of observations in the sample minus the number of parameters. In this study, df equals 14 (i.e., 16 minus 2).

Therefore, $\delta_{\alpha\alpha}^2$, δ_{nn}^2, $\delta_{\alpha n}^2$ (or $\delta_{n\alpha}^2$) in Eq. (13) are the estimated variance of α, variance of n, and covariance of α and n, respectively. Specifically, $\delta_{\alpha\alpha}$ and δ_{nn} are the standard errors used to characterize the uncertainties of α and n, respectively. At 95% confidence, the intervals of α and n are $\hat{\alpha} \pm 1.96\, \delta_{\alpha\alpha}$, $\hat{n} \pm 1.96\, \delta_{nn}$, respectively. SigmaPlot 10.0 is used to estimate the parameters and associated standard errors.

Monte Carlo method to estimate parameter uncertainty

Monte Carlo method is an analytical technique for solving a problem by performing a large number of simulations and inferring a solution from the collective results of the simulations. It is a method to calculate the probability distribution of possible outcomes.

In this paper, Monte Carlo simulation is performed to obtain residues of dependent variable θ. The residues follow a specified distribution with a mean of zero and standard deviation of $\sqrt{SSE/df}$. The simulated residues are added to the predicted θ ($\hat{\theta}$) to reconstruct new observations for dependent variable θ. The expression for obtaining new observations for dependent variable θ in Excel is:

$$\theta = \hat{\theta} + \text{NORM.INV}\left(\text{RAND}(), 0, \text{SQRT}\left(\frac{SSE}{df} \right) \right) \tag{15}$$

where function NORM.INV gives a value which follow a normal distribution with a mean of zero and standard deviation of $\sqrt{SSE/df}$ at a probability of RAND(). Therefore, normal distribution on the θ is assumed for Monte Carlo method. Excel function RAND produces a random value that is greater than or equal to 0 and less than 1. SQRT is a function to obtain the square root of a variable.

Monte Carlo simulations are performed 2000 times. Nonlinear regression is made on the simulated θ values versus $|\psi|$ to obtain 2000 values for parameters α and n. The fitted values with different numbers (from 100 to 2000 with intervals of 100) of simulation is analyzed separately to determine the influences of number of simulation on uncertainty estimates. For each dataset, the

probability distribution of α and n will be determined by the Shapiro-Wilk test using SPSS 16.0, and the 95% confidence intervals of α and n will be calculated to represent their uncertainties. For simplification, only 200 simulations are shown as an example. Readers can run different numbers of simulation by analogy.

Bootstrap method to estimate parameter uncertainty

Bootstrap method is an alternative method first introduced by Efron (1979) for determining uncertainty in any statistic caused by sampling error. The main idea of this method is to resample with replacement from the sample data at hand and create a large number of "phantom samples" known as bootstrap samples (Singh and Xie 2013). Bootstrap method is a nonparametric method which requires no assumptions about the data distribution.

Residues of θ are calculated by subtracting the $\hat{\theta}$ from the original θ measurements. Bootstrap method is used to resample the residues with replacement for each θ from the calculated residues. The re-sampled residues are added to the $\hat{\theta}$ to reconstruct new observations for dependent variable θ. The expression for obtaining new observations for dependent variable θ using bootstrap method in Excel is:

$$\theta = \hat{\theta} + \text{INDEX}(\text{Range of residual}, \text{INT}(\text{RAND}()) * \text{Row number}) \tag{16}$$

where function INDEX is used to randomly return a calculated residual from a certain array. Range of residual refers to the calculated residues. INT is a function to round a given number, which is randomly produced by RAND() multiplied by row number.

The non-parametric bootstrap method is a special case of Monte Carlo method used for obtaining the distribution of residues of θ which can be representative of the population. The idea behind the bootstrap method is that the calculated residues can be an estimate of the population, so the distribution of the residues can be obtained by drawing many samples with replacement from the calculated residues. For the Monte Carlo method, however, it creates the distribution of residues of θ with a theoretical (i.e., normal) distribution. From this aspect, the bootstrap method is more empirically based and the Monte Carlo method is more theoretically based.

Similar to the Monte Carlo method, bootstrap simulations are performed 2000 times. Distribution type and 95% confidence intervals of α and n will also be determined for fitted datasets with different numbers (from 100 to 2000 with intervals of 100) of simulation. For simplification, only 200 simulations are shown as an example.

Competing interests
The author declares that there are no competing interests associated with this research work.

Authors' contributions
WH analyzed the data and wrote the draft. JX and HC participated in the data analysis. BS designed the study. All authors read and approved the final manuscript.

Authors' information
Wei Hu is a professional research associate at the University of Saskatchewan and specialist for soil hydrology. Jing Xie is a PhD student in University of Saskatchewan who is investigating legume fertilization. Henry Wai Chau is a lecturer in environmental physics in Lincoln University. Bing Cheng Si is a full professor in University of Saskatchewan and specializes in soil physics.

Acknowledgements
The project was funded by the Natural Sciences and Engineering Research Council (NSERC) of Canada.

Author details
[1]Department of Soil Science, University of Saskatchewan, Saskatoon, SK S7N 5A8, Canada. [2]Department of Soil and Physical Science, Lincoln University, PO Box 84Lincoln, Christchurch 7647, New Zealand.

References
Berger RL (2007) Nonstandard operator precedence in Excel. Comput Stat Data An 51:2788–2791

Brown AM (2001) A step-by-step guide to non-linear regression analysis of experimental data using a Microsoft Excel spreadsheet. Comp Meth Prog Bio 65:191–200

Conedera M, Torriani D, Neff C, Ricotta C, Bajocco S, Pezzatti GB (2011) Using Monte Carlo simulations to estimate relative fire ignition danger in a low-to-medium fire-prone region. Forest Ecol Manag 26:2179–2187

Cwiertny DM, Roberts AL (2005) On the nonlinear relationship between k(obs) and reductant mass loading in iron batch systems. Environ Sci Technol 39:8948–8957

Delboy H (1994) A non-linear fitting program in pharmacokinetics with Microsoft® Excel spreadsheet. Int J Biomed Comput 37:1–14

Efron B (1979) Bootstrap method: another look at the Jackknife. Ann Stat 7:1–26

Efron B, Tibshirani R (1993) An introduction to the Bootstrap. Champman & Hall, London, UK

Fox J, Weisberg S (2010) Nonlinear regression and nonlinear least squares in R: An appendix to an R companion to applied regression, second edition. http://socserv.socsci.mcmaster.ca/jfox/Books/Companion/appendix/Appendix-Nonlinear-Regression.pdf.

Fredlund DG, Xing AQ, Huang SY (1994) Predicting the permeability function for unsaturated soils using the soil-water characteristic curve. Can Geotech J 31:533–546

Han XW, Shao MA, Hortaon R (2010) Estimating van Genuchten model parameters of undisturbed soils using an integral method. Pedosphere 20:55–62

Harris DC (1998) Nonlinear least-squares curve fitting with Microsoft Excel Solver. J Chem Educ 75:119–121

Luo B, Maqsood I, Yin YY, Huang GH, Cohen SJ (2003) Adaption to climate change through water trading under uncertainty - An inexact two-stage nonlinear programming approach. J Environ Inform 2:58–68

Singh K, Xie M (2013) Bootstrap: A statistical method. From Rutgers University. http://www.stat.rutgers.edu/home/mxie/rcpapers/bootstrap.pdf.

Smith LH, McCarty PL, Kitanidis PK (1998) Spreadsheet method for evaluation of biochemical reaction rate coefficients and their uncertainties by weighted nonlinear least-squares analysis of the integrated Monod equation. Appl Environ Microbiol 64:2044–2050

Tong L, Chang C, Jin S, Saminathan R (2012) Quantifying uncertainty of emission estimates in National Greenhouse Gas Inventories using bootstrap confidence intervals. Atmos Environ 56:80–87

van Genuchten MT (1980) A closed-form equation for predicting the hydraulic conductivity of unsaturated soils. Soil Sci Soc Am J 44:892–898

Wang QJ, Horton R, Shao MA (2002) Horizontal infiltration method for determining Brooks-Corey model parameters. Soil Sci Soc Am J 66:1733–1739

Wraith JM, Or D (1998) Nonlinear parameter estimation using spreadsheet software. J Nat Resour Life Sci Educ 27:13–19

6

Confidence modeling with reliability: a systems approach to sustainable energy planning

Xiaosheng Qin[1,2*], Ye Xu[3] and Jianjun Yu[4]

Abstract

Background: Energy systems planning has played a strong role in setting up the framework for developing long-term policies of energy activities to help guide the future of a local, regional or national energy system. However, the planning process is complicated with a variety of uncertainties and complexities. In this study, a fuzzy confidence model coupled with mixed-integer programming was proposed for regional energy systems planning.

Results: Application of the model to a hypothetical case indicated that the model was capable of handling uncertainties expressed as fuzzy sets and taking capacity-expansion issues of energy facilities into consideration. The solutions from the proposed model could meet system constraints at different confidence levels, where each confidence level was further associated with different reliability scenarios.

Conclusions: The proposed model could help decision makers analyze the trade-offs between system economy and reliability, and explore cost-effective energy systems planning strategies under uncertainty.

Keywords: Fuzzy programming, Mixed-integer programming, Energy, Uncertainty

Background

It has been widely accepted that the task of energy systems planning process involves a variety of social, economic, environmental, technical, and political factors that are characterized with temporal and spatial variabilities (Cai et al. 2009). The system is further complicated by the existence of uncertainties that may be associated with the planning processes. Previously, a number of inexact optimization techniques were developed for assisting in the formulation of energy management plans and generation of optimal decision schemes (Liu et al. 2000; Cai et al. 2009; Lin and Huang, 2009). Among various alternatives, the fuzzy chance-constrained programming (FCCP) was advantageous in dealing with optimization problems subject to fuzzy constraints at prescribed confidence levels (Liu and Iwamura, 1998). In recent years, FCCP was successfully used in many environmental management fields (Cao et al. 2009). However, its application in energy systems planning field was very limited. Moreover, a FCCP

model is incapable of handling the binary-decision (i.e. yes/no) problems which is important in energy systems planning for seeking solutions to capacity-expansion or operation-scheduling issues. To fill this gap, this study aims to develop a double-sided fuzzy chance-constrained mixed-integer programming (DFCCMIP) model for supporting regional energy systems planning under uncertainty. A hypothetical case will be used for demonstration.

Energy systems planning model

A regional energy system planning system, modified from Li et al. (2010), is to be investigated. Within the energy system, multiple energy sources are considered and various power conversion technologies are applied to generate the electricity from the provided energy sources. Generally, large-scale conversion technologies are responsible for conventional energy resources, and small-scale plants are based on local availability of renewable resources (Li et al. 2010). The generated electricity will be used to meet the requirement of multiple end-users including industrial, commercial, agricultural, transportational and residential sectors. For environmental protection, the generated air pollutants from power conversion plants should be mitigated by different treatment technologies in order to meet the related emission standards. The decision maker is

* Correspondence: xsqin@ntu.edu.sg
[1]School of Civil & Environmental Engineering, Nanyang Technological University, 50 Nanyang Avenue 639798 Singapore
[2]Earth Observatory of Singapore (EOS), Nanyang Technological University, 50 Nanyang Avenue 639798 Singapore
Full list of author information is available at the end of the article

responsible for allocating energy resources/services from multiple facilities to multiple end-users at a minimum system cost within a multi-period time horizon in light of environmental constraints and uncertainties.

It is assumed that the end-users' electricity demands can be described as triangular fuzzy sets. Over the three planning periods (each one has 5 years), the demand amounts are (50, 70, 96), (85, 112, 147) and (135, 170, 200) $\times 10^3$ GWh, respectively. The peak load demands are considered as deterministic, being 1.5, 2.0, 2.5 GW in three periods, respectively. Totally, five energy sources (i.e., coal, natural gas, hydropower, wind, solar and nuclear) are used for power generation. Over three planning periods, the supplied costs of coal are 2.5, 3 and 3.5 ($\times 10^3$ \$/TJ), respectively; those for natural gas are 5, 5.5 and 6 ($\times 10^3$ \$/TJ), respectively; those for electricity are 900, 1000 and 1100 ($\times 10^3$ \$/GWh), respectively. In order to meet the increasing energy demand from the end-users, capacity expansions of energy-supply facilities are necessary. Based on Li et al. (2010), the existing capacities of the coal-fired, gas-fired, hydropower, wind power, solar power and nuclear power conversion technologies are set as 10, 2.2, 2.8, 0, 0, 0 (GW), respectively. Other related parameters associated with the expansion options are listed in Table 1.

Sulfur dioxide (SO_2), nitrogen oxides (NO_x) and particulate matter (PM) are the main pollutants emitted from power plants. To achieve the related environmental targets, each emission source has installed various mitigation measures to avoid penalties from government. The applied control techniques mainly include: (i) soda ash scrubber (SAS), wet limestone scrubber (WLS) and lime spray dryer (LSD) for reducing the SO_2 amounts; (ii) Selective catalytic reduction (SCR) and selective non-catalytic reduction (SNCR) for controlling NO_x emissions; (iii) Fabric filiter/baghouse (BH), electrostatic precipitator (ESP) and wet collector (WC) for mitigating the PM emissions (Li et al. 2010). The treatment efficiencies of various technologies, the allowable emission amounts of power plants, and design safety coefficients for energy supply are also described by triangular fuzzy sets (see Table 2).

Model Formulation

Double-sided fuzzy chance-constrained programming (DFCCP) was firstly proposed by Fiedler et al. (2006). In a DFCCP model, constraints with fuzzy variables can be satisfied at a series of predetermined confidence levels with two reliability scenarios, i.e. the minimum and maximum reliabilities. The model could eventually be converted into two crisp equivalents for solution. The related solution algorithms could be referred to Fiedler et al. (2006). In addition, expansion of facility capacity in energy systems planning is necessary in order to meet the increasing energy demand over the planning horizons. Mixed integer linear

Table 1 Parameters related to power conversion technologies

Conversion technology	Time period		
	t = 1	t = 2	t = 3
Power generation cost (\$10³/GWh) and operating time (h) of conversion technology			
Coal-fired power	5.0 (24900*)	5.5 (24900)	6.0 (24900)
Gas-fired power	4.5 (24600)	5.0 (24600)	5.5 (24600)
Hydropower	4.0 (21000)	4.5 (21000)	5.0 (21000)
Wind power	2.5 (15000)	3.0 (15000)	3.5 (15000)
Solar power	2.0 (15000)	2.5 (15000)	3.0 (15000)
Nuclear power	10.0 (24600)	11.0	12.0 (24600)
Fixed (\$10⁶) and variable (\$10⁶/GW) costs for capacity expansion			
Coal-fired power	325 (700**)	385 (750)	445 (800)
Gas-fired power	300 (650)	350 (700)	400 (750)
Hydropower	700 (1800)	770 (1900)	840 (2000)
Wind power	800 (1900)	880 (1950)	960 (2000)
Solar power	900 (2000)	990 (2100)	1080 (2200)
Nuclear power	1000 (1950)	1100 (2100)	1200 (2250)
Variable upper bounds for capacity expansion (GW)			
Coal-fired power	6.5	4.5	2.5
Gas-fired power	4.8	5.8	6.8
Hydropower	2.5	3.5	4.5
Wind power	0.8	1.8	2.8
Solar power	1.8	2.8	3.8
Nuclear power	2.5	3.5	4.5
Energy consumption per units of electricity production (TJ/GWh)			
Coal-fired power	12.5	12.4	12.3
Gas-fired power	11.5	11.4	11.3
Hydropower	4.0	3.95	3.9
Wind power	0.13	0.12	0.11
Solar power	5.0	4.9	4.8
Nuclear power	13.0	12.8	12.6
Available amounts of renewable energy (10³ TJ)			
Hydropower	90000	90000	90000
Wind power	15000	15000	15000
Solar power	20000	20000	20000
Nuclear power	150000	150000	150000

Notes: data are modified from Li et al. (2010); * is the operating time for conversion technology; ** is the variable costs for capacity expansion.

programming (MILP) is a useful tool (through using 0–1 integer variables) to help determine whether or not a particular facility development or an expansion option needs to be undertaken (Huang et al. 1995). Coupling DFCCP and MILP into a general framework, a double-sided fuzzy chance-constrained mixed-integer

Table 2 Parameters related to pollution control technologies

Pollution control technology	Time period		
	$t=1$	$t=2$	$t=3$
Treatment cost of SO_2 emission ($/tonne)			
SAS	55	57	59
WLS	45	48	51
LSD	30	33	36
Treatment cost of NO_x emission ($/tonne)			
SCR	55	59	62
SNCR	35	38	40
Treatment cost of PM emission ($/tonne)			
BH	135	140	145
ESP	125	133	140
WC	115	125	135
Allowable emission amounts of pollutants (tonne)			
SO_2	(40, 44, 50)*	(61, 72, 81)	(69, 82, 93)
NO_x	(20, 32, 45)	(31, 47, 61)	(35, 57, 80)
PM	(0.3, 0.52, 0.7)	(0.45, 0.8, 1.1)	(0.49, 0.89, 1.2)
Treatment efficiency of pollutants (%)			
SAS	(0.85, 0.91, 0.99)		
WLS	(0.76, 0.82, 0.9)		
LSD	(0.7, 0.77, 0.85)		
SCR	(0.8, 0.86, 0.9)		
SNCR	(0.5, 0.62, 0.7)		
BH	(0.96, 0.975, 0.99)		
ESP	(0.95, 0.964, 0.98)		
WC	(0.94, 0.958, 0.97)		

Notes: data are modified from Li et al. (2010); *(a, b, c) represents a triangular fuzzy set, where *a* and *c* are the minimum and the maximum possible values, and *b* is the most likely value.

programming (DFCCMIP) model can be formulated as follows (Li et al. 2010):

$$Min f = \sum_{t=1}^{T}(CEC_t*XC_t + CEN_t*XG_t) + \sum_{t=1}^{T}CIE_t*XE_t$$

$$+\sum_{i=1}^{I}\sum_{t=1}^{T}CV_{it}*XW_{it} + \sum_{i=1}^{I}\sum_{t=1}^{T}(Y_{it}*A_{it} + B_{it}*X_{it})$$

$$+\sum_{i=1}^{I}\sum_{o=1}^{O}\sum_{t=1}^{T}CS_{ot}*XS_{iot} + \sum_{i=1}^{I}\sum_{p=1}^{P}\sum_{t=1}^{T}CN_{pt}*XN_{ipt}$$

$$+\sum_{i=1}^{I}\sum_{q=1}^{Q}\sum_{t=1}^{T}CP_{qt}*XP_{iqt} \qquad (1a)$$

Subject to:

(1) Constraints for mass balance of fossil fuels:

$$XW_{1t}*FE_{1t} \le XC_t, \quad \forall t \qquad (1b)$$

$$XW_{2t}*FE_{2t} \le XG_t, \quad \forall t \qquad (1c)$$

(2) Constraints for availabilities of energy resources:

$$XW_{it}*FE_{it} \le UP_{it}, \quad \forall t \text{ for } i \ge 3 \qquad (1d)$$

(3) Constraints for electricity supply and demand balance:

$$Pos\left\{\tilde{\alpha}_l, \tilde{d}_t \middle| \tilde{\alpha}_l \sum_{i=1}^{I}(XW_{it} + XE_t) \ge \tilde{d}_t\right\} \ge \beta_l, \quad \forall t \qquad (1e)$$

(4) Constraints for electricity generation of every power conversion technology:

$$\left(\sum_{t'=1}^{t}X_{it'} + RC_i\right)*ST_{it} \ge XW_{it}, \quad \forall i, t \qquad (1f)$$

(5) Constraints for electricity peak load demand:

$$\sum_{i=1}^{I}RC_i + \sum_{i=1}^{I}\sum_{t'=1}^{t}X_{it'} \ge V_t, \quad \forall t \qquad (1g)$$

(6) Constraints for capacity expansion of electricity-generation facilities:

$$Y_{it}\begin{cases} = 1, \text{ if capacity expansion is undertaken} \\ = 0, \text{ if otherwise} \end{cases}, \quad \forall i, t \qquad (1h)$$

$$X_{it} \le M_{it}*Y_{it}, \quad \forall i, t \qquad (1i)$$

(7) Constraints for air pollution control demand:

$$\sum_{o=1}^{O} XS_{iot} = W_{it}*INS_{it}, \ \forall i, \ t \tag{1j}$$

$$\sum_{p=1}^{P} XN_{ipt} = W_{it}*INN_{it}, \ \forall i, \ t \tag{1k}$$

$$\sum_{q=1}^{Q} XP_{iqt} = W_{it}*INP_{it}, \ \forall i, \ t \tag{1l}$$

(8) Constraints for air pollutants emissions:

$$Pos\left\{\tilde{\eta}_o, \ \widetilde{ES}_t \left| \sum_{i=1}^{I}\sum_{o=1}^{O} \left(1 - \tilde{\eta}_o\right)XS_{iot} \le \widetilde{ES}_t \right.\right\} \ge \beta_l, \ \forall t \tag{1m}$$

$$Pos\left\{\tilde{\eta}_P, \ \widetilde{EN}_t \left| \sum_{i=1}^{I}\sum_{p=1}^{P} \left(1 - \tilde{\eta}_p\right)XN_{ipt} \le \widetilde{EN}_t \right.\right\} \ge \beta_l, \ \forall t \tag{1n}$$

$$Pos\left\{\tilde{\eta}_q, \ \widetilde{EP}_t \left| \sum_{i=1}^{I}\sum_{q=1}^{Q} \left(1 - \tilde{\eta}_q\right)XP_{iqt} \le \widetilde{EP}_t \right.\right\} \ge \beta_l, \ \forall t \tag{1o}$$

(9) Non-negative constraints:

$$XC_t, \ XG_t, \ XE_t, \ XW_{it}, X_{it} \ge 0, \ \forall i, \ t \tag{1p}$$

where f is expected system cost for energy system management over the planning horizon ($\$10^9$); i is type of power conversion technology, $i = 1,2,.., I$ (in this study, I is considered as 6, where $i = 1, 2,...,6$ means the coal, natural gas, hydropower, wind power, solar power and nuclear power, respectively); o is type of SO_2 control measure, $o = 1, 2,..., O$; p is type of NO_x control measure, $p = 1, 2,..., P$; q is type of PM control measure, $q = 1, 2,..., Q$; O, P and Q are numbers of control measure of the pollutants, respectively; t is time period, $t = 1,2,.., T$; t' is an intermediate index satisfying $1 \le t' \le t$; CEC_t and CEN_t are cost for coal and nature gas supply in period t ($\$10^3$/TJ), respectively; CIE_t are cost for imported electricity supply in period t ($\$10^3$/GWh), respectively; UP_{it} $(i \ge 3)$ are available amounts of

hydropower, wind power, solar power and nuclear power in period t (10^3 TJ), respectively; CV_{it} is operating cost of power conversion technology i for electricity generation in period t ($\$10^3$/GWh); CS_{ot}, CN_{pt} and CP_{qt} are unit operating cost of controlling SO_2, NO_x and PM emissions during period t, ($\$$/tonne), respectively; ST_{it} is average service time of power conversion technology i in period t (h); V_t is peak load demand in period t (GW); A_{it} and B_{it} are fixed-charge and variable cost for capacity expansion of power conversion technology i in period t ($\$10^6$, $\$10^6$/GW), respectively; RC_i is the existing capacity of conversion technology i (GW); FE_{it} is the units of energy consumption per units of electricity production for power conversion technology i in period t (TJ/GWh); M_{it} is variable upper bounds for capacity expansion of power conversion technology i in period t (GW); INS_{it}, INN_{it} and INP_{it} are units of SO_2, NO_x and PM emission per unit of electricity production for power conversion technology i in period t (tonne/GWh), respectively; $\tilde{\eta}_o$, $\tilde{\eta}_p$ and $\tilde{\eta}_q$ are the average efficiency of SO_2, NO_x and PM control measure (%), which are expressed as the triangular fuzzy sets, respectively; ES_t, EN_t and EP_t are the emission allowance of SO_2, NO_x and PM in period t (tonne), which are expressed as the triangular fuzzy sets, respectively; $\tilde{\alpha}_l$ is design safety factors assuring the electricity demand can be satisfied completely during period t, which are expressed as the triangular fuzzy sets; $\tilde{d}_t$ is the total electricity demand during period t (10^3 GWh), which are expressed as the triangular fuzzy sets; $\tilde{d}_t$ is the total electricity demand during period t (10^3 GWh), which are expressed as the triangular fuzzy sets; $Pos\{\cdot\}$ denotes possibility of events in $\{\cdot\}$ where β_l is a predetermined confidence level and l is type of confidence levels; XC_t and XG_t are supply amounts of the coal and natural gas in period t (TJ), respectively; XE_t is the imported electricity supply in period t (10^3 GWh); XW_{it} is electricity generation amounts of power conversion technology i during period t (10^3GWh); X_{it} is continuous variables about the amount of capacity expansion of power conversion technology i in period t (GW); Y_{it} is the binary variables for identifying whether or not a capacity expansion action of power conversion technology i needs to be undertaken in period t; XS_{iot}, XN_{ipt} and XP_{iqt} are the SO_2, NO_x and PM amount generated from power conversion technology i to be treated by control measure o, p and q in period t (tonne), respectively.

Results and discussion

Figure 1 shows the model solutions for the supplied amounts of coal and gas. It is indicated that the temporal and spatial variations of electricity demand may result in varied energy supply schemes. As the electricity demand increases, the supplied amounts of the coal and natural gas would increase over the three planning periods, and the supplied amounts of the coal are higher than those of the natural gas. It is also found that there is no need to import

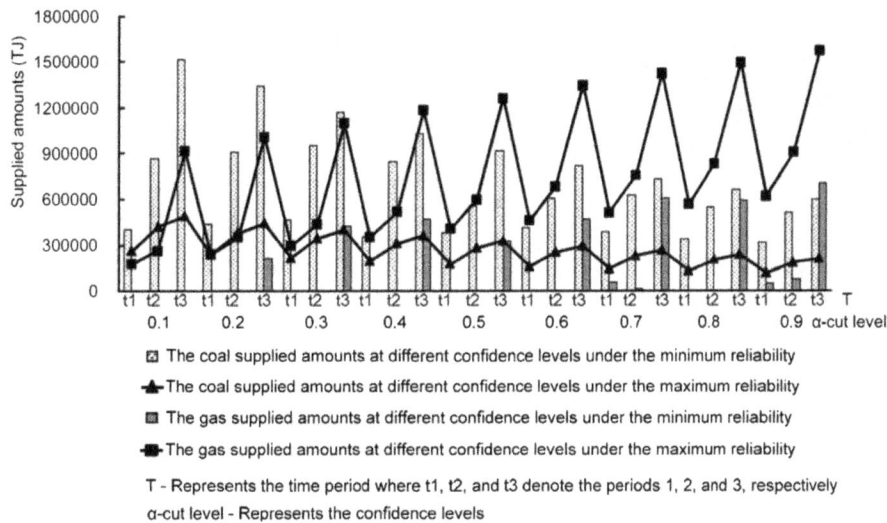

Figure 1 Solutions of coal and gas supplied amounts.

electricity from other regions. For example, at a confidence level of 0.4 with the minimum reliability, the amounts of coal supply are 351.69, 851.52 and 1033.97 × 10³ TJ over the three planning periods, respectively; the supplied amounts of natural gas are 0, 0 and 471.98 × 10³ TJ, respectively. This is because the coal owns the lowest unit supply cost (i.e. 31.25, 37.20 and 43.05 × 10³ $/TJ), the natural gas ranked in the middle (i.e. 57.50, 62.70 and 67.80 × 10³ $/TJ), and the imported electricity is the highest (i.e. 900, 1000 and 1100 × 10³ $/GWh). Figure 2 shows the solution of electricity generation amounts. For the renewable energy, the electricity generation amounts by the hydropower are 22.50, 22.78 and 23.08 GWh over the three planning periods, respectively. Those by the windpower are 12.00, 12.00 and 12.00 GWh over the three planning periods, respectively. This is due to the fact that the hydropower has the highest existing capacity, and the lowest fixed and variable costs for capacity expansion among all renewable energy sources.

From Figures 1 and 2, the solutions of decision variables have notable variations at different α-cut levels with two reliability scenarios. The supplied amounts of the coal with the maximum reliability are in a decreasing trend when the α-cut level is increasing. Compared with those of the coal, the supplied amounts of the natural gas would increase with the increase of the α-cut level. This is because, when the confidence level increases, the electricity demand would increase according to fuzzy algorithm rule (Fiedler et al. 2006); meanwhile, the constraints of the environmental emission standards would become stricter. This will lead to the decrease of the coal supplied amounts, which are deciding factor for pollutant emissions. At the same time, the supplied amounts of the natural gas with low pollutant emissions would increase. This reflects the trade-off

between system economy and reliability. A lower system cost would be incurred if a larger quantity of pollutant emission is allowable; meanwhile, the planning scheme with a higher cost would urge the environmental quality maintain at a higher level.

The variation trend of the solutions under the minimum reliability is generally similar to those under the maximum one, except for a few inconsistent points. For example, in period 3, the supplied amounts of the coal are 1520.76, 1345.90, 1173.80, 1033.97, 918.11, 820.54, 737.25, 665.32 and 602.57 × 10³ TJ, respectively; those of the natural gas are 0, 213.41, 423.83, 471.98, 328.16, 473.37, 606.19, 594.80 and 710.26 × 10³ TJ, respectively. The supplied amounts of the natural gas at the confidence levels of 0.5 and 0.8 in period 3 are not consistent with the general changing trend, as are the cases for a number of solutions of coal in periods 1 and 2. This is mainly due to the complex interactive relationships among various components of the planning system. Another reason may be the loose constraints incurred by the setting of minimum reliability, making the effect of confidence levels become less significant. In addition, at the same α-cut levels, the supplied amounts of the coal under the minimum reliability are higher than those under the maximum one; for natural gas, the varying trend is opposite. This is due to the fact that the minimum reliability prefers looser environmental emission standards and lower electricity demand than the maximum reliability does. Therefore, the energy source of coal would become more popular than the natural gas.

The optimal treatment amounts of pollutants can be generated from various conversion technologies under two reliability levels. The solution indicates that the increase of the coal and natural gas amounts would result in

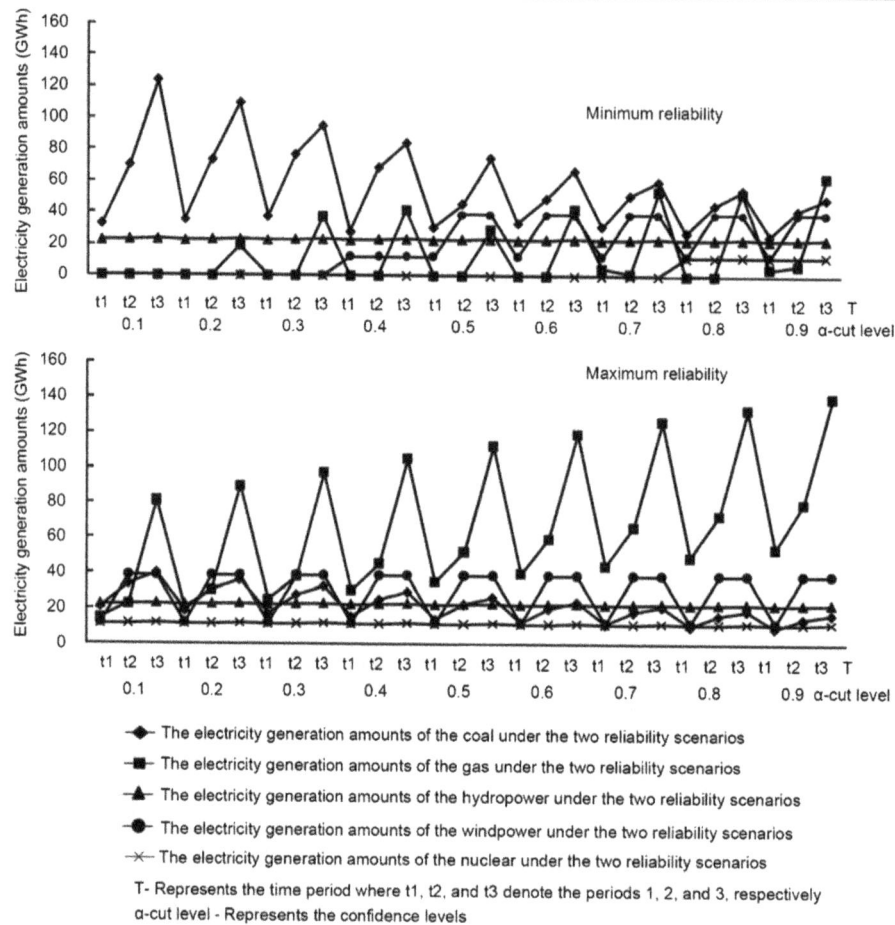

Figure 2 Solutions of electricity generation amounts.

an increased amount of pollutant reduction. For example, the disposal amounts of SO_2 generated from the coal-fired power conversion technology by SAS over the three planning periods are 0, 170.25 and 243.56 t, respectively; the allocated amounts of NO_x to the SCR technology are 0, 49.61 and 62.89 t, respectively; the treated amounts of PM to the BH technology are 13.47, 55.92 and 67.25 t, respectively. Generally, the varying trends of the pollutant treatment amounts were similar to those of the supplied amounts for the emission sources. When the confidence level increases, the treated amounts of SO_2 from coal would generally decrease. For example, under the minimum reliability, at the period 3, the treated amounts by SAS at the different confidence-levels are 459.86, 390.28, 310.01, 243.56, 186.98, 140.01, 99.44, 61.14 and 31.88 t, respectively; those by the LSD are 529.26, 485.10, 453.44, 428.94, 410.16, 393.68, 380.07, 371.59 and 360.03 t, respectively. The reason is that, when the confidence level goes higher, the constraints of the environmental standards would become stricter; this would lead to reduced supplied amounts of the coal, and consequently decreased emission levels of the pollutants. Meanwhile, over the

three planning periods, the treated amounts of the coal by SAS are lower than those by LSD, due to its higher operational cost. Similar trends can also be found for the pollutants generated from the natural gas.

The total costs at different α-cut levels also vary with reliabilities. As the α-cut level increases, the electricity supply would increase, leading to better environmental quality; however, the related system cost would increase. The system cost with the minimum reliability would be lower than that with the maximum one, indicating that a lower system cost is related to a higher environmental risk. A conservative alternative is more effective to meet the electricity requirement and maintain the environmental quality. A trade-off between the total system cost and reliability of satisfying model constraints needs to be analyzed in order to gain an in-depth insight into the characteristics of energy planning systems.

Conclusions

A double-sided fuzzy chance-constrained mixed-integer programming (DFCCMIP) model was developed in this study and applied to a regional energy systems planning problem. The model coupled DFCCP and MIP models into a general framework, and could help deal with uncertainties expressed as fuzzy sets associated with both the left- and right-hand-side components of constraints, and the capacity expansion issue of energy-production facilities. The study results indicated that DFCCMIP allowed violation of system constraints at specified confidence levels with various reliability scenarios. The solutions of continuous and binary variables could help decision makers establish various energy production patterns and capacity-expansion plans under complex uncertainties, and gain in-depth insights into the trade-offs between system economy and reliability.

Competing interests
The authors declared that they have no competing interest.

Acknowledgement
This research was supported by Earth Observatory of Singapore (EOS) Project (M4080891.B50) and Singapore's Ministry of Education (MOM) AcRF Tier 1 Project (M4010973.030).

Author details
[1]School of Civil & Environmental Engineering, Nanyang Technological University, 50 Nanyang Avenue 639798 Singapore. [2]Earth Observatory of Singapore (EOS), Nanyang Technological University, 50 Nanyang Avenue 639798 Singapore. [3]S-C Research Academy of Energy & Environmental Studies, North China Electric Power University, Beijing 102206, China. [4]DHI-NTU Water & Environment Research Centre and Education Hub, Nanyang Technological University, 50 Nanyang Avenue 639798, Singapore.

Authors' contributions
QXS is responsible for the development of the energy systems planning model and drafted the manuscript. XY participated in model computation, and involved in result analysis. YJJ participated in the model development, and helped polish the manuscript. All authors read and approved the final manuscript.

References
Cai YP, Huang GH, Yang ZF, Lin QG, Tan Q (2009) Community-scale renewable energy systems planning under uncertainty-An interval chance-constrained programming approach. Renew Sustain Energy Rev 13:721–735

Cao CW, Gu XS, Xin Z (2009) Chance-constrained programming models for refinery short-term crude oil scheduling problem. Appl Math Model 33:1696–1707

Fiedler M, Nedoma J, Ramík J, Rohn J, Zimmermann K (2006) Linear optimization problems with inexact data. Springer-Verlag, New York

Huang GH, Baetz BW, Patry GG (1995) Grey integer programming: an application to waste management planning under uncertainty. Eur J Oper Res 83:594–620

Li YF, Li YP, Huang GH, Chen X (2010) Energy and environmental systems planning under uncertainty - An inexact fuzzy-stochastic programming approach. Appl Energ 87:3189–3211

Lin QG, Huang GH (2009) Planning of energy system management and GHG-emission control in the Municipality of Beijing-An inexact-dynamic stochastic programming model. Energy Policy 37:4463–4473

Liu BD, Iwamura K (1998) Chance-constrained programming with fuzzy parameters. Fuzzy Set Syst 94:227–237

Liu L, Huang GH, Fuller GA, Chakma A, Guo HC (2000) A dynamic optimization approach for nonrenewable energy resources management under uncertainty. J Petrol Sci Eng 26:301–309

7

Endogenous social discount rate, proportional carbon tax, and sustainability: Do we have the right to discount future generations' utility?

Masayuki Otaki

Abstract

Background: This paper examines a serious issue - whether future generations of utility should be discounted. The issue is of vital importance because future generations will never have the opportunity to reveal their preference regarding the current resource allocation and yet this will ultimately affect their utility. This paper addresses with this issue in the context of the phenomenon of global warming that is crucially connected with the emission and accumulation of CO_2.

Results: The analysis focuses on how the social optimum is attained under the constraint of *sustainability* and reveals the following relationship between the optimal policies: not discounting utility in social planning corresponds to adopting the socially optimal carbon tax rate in a decentralized economy.
We also prove that the optimal carbon tax regime satisfies time consistency, indicating that policy is Pareto efficient for every generation given the *sustainability* constraint. In addition, it is shown that the theory can be extended to apply to an infinite horizon.
Finally, the second-best proportional carbon tax rates are calculated using available data. The result astonishingly reveals that even if we apply a social discount rate of 5 per cent to annum in the planning economy, it is still equivalent to levying 32 per cent proportional carbon tax rate.

Conclusions: Considering the actual absorption capacity of oceans concerning CO_2, we can never be too prudent in discounting the utility of future generations with regard global climate change. This fact indicates the need for urgent introduction of a proportional carbon tax.

Keywords: Endogenous social discount rate, Proportional carbon tax, Ordinal utilitarian, Sustainability, Time consistency

Background

Evaluation of future generations' utility is a difficult issue, as they cannot question their ancestors' behavior. In other words, the economic activity of the current generation affects voiceless future generations. The most prominent example of such a problem is global warming which is inseparable from the emission and accumulation of CO_2.

There is a huge volume of research accumulation that concerns the relationship between the concentration of CO_2 and global warming. Hulme (2009, Ch.2) concisely reviews theory of the anthropogenic climate change that

originates from Tyndal's experiment in 1859. Calendar (1938) estimated that the global temperature would rise at the rate of 0.3°C per century. On the other hand, Keeling's measurements at the South Pole (from 1957) and Mt. Mauna Loa (from 1958) revealed that the CO_2 concentration was rising by 0.5 to 1.3 ppm per year at both sites.

According to Pearson (2000, p.385), there is evidence that the CO_2 concentration leads to the global warming. Pearson and Pryor (1994) found that the CO_2 concentration increases from 315ppm to 331ppm between 1958 and 1975, furthermore, their observations revealed that concentrations would reach 358ppm by 1994. On the other hand, the concentration during the preindustrial era is estimated at 280ppm. The global average temperature increased between 0.5-1.7°F. Ramanathan et al. (1985)

Correspondence: ohtaki@iss.u-tokyo.ac.jp
Institute of Social Science, University of Tokyo, 7-3-1 Hongo, Bunkyo, Tokyo, Japan

estimated that the surface temperature will rise within 1.5-2.5°K from 1980 to 2030 (0.5-1.7°F). In addition, Houghton et al. (1996) estimated that due to a discernible human influence on the global climate, temperature will increase by 1.1-3.3°K (2-6°F) over the next 100 years. Uzawa (2003, pp.1-21) gives a precise survey of more specialized articles concerning global warming as it relates to economic theory.

Much research within resource and environmental economics discounts the utility of future generations a priori.[a] However, no solid logical foundation is given for the reason permitting such discounting. As such, the main purpose of this article is to determine endogenously the optimal discount rate incurred in the CO_2 emission/accumulation problem based on *ordinal* utility. We solve the problem using the constraint of "sustainability" proposed by Pezzey (1997) in a planning economy.

As Dasgupta (2008) summarizes, whether discounting is permitted is can be attributed to the problem of "ethics". In this context, "ethics" indicates a method that introduces additional, exogenous, and stronger axioms concerning the comparison of utility streams. Koopmans (1960) classifies the case in which discounting is permitted. Diamond (1965) explains that no utility-discounting leads to a contradiction when some assumptions are imposed in addition to plausible standards of utility-stream comparison. Both studies negate the possibility of non-discounting.

On the other hand, Cowen (1992) and Blackorby et al. (1995) assume *cardinal* utility, in which utilities are comparable between individuals, and insist that discounting is not permissible. However, the cardinality of individual utility is quite a strong assumption regarding the current welfare economics criterion.

Although arriving at opposite conclusions, such studies have common characteristics. That is, in addition to the usual assumptions concerning the utility function of each generation, they all give exogenous and restrictive value judgments in attaining their results.

This article is based on a solely *ordinal* utilitarian definition of sustainability proposed by Pezzey (1997). That is, we call an economy sustainable when each generation j can enjoy the minimal utility level $\bar{U}$.[b] By using *ordinal* utility, we entirely exclude any exogenous and transcendent utility comparisons between generations. Such a broad value judgment is not adopted in the preceding research.

This paper determines the optimal social discount rate using the steps outlined below, while avoiding Dasgupta's "ethical" problem. That is, while never imposing any additional restrictions on individuals' ordinal utility function, we arrive at the optimal discount rate using the following two steps.

As the first-best benchmark case, we calculate the optimal tax that is proportional to emissions of CO_2, based on the given definition of sustainability. Since, for simplicity, we assume that current generations are not concerned with the utility of generations thereafter, and only a part of CO_2 emission is absorbed by oceans, biomass, and etc., new emissions will accumulate in the future. A proportional carbon tax is desirable because the tax fully reflect the true price of CO_2, and thus the price mechanism use to deter excess emissions. In other words, when a higher tax is incurred by the CO_2 emissions, individuals reduce their emissions, which consequently reduce CO_2 carried over to future generations.

Second, we solve the social planning problem to attain the same utility level as that which occurs under the optimal proportional carbon tax (the first-best solution). This procedure, also known as the method of Negishi (1960), reveals the relation between the optimal carbon tax rate in a market economy and the endogenously determined social discount rate in a planning economy.

This gives the result that conventional discounting in social planning cannot achieve the optimal taxation in a market economy, if each generation has a right to enjoy some fixed utility, utility should be equally weighed.[c]

It is noteworthy that utility discounting is compatible with our *ordinal* utilitarian definition of sustainability. A positive discount rate in the economy implies that every generation prefers more consumption to the amenity acquired from a reduction in CO_2 emissions. As such, even in the stationary state, the economy will not necessarily reach the first-best resource allocation. Thus, the optimal social discount rate is not self-evident, even with the *ordinal* utilitarian view of sustainability in stationary state.

Results and Discussion
The Model
Dynamics of CO_2

We assume the dynamics of CO_2 accumulation is as follows:

$$e_t = \alpha e_{t-1} + c_t, \quad 0 < \alpha < 1, \qquad (1)$$

where e_t is the stock of CO_2 measured by its weight, c_t is current consumption, and α denotes the proportion of CO_2 remaining per period. We further assume that a unit production/consumption emits a unit weight CO_2. This assumption is permitted by defining the unit of production/consumption volume so as to correspond to the unit weight emission of CO_2.

The economic meaning of equation (1) is as follows: The sum of CO_2 that remains within the atmosphere at the end of period t consists of CO_2 carried over from the previous period, which was not absorbed by oceans, forests, and etc., that is, αe_{t-1}, and the newly emitted CO_2 by generation t's economic activity, that is, c_t. Note that although the linear relationship in (1) seems to be oversimplified, it is not difficult to extend to the non-linear feasible set as long as the set is strictly convex by using the separation theorem of convex sets (see Figure 1). Nevertheless, in such a case, it becomes necessary to have information concerning the utility function to determine the stationary state in which the highest utility is attained.

Another way to consider the difference equation (1) is as follows: The current total CO_2 stock e_t consists of two parts. One is the accumulated stock of CO_2 carried over from the previous period αe_{t-1}. The other is the flow emitted due to current-period consumption c_t. Thus, the dimension of equation (1) is the weight of CO_2 (giga-ton).

Takahashi et al. (1980) estimate that the portion of remaining CO_2, α is approximately 40% per annum although this dissociation rate is very sensitive to the depth, temperature, salinity, and alkalinity of oceans, etc..[d]

Individuals

Individuals live for one period. Their identical, strictly concave, utility function U_t is as follows:

$$U_t \equiv u(c_t, e_t), \quad \frac{\partial u}{\partial c_t} > 0, \quad \frac{\partial u}{\partial e_t} < 0. \tag{2}$$

The individuals' feasibility constraint is

$$e_t = \alpha e_{t-1} + [1 + \theta] c_t - \tau_t, \tag{3}$$

where θ denotes the proportional carbon tax rate and τ_t is a transfer from the government.

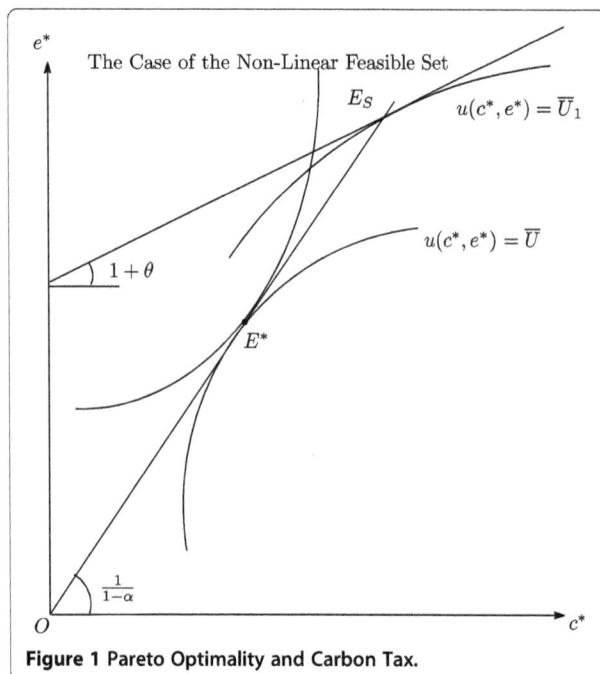

Figure 1 Pareto Optimality and Carbon Tax.

The meaning of (3) is as follows: Since θ additional units of the consumption good are necessary due to the carbon tax incurred, the effective price of the consumption good/CO_2 price becomes $1 + \theta$, as in the second term of (3) (note that from (1), a unit of consumption is assumed to emit a unit of CO_2). On the other hand, the transfer from the government reduces the production and mitigates emissions. This is why τ_t has a negative sign in the right-hand side of (3). It reduces to the difference equation (1) that represents the dynamic of CO_2 accumulation, if we take the budget constraint of the government.

Government

The government transfers the collected tax to individuals equally. [e] Namely, the budget constraint of the government is as follows:

$$\theta c_t = \tau_t. \tag{4}$$

Market economy and social planning

In this section, we first solve for the optimal carbon tax rate in a stationary market economy. Second, we consider the optimal social planning regarding CO_2 emission that has the same effect as the optimal proportional carbon tax. Based on these considerations, we answer the question of whether the utility of future generations should be discounted in a planning economy.

Optimal tax rate in a stationary market economy

In a market economy, each individual maximizes her/his utility (2) subject to the feasibility constraint (3). Hence, the following first-order condition holds.

$$\frac{1}{1 + \theta} \frac{\partial u}{\partial c_t} + \frac{\partial u}{\partial e_t} = 0, \quad \forall t. \tag{5}$$

The dynamics of the market economy is fully described by two difference equations: (1) and (5).

For simplicity, we assume that the economy is initially located at some stationary state (c^*, e^*). Then, it is clear from Figure 1 that the optimal tax rate θ^* in the stationary state E^* is $\frac{\alpha}{1-\alpha}$.

Optimal social planning and utility discounting

We now focus on the social planning problem under an ordinal utilitarian-egalitarian definition of sustainability. [f] Namely, we have the following.

Definition 1 *An economy is sustainable iff*

$$u(c_t, e_t) \geq \bar{U}, \quad \forall t \tag{6}$$

holds for some given $\bar{U}$.

With this definition, we can express the maximization problem of the government as follows:

$$\max_{\{c_t, e_t\}_{t=0}^T} u(c_0, e_0), \quad s.t. e_{-1} = \bar{e}, \quad u(c_t, e_t) \geq \bar{U} \quad \forall t.$$

$$(7)$$

The corresponding Lagrangean L is

$$L \equiv u(e_0 - \alpha e_{-1}, e_0) + \sum_{t=1}^T \lambda_t [u(e_t - \alpha e_{t-1}, e_t) - \bar{U}].$$

$$(8)$$

The first-order condition yields

$$\lambda_t \left[\frac{\partial u}{\partial c_t} + \frac{\partial u}{\partial e_t} \right] - \alpha \lambda_{t+1} \frac{\partial u}{\partial c_{t+1}} = 0. \tag{9}$$

Note that the constraints (6) always bind at the point of the optimal plan (c_t^*, e_t^*). That is,

$$u(c_t^*, e_t^*) = \bar{U}, \quad \forall t. \tag{10}$$

This property can easily be shown as follows: By the Kuhn-Tucker theorem, the necessary-sufficient condition in a convex environment is to find the saddle point of the Lagrangean, and thus

$$L(c, e, \lambda^*) \leq L(c^*, e^*, \lambda^*) \leq L(c^*, e^*, \lambda) \tag{11}$$

holds. (c, e, λ) is the $T+1$ dimensional vector of consumption, CO_2 emission, and the Lagrangean multiplier. *indicates optimal values.

From the right-half of inequality (11),

$$[\lambda - \lambda^*]^{\cdot} [u(c^*, e^*) - \bar{U}] \geq 0, \quad \forall \lambda \geq 0 \tag{12}$$

must hold. Here $^{\cdot}$ denotes the inner product. Since we can select λ to be smaller than λ^*, (10) must hold for inequality (12) to hold.[g]

An important property of the stationary state emerges in (9). The stream of Lagrangean multipliers, $\{\lambda_t^*\}_{t=1}^T$, satisfies the following difference equation:

$$\lambda_t^* \left[1 - \left[-\frac{\frac{\partial u}{\partial e^*}}{\frac{\partial u}{\partial c^*}} \right] \right] = \alpha \lambda_{t+1}^* \quad \Leftrightarrow \quad \lambda_{t+1}^*$$
$$= \frac{1}{\alpha} \left[1 - \left[-\frac{\frac{\partial u}{\partial e^*}}{\frac{\partial u}{\partial c^*}} \right] \right] \cdot \lambda_t^*, \quad \lambda_0^* = 1,$$

$$(13)$$

where (c^*, e^*) is a stationary state of equation (1) and $-\frac{\frac{\partial u}{\partial e^*}}{\frac{\partial u}{\partial c^*}}$ is the marginal rate of substitution between consumption and the stock of CO_2 in the stationary state.

Using the method of Negishi (1960), we can easily prove that the maximization problem (8) with the initial condition, $e_{-1} = e^*$, is equivalent to the maximization of the following social welfare function:

$$\max_{e_t} SW \equiv \max_{e_t} \sum_{t=0}^T \lambda_t^* u(e_t - \alpha e_{t-1}, e_t), \quad e_{-1} = e^*.$$

$$(14)$$

Thus, the endogenously determined Lagrangean multipliers $\{\lambda_t^*\}_{t \geq 0}$ correspond to the optimal social discount rates.[h]

The first-order condition of (14) in the steady state is

$$\frac{\partial SW}{\partial e_t} = [\lambda_t^* - \alpha \lambda_{t+1}^*] \frac{\partial u}{\partial c^*} + \lambda_t^* \frac{\partial u}{\partial e^*} = 0.$$

It is clear from (13) that such conditions are satisfied due to the definition of $\{\lambda_t^*\}_{t \geq 0}$.

It is also clear from (13) that utility discounting is permitted only when

$$\frac{\lambda_{t+1}^*}{\lambda_t^*} < 1 \quad \Leftrightarrow \quad 1 - \alpha = \frac{1}{1+\theta^*} < -\frac{\frac{\partial u}{\partial e^*}}{\frac{\partial u}{\partial c^*}}.$$

Whenever the same allocation is also attained by a market economy, (5) will hold. Consequently, the necessary and sufficient condition for permitting discounting-programming is

$$\frac{1}{1+\theta} > \frac{1}{1+\theta^*} \quad \Leftrightarrow \quad \theta^* > \theta, \tag{15}$$

where θ is the existing carbon tax rate. That is, the social programming with discounting corresponds to lowering the proportional tax to a level lower than that in the market economy optimum. Such a steady state is illustrated by point E_S in Figure 1. This implies that utility-discounting social programming leads to excess emissions of CO_2, even from the *ordinal* utilitarian viewpoint. To sum up, we have the following theorem.

Theorem 1 *As long as the economy is sustainable, the weight of each generation's utility in the social welfare function should be allotted equally in a planning economy. Utility discounting corresponds to the carbon tax rate being lower than the optimum in a market economy. That is,*

$$\lambda_t^* \leq 1 \quad \forall t \quad \Leftrightarrow \quad \theta \leq \theta^* = \frac{\alpha}{1-\alpha}$$

holds with equality only when $\lambda_t^* = 1$.

Sustainability and discounting

In this subsection, we deal with the relationship between the sustainable utility $\bar{U}_j$ and utility discounting.

We have the following representation that is equivalent to Theorem 1.

Theorem 2 *The social welfare function (14) is maximized when the Lagrangean (8) evaluated at the optimal solution is maximized on the feasible sustainable utility level $\bar{U}$.*[i]

Proof.

According to Theorem 1, applying the optimal carbon tax rate θ^* to a market economy is equivalent to setting the discount rate as zero in the social welfare function. Hence, we can sufficiently prove the theorem by showing that maximizing the Lagrangean evaluated at the optimal solution concerning feasible sustainable utility level $\bar{U}$ is equivalent to applying the optimal tax rate $\theta^* = \frac{\alpha}{1-\alpha}$.

The Lagrangean evaluated at the optimal solution L^* is

$$L^*(\bar{U}) = u\big(c_t^*(\bar{U}), e_t^*(\bar{U})\big), \qquad (16)$$

where * indicates optimal values. Since all constraints are binding for the optimal solution, all terms of the Lagrangean after the current period t vanish.

Differentiating (16), we obtain

$$\frac{dL^*}{d\bar{U}} = \frac{\partial u}{\partial c}\frac{dc_t^*}{d\bar{U}} + \frac{\partial u}{\partial e}\frac{de_t^*}{d\bar{U}}$$

$$= \left[1 - \frac{\dfrac{\partial u}{\partial e_t^*}\dfrac{de_t^*}{d\bar{U}}}{\dfrac{\partial u}{\partial c_t^*}\dfrac{dc_t^*}{d\bar{U}}}\right]\frac{\partial u}{\partial c}\frac{dc_t^*}{d\bar{U}}$$

$$= \left[1 - \frac{\dfrac{\partial u}{\partial e_t^*}}{\dfrac{\partial u}{\partial c_t^*}}\frac{de_t^*}{dc_t^*}\right]\frac{\partial u}{\partial c}\frac{dc_t^*}{d\bar{U}}.$$

$$(17)$$

From (3), it is clear that $\frac{de_t^*}{dc_t^*} = \frac{1}{1-\alpha}$ in the stationary state. Hence (17) can be rewritten as

$$\frac{dL^*}{d\bar{U}} = \left[1 - \frac{1}{1-\alpha}\frac{\frac{\partial u}{\partial e_t^*}}{\frac{\partial u}{\partial c_t^*}}\right]\frac{\partial u}{\partial c}\frac{dc_t^*}{d\bar{U}}. \qquad (18)$$

The optimal solutions satisfy (5). Furthermore, it is clear from Figure 1 that

$$\frac{dc_t^*}{d\bar{U}} < 0, \quad if \quad \frac{\frac{\partial u}{\partial e_t^*}}{\frac{\partial u}{\partial c_t^*}} > \frac{1}{1+\theta^*}, \quad and \quad \frac{dc_t^*}{d\bar{U}}$$

$$> 0, \quad if \quad \frac{\frac{\partial u}{\partial e_t^*}}{\frac{\partial u}{\partial c_t^*}} < \frac{1}{1+\theta^*}.$$

Thus, this gives Figure 2. We can clearly see that (16) is maximized at

$$\frac{\frac{\partial u}{\partial e_t^*}}{\frac{\partial u}{\partial c_t^*}} = \frac{1}{1+\theta^*}.$$

Accordingly, L^* is maximized on $\bar{U}$ when a market economy adopts the optimal carbon tax rate θ^*.

Since Theorem 2 implies that

$$\bar{U}^* = u\big(c^*(\bar{U}^*), e^*(\bar{U}^*)\big), \quad \bar{U}^* \equiv \arg\max_{\bar{U}} u(c^*(\bar{U}), e^*(\bar{U})),$$

$$(19)$$

in principle, there is no conflict between generations regarding the optimal carbon taxation. Equation (19) indicates that the optimal social planning by the current generation is also optimal from the viewpoint of future generations. In this sense, the social planning is *time consistent*, and hence, the utility level required for sustainability is set at the most desired level (see Figure 1).[j]

Infinite horizon case

Finally, we discuss the case of an infinite time horizon for the social planning. According to Theorem 1, equal weighting is not constrained to unity. As such, we set $\frac{1}{T}$ as the weight and take the limit $T \to +\infty$. Then, we have the social welfare function for the infinite horizon case:

$$\lim_{T \to +\infty} \frac{1}{T}\sum_{t=0}^{T} u(c_t, e_t). \qquad (20)$$

Thus, the divergence problem for the sum of utility can be avoided, even if the social discount rate is unity. Dutta (1991), Broome (1992), and Cline (1992) apply this

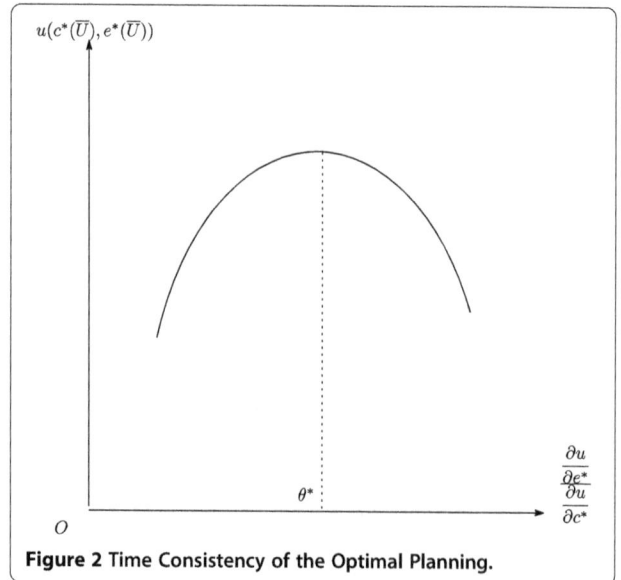

Figure 2 Time Consistency of the Optimal Planning.

type of utility function to the global warming problem. Our paper provides a microeconomic foundation for their research, using Pareto efficiency and stationary equilibrium as a basis.

Case Study: Second-best proportional carbon tax and the social discount rate

In this section, as a case study, we calculate the relationship between the second-best carbon tax rate in a market economy and the corresponding social discount rate in a planning economy.

From equations (5) and (13), we obtain

$$\frac{\lambda_{t+1}^*}{\lambda_t^*} = \frac{1}{\alpha} \left[1 - \left[-\frac{\frac{\partial u}{\partial e^*}}{\frac{\partial u}{\partial c^*}} \right] \right] = \frac{1}{\alpha} \left[1 - \frac{1}{1+\theta} \right]. \quad (21)$$

The left-hand side of (21) is the second-best social discount rate $SDR(\theta)$, which corresponds to the carbon tax rate θ. We consider the length of one generation as approximately twenty years. Relying on Takahashi et al. (1980), we assume that emitted CO_2 remains at a uniform 40 percent annually. That is, α is estimated as

$$\alpha = \sum_{k=1}^{20} 0.4^k = \frac{0.4 \times [1 - 0.4^{20}]}{1 - 0.4} \approx .67.$$

Thus, we obtain Table 1.

This estimation is robust, as the calculation does not use any information concerning the utility function. In other words, any well-behaved utility function will obey equation (21). In addition, although the portion remaining α is probably underestimated, the tax rate corresponding to the endogenously determined social discount rate is never particularly low; hence, we should never be optimistic about the problem of CO_2 emissions.

Conclusions

This paper analyzed the theoretical relationship between the social discount rate in a planning economy and the tax rate on CO_2 emissions in a market economy. If a social planner discounts the utility of future generations, this corresponds to lowering the carbon tax rate below the optimum level in a market economy. Given the ordinal utilitarian view of

sustainability, it is desirable to pay the true price of CO_2, which is

$$\alpha + \alpha^2 + \cdots = \frac{\alpha}{1-\alpha}.$$

This implies that a planner should not discount the utility of any generation in a centralized economy, and that the optimal tax rate must be equal to the total portion of CO_2 remaining $\frac{\alpha}{1-\alpha}$ in a decentralized economy.

Finally, we must note that some important issues are not analyzed in this paper. First, as we concentrated on the properties of the stationary state, we have neglected the issues of the transition process. While such an analysis is beyond the analytical procedure used here, it would be worthwhile to try a numerical analysis or simulation to investigate how the optimal discount (possibly premium) rate varies through such adjustment periods.

Second, we have completely neglected the existence of uncertainty. Specifically, it is natural to assume that governments do not accumulate information reliable enough to determine how the CO_2 emissions and economic activities correlate. This fact makes it very difficult to formulate the agreement necessary to introduce a proportional carbon tax. It also implies that governments tend to underestimate the optimal discount rate. The optimal decision under such uncertainty would be an important topic for future research.

Methods

The tradeoff between consumption utility and the disutility from the climate change which is summarized by the total stock of carbon dioxide is expressed by a utility function. Each generation enjoys consumption, but instead, emits carbon dioxide, and thus the concentration of carbon dioxide occurs. We calculate the optimal proportional carbon tax that prevents the excess emission by using only the information concerning the feasibility condition between consumption and emission independent of the form of utility function. The optimal social discount rates in a planning economy are calculated as the rates by which the economy can attain the same intergenerational resource allocation as that in a market economy where the optimal proportional carbon tax is adopted. We find such an endogenously determined optimal discount rate is zero.

Endnotes

[a]For example, see Dasgupta and Heal (1974), Hartwick (1977), and Solow (1986). Weitzman (1998) finds that the lowest discount rate should be applied to the far-distant future. However, he does not clarify why future projects per se should be discounted.

[b]We use the term "ordinal utilitarian" in the following sense. Ordinal utilitarian implies that an individual welfare is measured only by their consumption and CO_2

Table 1 The Social discount rate and the second-best proportional carbon tax rate

SDR	1	0.99	0.98	0.97	0.96	0.95	(per annum)
$\frac{\lambda_{t+1}^*}{\lambda_t^*}$	1	0.98	0.67	0.55	0.45	0.37	(per twenty years)
θ	200	191	122	58	43	32	(%)

level, which is considered a cause of global climate change, and assumes that the comparison of utilities between different individuals is not feasible.

[c]We never comment on whether a market economy or a planning economy is preferable for preventing excess emissions problem because both systems potentially give the best outcome. Our main interest is in investigating how the best outcome is attained by these alternative systems.

[d]Takahashi et al. (1980) consider two cases: First, the case that there is a homogenous 75m thick surface layer, which is isolated from the underlying deep water. The other case deals with a 4000m deep homogeneous ocean. The former case shows that about 40% of emitted CO_2 remains in the air although in the latter case only 2% of CO_2 remains. Weiss (1974) deals with the issue of how the solubility of CO_2 varies with salinity.

As Tanaka (1993) discusses, there is a serious problem of the 'missing sink' concerning the absorption of CO_2. According to Houghton et al. (1990), a discrepancy exists in the emission/absorption of CO_2 of the order of 1.6 ± 1.4 giga-ton. Since the emission from fossil fuel combusting is more accurately estimated at 5.4 ± 0.5 giga-ton, such a discrepancy cannot be neglected. Generally, the current absorption ability of oceans is estimated at 2.0 ± 0.8 giga-ton.

[e]Note that although individuals know the *values* of θ and τ, they do not consider the government's budget constraint *relationship* in their decision process. It never contradicts their rational economic behavior.

[f]This definition of sustainability is identical to that in Pezzey (1997), although we do not regard there is contradiction between sustainability and "optimality" as Pezzey (1997) insists.

[g]This concise proof is based on Uzawa's proof. For more details, see Uzawa (1958).

[h]Negishi (1960) shows that the Pareto-efficient equilibrium in a market economy is equivalent to maximizing *as if* considering an additive-separable social utility function in a planning economy wherein the weight of each individual's utility is equal to his/her inverse of the marginal utility of income.

[i]Note that whenever we predetermine the optimal social discount rate λ^*, the minimum required utility $\bar{U}_t$ is determined to be consistent with the discount rate.

[j]Dasgupta and Heal (1974), Hartwick (1977), and Solow (1986) define sustainability as maintaining a constant consumption level over time. However, their definition of sustainability has an ambiguous welfare economics foundation. Arrow et al. (2003) define sustainability as a utility integral that increases over time. Besides the ambiguity in its welfare economics foundation, this utility integral analysis may lack the ability to deal with the conflict between generations easily.

Competing interests

The author declares that he has no competing interests.

Author's contribution

MO carried out the modeling and analyzed the model. He also collected data that are necessary for estimating the first and second best carbon-tax ratios and corresponding social discount rates and drafted the manuscript. All authors read and approved the final manuscript.

Acknowledgement

The author is thankful to Susumu Cato and anonymous reviewers for their constructive and incisive comments and suggestions.

References

Arrow K, Dasgupta P, Mäler K-G (2003) Evaluating projects and assessing sustainable development in imperfect economies. Environ Resour Econ 26:647–685

Blackorby C, Bossert W, Donaldson D (1995) Intertemporal population ethics: Critical-level utilitarian principle. Econometrica 63:1303–1320

Broome J (1992) Counting the Cost of Global Warming. White House Press, London

Cline W (1992) The Economics of Global Warming. Institute for International Economics, Washington D.C

Cowen T (1992) Consequentialism implies a zero rate of intergenerational discount. In: Laslett P, Fishkin J (eds) Justice Between Age Groups and Generations. Yale University Press, New Haven

Dasgupta P (2008) Discounting climate change. J Risk Uncertain 37:141–169

Dasgupta P, Heal G (1974) The optimal depletion of exhaustible resources. Rev Econ Stud :3–28, Symposium

Diamond P (1965) The evaluation of infinite utility stream. Econometrica 33:170–177

Dutta P (1991) What do discounted optima converge to? J Econ Theory 55:64–94

Hartwick J (1977) Intergenerational equity and the investing rents from exhaustible resources. Am Econ Rev 67:972–974

Hulme M (2009) Why We Disagree about Climate Change: Understanding Controversy, Inaction and Opportunity. Cambridge University Press, Cambridge

Houghton J et al (1990) Climate Change. Intergovernmental Panel on Climate Change. Cambridge University Press, Cambridge, UK

Houghton J et al (1996) Climate Change 1995: The Science of Climate Change. Intergovernmental Panel on Climate Change, Working Group I. Cambridge University Press, Cambridge, UK

Koopmans T (1960) Stationary ordinal utility and impatience. Econometrica 28:286–309

Negishi T (1960) Welfare economics and existence of an equilibrium for a competitive economy. Metroeconomica 12:92–97

Pearson C (2000) Economics and the Global Environment. Cambridge University Press, Cambridge, UK

Pearson C, Pryor A (1994) Environment: North and South: An Economic Interpretation. John Wiley & Sons, New York, USA

Pezzey J (1997) Sustainability constraints versus "optimality" versus intertemporal concern, and axiom versus data. Land Econ 73:448–466

Ramanathan V, Cicerone R, Singh H, Kiehl J (1985) Trace gas trends and their potential role in climate change. J Geophys Res 90:5547–5566

Solow R (1986) On the intergenerational allocation of natural resources. Scand J Econ 88:141–149

Takahashi T, Wallas B, Werner S (1980) Carbonate chemistry of the surface waters of the world oceans. In: Goldberg ED, Horibe Y, Saruhashi K (eds) Isotope Marine Chemistry. Uchida-Rokakuho, Tokyo, Japan

Tanaka M (1993) The mechanism of global warming (in Japanese). In: Uzawa H, Kuninori M (eds) Economic Analysis of Global Warming. University of Tokyo Press, Tokyo, Japan

Uzawa H (1958) The Kuhn-Tucker theorem in concave programming. In: Arrow KJ, Hurwicz L, Uzawa H (eds) Studies in Linear and Non-Linear Programming. Stanford University Press, California

Uzawa H (2003) Economic Theory and Global Warming. Cambridge University Press, Cambridge, UK

Weiss RF (1974) Carbon dioxide in water and sea water: The solubility of a non-ideal gas. Mar Chem 2:203–215

Weitzman M (1998) Why the far-distant future should be discounted at its lowest possible rate. J Environ Econ Manag 36:201–208

8

Analyzing the impact of environmental variables on the repayment time for solar farms under feed-in tariff

Bin Lu[1][*] and Matt Davison[2]

Abstract

Background: Environmental concerns have promoted the rise of low emissions "green" power technologies such as solar power. In part to make these technologies of economic interest to investors, many green energy policies have been proposed, and a wide variety of green energy developments have been launched which take advantage of these policies. This paper studies the impact of the unpredictable solar insolation on two variables of key interest to solar plant developers: the repayment time and the cash flow at risk.

Results: Using a bootstrap analysis of solar irradiation time series, we model solar farms which sell their power output at a Feed-In Tariff (FIT) rate motivated by one used in the province of Ontario, Canada. We show that the feed-in tariff level which existed in Ontario in March 2012 was more than sufficient to remove the financial risks inherent in financing a solar PV plant.

Conclusions: We conclude that the Ontario Canada FIT 2012 program was an effective tool to encourage investment in solar PV plants. We also find that repayment time is strongly sensitive to FIT rates. So FIT is a very efficient tool to impact/control the volume risk.

Keywords: Solar PV, Feed-In Tariff, Repayment time, Bootstrap

Background

The renewable energy market has seen rapid growth during the past few years. According to the Renewables 2010 Global Status Report (REN21 2010), investment in clean energy assets (not including large hydro) was $29.5 billion in the first quarter of 2010, 63% above that in the same period of 2009. The global capacity of many renewable technologies increased at rates of 10 - 60% annually during the period from the end of 2004 through 2009. In the power sector, though conventional fuels (fossil fuels and nuclear) remain the primary suppliers of global energy, power production from renewable energy (excluding large hydro) increased by 22% in 2009. Worldwide among all types of renewable power generating technologies, solar photovoltaic (PV) power continues to be the fastest growing power generation technology. Cumulative global PV installed capacity was almost six times larger in 2012 than in 2004 (REN21 2010). In 2009, about 16% of all new electric power capacity additions in Europe were credited to Solar PV (REN21 2010). In North America, an estimated 470 MW of solar PV was installed in 2009 in the United States (REN21 2010) where 1800 MW of PV is expected to be installed on the power grid by 2013. Over 1600 MW of PV was under development in Ontario, Canada at the end of 2011 (Ontario Power Authority 2011).

Some facts of solar PV power are particularly favorable for investors: 1) The source of solar photo-voltaic (PV) is free and clean; 2) Solar PV power is easier to predict and more reliable/stable than wind power. Sunlight levels, while still at the mercy of weather patterns, are not as unpredictable as wind speeds; the fact that solar cells don't work at night is at least a predictable feature of their design; 3) Research (Rowlands 2005; Perez et al. 2012) conducted in the United States and Canadian electricity markets finds that solar PV power is highly associated with peak market demand and to somewhat lesser extent associated with high power prices. It also points out that

*Correspondence: blu7@uwo.ca
[1] Department of Applied Mathematics, University of Western Ontario, London, ON N6A 5B7, Canada
Full list of author information is available at the end of the article

the PV power is a potential solution to provide dependable peak power to meet growing summertime demand. However, investors must balance these desirable features against the high capital cost of solar PV power. Although a great deal of analysis has been done on the scientific and engineering development of solar cells, much less literature exists on the economic analysis of these cells. The paper by (Powell et al. 2009) calculates various financial indicators for an "organic" solar cell. Their work uses an approach (complementary to the one taken here) for simulating ground level insolation measurements for which historical data does not exist. The repayment time metric is calculated in the (Powell et al. 2009) paper, but assuming market electricity prices, and finds that payback periods are too large to be economically viable. In a later work, (Azzopardi et al. 2011) compute the average cost of generated power metric for such organic solar cells.

The dramatic growth in the solar PV industry has come in large part because of substantial government support. In the first decade of the 21st century, the world's major governments launched, updated or modified several programs to ensure that financial and administrative instruments are available to aid the development of renewable energy. Common policy measures for promoting solar PV power generation are feed-in tariffs, capital subsidies or grants, tax credits, net metering and direct public investment or financing. The most common policy used to encourage solar PV power is the feed-in tariff (FIT). A FIT offers stable prices under long-term contracts for energy generated from renewable sources. Germany is a pioneer and advocate for feed-in tariff policy among European countries. In 2000, Germany adopted the Renewable Energy Sources Act (Germany 2000), which is a replacement of the previous Electricity Feed Act launched in the 1990's. The German Renewable Energy Sources Act turned out to be a great success and has since been amended several times. Following Germany's success, between 2005 and 2010 at least 50 countries and 25 states or provinces adopted feed-in tariffs (REN21 2010). For instance, France adopted a feed-in tariff of EUR 42-58 cents/kWh for ground-mounted PV systems in 2009 (REN21 2010). Japan also implemented its first feed-in tariff of JPY 48/kWh for residential PV systems in 2009 (REN21 2010). In the Province of Ontario Canada, the current (2012) FIT program provides much higher rates than the market price for the electricity generated from solar PV. In addition, the rates are fixed even though the market price is variable. The Ontario FIT offers CAD 44 cents/kWh for large scale solar PV plants (Ontario Power Authority 2012). On the other hand, the monthly volume weighted average Hourly Ontario Energy Price (HOEP) between 2003 and 2011 has always been below CAD 10 cents/kWh (IESO 2012).

An investor in solar PV projects faces high risks, of which the three largest are: high capital costs, price risk, and volume risk. As mentioned above, high capital cost is indeed a concern but the trend of such costs is falling and capital subsidies may be available in local jurisdictions. Price risk arising from highly volatile electricity prices is another big issue in power plant financing. Even though solar PV power, which is not generated during the low demand night time hours, is associated with peak electricity prices (Rowlands 2005), this does not suffice to remove all price risk. The goal of a FIT program is to provide constant power rates, thereby removing price risk, and to make these rates sufficiently large that sufficient funds may be generated by the developer even in years which are not very sunny, thereby vastly reducing the impact of volume risk. This paper presents a statistical framework for answering the question of whether a given feed-in tariff is high enough to effectively eliminate both price and volume risk. The statistical framework is applied to show that the FIT tariff level in 2012 Ontario sufficed to eliminate solar farm volume risk, as measured by two financial metrics.

Risks other than volume risk include weather damage to panels from hail or snow, faster than expected degradation in panel performance due to extreme cold or heat and transmission line failure. This paper does not consider these risks, which may be hedged using insurance or product warranties.

This paper presents a representative case study from the city of London in the Canadian province of Ontario. The case study demonstrates how FIT performs as a financial inducement to promote solar PV power generation. We seek to answer the question: is the Ontario FIT price at the correct level? This question is answered using two types of financial metric. The first is based on the repayment time. This is the length of time a developer would take to repay the loan taken out to construct the solar farm. This repayment time will fluctuate according to future sunshine level patterns, and so will be a random variable; histograms of the outcomes of this random variable will be generated and reported. The repayment time conclusions are then reinforced using calculations of the Cash Flow at Risk (CFaR) metric (RiskMetrics Group 1999). The relatively new CFaR metric captures some of the risks due to uncertain cash flows in a way that is impossible for a more traditional engineering economic analysis performed using a Discounted Cash Flow approach (White et al. 2010). The CFaR metric reports the worst cash flow that can happen at a given materiality threshold; for instance, that 19 times out 20 the worst lowest cash flow that would be realized in a given year is X. Minimal work has focused on the application of the CFaR metric in the solar PV industry; this new metric will introduce an important new perspective for solar investors. The results of the repayment time and CFaR studies both show that the Ontario FIT levels were, in 2012, more than sufficient to cover the variable insolation risk considered here.

The rest of this paper is organized as follows. Section 'Solar PV financial basics' describes the financial basics of a solar PV plant. Section 'Impact of volume risk on repayment time' describes the methodology used to analyze repayment time of a solar PV plant and discusses the results. Section 'Sensitivity analysis' presents a sensitivity analysis. Section 'Cash flow at risk' presents a Cash Flow at Risk analysis. Section 'Conclusions' concludes the paper.

Solar PV financial basics

Solar energy comes from incoming solar radiation (insolation). A solar PV plant employs solar panels to directly convert sunlight into electricity. A solar PV power plant is characterized both by its geographical location and by its installed capacity. Installed capacity, also known as nominal capacity or nameplate capacity, refers to intended sustained power output levels under ideal conditions. The ideal working condition for a PV array is at a cell temperature of 25°C and at solar irradiance of 1000 W/m^2. This is called the standard test condition (STC) of a solar cell. However, PV arrays rarely operate under these conditions. The actual output of the plant varies with orientation and efficiency of PV arrays, time of the day, season of the year and state of the atmosphere. The capacity factor is a measurement of the efficiency of the plant's actual output power. Solar PV capacity factors are typically under 25% (Bellemare 2003).

DCF analysis

It is common in engineering economic analysis (White et al. 2010) to employ discounted cash flow calculations. This methodology is also known as net present value (NPV) analysis. Future net cash receipts of a project are projected and then discounted with the appropriate time value of money and then summed to decide if a project is worthwhile. Within the DCF framework, the repayment time is a common metric to characterize the economic performance of industrial projects. Some papers (Drury et al. 2011; Sidira and Koukios 2005) use time-to-net-positive-cash-flow (TNP) payback time which is similar to the repayment time idea used in this paper. The repayment time is the earliest time at which a project is able to repay all of its debt. In other words, for our example, it is the time horizon for which the solar PV plant has zero net present value. NPV or DCF analyses are standard ways to evaluate investment profitability. For example, both (Drury et al. 2011; Pappas et al. 2012; Rehman et al. 2007; Borenstein 2008) used NPV as an indicator in their economic analysis.

The cash outflow in a solar PV plant contains initial investment (the cost of purchasing and installing solar modules, land cost etc.) as well as annual operation and maintenance (O&M) costs. The annual cash inflow of a solar PV plant is simply the sales resulting from selling the electricity the panels generate from solar irradiation. Negative numbers denote cash outflows and positive numbers denote cash inflows. All annual inflows are discounted back to a common initial reference time t_0 using a common discount rate or internal rate of return. The first time at which the present value of all cash outflows exactly balances the present value of all cash inflows is termed the repayment time of the plant. In mathematical terms the result is equation (1).

$$V(t) = -C_0 - \sum_{i=1}^{t} \frac{OM_i}{(1+r_{OM})^i} - \sum_{i=0}^{t-1} \frac{L_i}{(1+r_L)^i} + \sum_{i=1}^{t} \frac{E_i}{(1+r_E)^i}.$$
(1)

Here:

$V(t)$ is the present value of solar PV plant [Units $],
C_0 is the initial capital cost [Units $],
t is the time, measured in years [Units years],
OM_i denotes the O&M costs for ith year [Units $][N.B. paid in arrears, see section 'Land cost'],
L_i denotes the land costs for ith year [Units $][N.B. paid in advance, see section 'Land cost'],
r_{OM} is the real interest rate for OM cost [Units %/year],
r_L is the real interest rate for land cost [Units %/year],
r_E is the real interest rate for energy price [Units %/year],
E_i denotes the earnings for ith year by selling electricity [Units $].

Parameter identification
Solar PV plant

Our case study assumes a PV plant of capacity 10MW is to be built in a rural area near London Ontario, Canada. We assume that the power plant will operate for 20 years in order to match the terms of Ontario's FIT contract. We assume the PV plant capacity factor to be 16% and that inverters of the PV system which convert direct current (DC) to alternating current (AC) are 90% efficient. To install a nominal capacity of 1kW modules, an area between $7m^2$ and $10m^2$ is required (Solar Server 2010). This plant requires an area of approximately $70,000m^2$ (approximately equivalent to 17.29 acres) for a solar PV plant with 10MW installed capacity.

Interest rates

We assume that the project is financed before construction and that the same interest rate applies for all cash flows. To estimate the applicable rate, we consider Clean Renewable Energy Bonds (CREBs), a federal loan program sponsored by US Internal Revenue Service (IRS) to finance eligible clean renewable energy projects. On March 1st

2012, the interest rate of a CREB with 20-year maturity was 4.79% (U.S. Treasury 2012). Therefore we believe it is appropriate to estimate an interest rate of 5% for our case study.

Module cost

Generally, PV module costs represent 40-60% of total PV system costs, and installation costs account for the remaining costs (REN21 2010). The larger the solar plant, the smaller the installation cost for each unit.

As of 2007, the unit installation cost for a typical 10kW residential system was $8.00/$W$ (Borenstein 2008).

By the end of 2009, solar PV module prices fell below $2.00/$W$ in some instances (REN21 2010). If we double that to include installation costs, we may assume a total cost of $4.00/$W$ for solar PV systems module and installation.

Combining the above information with the figures quoted in (International Energy Agency 2009) and (Drury et al. 2011), where average annual PV system costs was reported as $3.85/$W$ and $3.91/$W$ respectively, we believe it is reasonable to assume the costs of PV modules and installation are $4.00/$W$ in this case study.

Typical OM costs are 1% of total initial investment (PVResources 2012).

Land cost

One may purchase or rent the premise in order to run the power plant. In this case study, we assume the solar farm operator leases a premise in rural area. The land so rented is assumed to be otherwise unproductive except for relatively low value agricultural or recreational uses. According to a survey in (Weersink et al. 2011), the weighted average monthly rent for Middlesex County (which surrounds London, Ontario) was $200/acre in summer 2010. Therefore, we estimate a total land rent of $3,458/month. In Ontario, rent is typically paid in advance. The details of a lease will vary across rental agreements, but to be conservative we assume that the rent is paid one year in advance. In contrast, work done is often paid in response to the receipt of an invoice, which must be settled on normal business terms of for example 90 days in arrears, so it makes sense to model other costs of the project as being covered at the end of the year. The interest rates used in this analysis are low enough to make the impact of these choices rather small in the final results.

Insolation

We obtained hourly global insolation from 1955 to 2004 at London International Airport weather station from Environment Canada Canadian Weather Energy and Engineering Data Sets (CWEEDS) (Environment Canada 2012). our financial model requires annual cash flows, so we average the data set by year in order to get a constant daily global insolation for each year.

Ontario's FIT

The FIT in Ontario guarantees that all power generated from solar PV plants may be sold at $0.44/$kWh$ (Ontario Power Authority 2012).

The calculation

By listing all the details and calculations each year's cash-flow from equation (1), we obtain the following tables for DCF analysis. These tables may then be used to compute various financial metrics including repayment time.

Tables 1 and 2 include all parameters and the DCF calculations. Table 1 exhibits the values for input parameters and steps for carrying out annual income (not yet discounted) from selling solar-generated power. Table 2 lists each year's cash flow after applying the appropriate discount factors.

Methods

Tables 1 and 2 serves the purpose of demonstrating the DCF analysis. We denote the result of Table 1 the "constant insolation" scenario since we use constant daily global insolation for each operating year in the above DCF analysis. But in reality, insolation is random, fluctuating around some constant level. As financing a solar PV plant involves a huge amount of capital, decision makers must thoroughly understand all risks which impact the plant's repayment time. As mentioned in section 'Introduction', volume risk is the major remaining risk. In order to reflect the impact of volume risk on the repayment time, we introduce some dynamics to model the solar insolation.

Although cloud cover data is readily available, it turns out to be unsuitable for the current purpose, for reasons explained in Appendix A.

The next step is to apply both retrospective and bootstrap analysis to model randomness of solar insolation. Some data abnormality issues emerge. The next subsection explains these abnormalities and then return to retrospective and bootstrap analysis.

Data abnormality

Our data is the CWEEDS hourly global insolation at London International Airport from 1955 to 2004. We average every year's hourly global insolation to obtain annual total global insolation. The time series for annual average global insolation are plotted in Figure 1, which reveals some possible data inconsistency issues.

If we determine the maximum reading for each individual hour from each year, we observe two regimes of data from some of these time series. For example, we pick the annual maximum irradiance for 11AM, noon (12PM),

Table 1 DCF analysis input parameters, annual production etc

	Quantity	Unit	Total
INITIAL SYSTEM COST			
Installed Capacity	10	MW	
panel purchase and installation	4	$/W	
Initial System Cost Total		$	40,000,000
OTHER COSTS			
Land Area	17.29	acre	
Land Rent	200	$/acre/month	
Land Cost Total		$/yr	41,496
Application Cost		$	5,000
OM COSTS			
% of Installation Costs	1	%	
OM Annual Total		$	400,000
ANNUAL PRODUCTION			
Installed Capacity	10	MW	
Mean daily global insolation in Toronto Area	4	kWh/m^2	
Total Energy received annually		kWh/m^2/yr	1,394
PV moduel efficiency	16	%	
Total Annual DC output		kWh/m^2/yr	251
DC to AC conversion factor	90	%	
Estimated actual Annual output		kWh/m^2/yr	226
Area of modules needed to provide this rated capacity	70,000	m^2	
Estimated annual kilowatt hours		kWh/yr	14,069,261
Ontario FIT rates	0.44	$/kWh	
Estimated Annual Income		$/yr	6,232,683
Annual Interest rate	5.0	%	

Table 2 DCF calculation of each year's revenues, costs and net profit, in 2012 Canadian dollars

	Year 0	Year 1	Year 2	Year 3	Year 4
Initial System Cost	(40,005,000)				
OM Cost		(380,952)	(362,812)	(345,535)	(329,081)
Land Cost	(41,496)	(39,520)	(37,638)	(35,846)	(34,139)
Electricity Sales		5,935,888	5,653,227	5,384,026	5,127,643
Cumulative Income		5,935,888	11,589,115	16,973,140	22,100,784
Cashflow		(34,450,064)	(29,159,649)	(24,121,159)	(19,322,596)

Year 5	Year 6	Year 7	Year 8	Year 9	Year 10
(313,410)	(298,486)	(284,273)	(270,736)	(257,844)	(245,565)
(32,513)	(30,965)	(29,490)	(28,086)	(26,749)	(25,475)
4,883,470	4,650,924	4,429,451	4,218,525	4,017,643	3,826,326
26,984,254	31,635,177	36,064,628	40,283,153	44,300,796	48,127,122
(14,752,537)	(10,400,100)	(6,254,921)	(2,307,132)	1,452,667	5,033,428

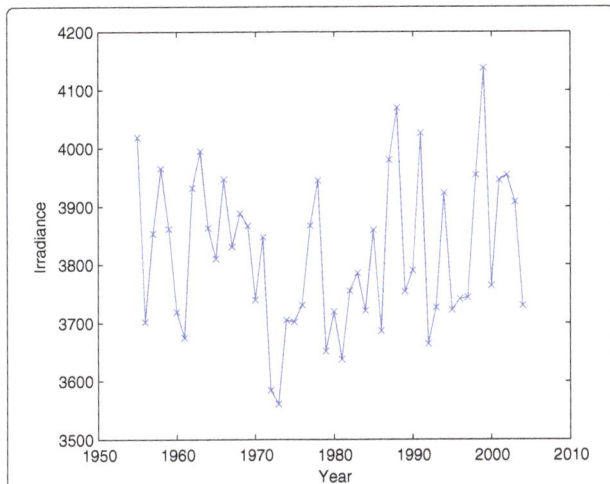

Figure 1 Annual average daily global horizontal irradiance from year 1955 to 2004, London International Airport, in W/m^2, data courtesy of Environment Canada CWEEDS.

1PM and 2PM separately. The time series plots for those hours are shown in Figure 2.

There clearly appears to be two sets of data with different average levels in each of these plots. The difference of these two average levels is about 10%. The reason for this is unknown although we suspect a change in measurement technique around 1978. Because of this issue, we choose to use only data from 1978 and after in order to remain consistent with measured recent data.

Data autocorrelation

Evidence fails to support the hypothesis that the total annual insolation at a point is meaningfully autocorrelated, although of course is likely autocorrelation at higher frequencies. The lack of annual autocorrelation is seen in the autocorrelation (acf) plot of Figure 3 which shows that the autocorrelations at all nonzero lags lie below the significance threshold which distinguishes them from zero. Further evidence is given by the related scatter plot of Figure 4 which plots insolation at year k+1 against insolation at year k. The regression line fitted to the Figure 4 data shows little evidence of autoregressions even at time lag 1.

Retrospective and bootstrap analysis

Now we are ready to employ retrospective and bootstrap analysis on the data sample. The retrospective and bootstrap analyses use identical settings to the constant insolation scenario, with the one change being that they model fluctuations in the annual global insolation. We no longer assume constant insolation. We specify the contract life to be 20 years. So each time we need to generate 20 random numbers to give the insolation for each year in order to work out the repayment time problem.

Retrospective analysis means evaluating what would have happened with actual historical results. We apply the actual observed sequence of historical insolation. Each sequence contains 20 observations. In other words, retrospective analysis shows the repayment time for a solar

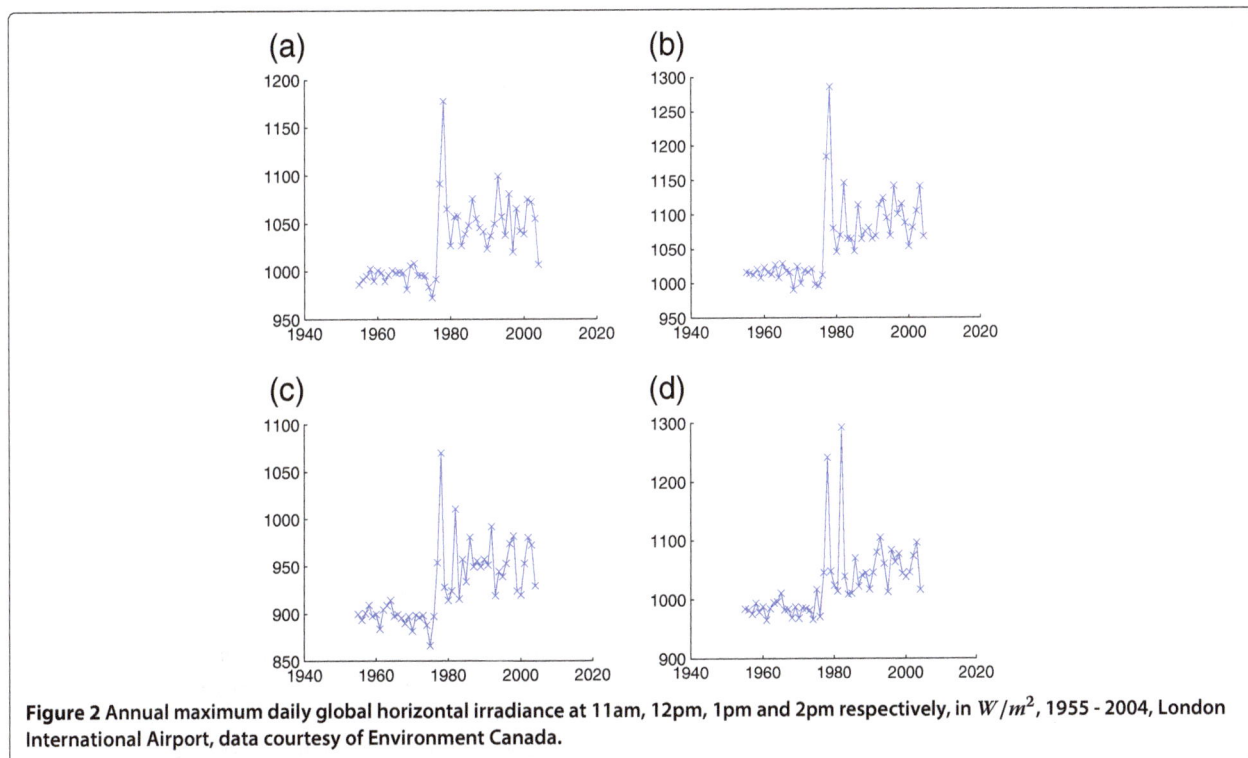

Figure 2 Annual maximum daily global horizontal irradiance at 11am, 12pm, 1pm and 2pm respectively, in W/m^2, 1955 - 2004, London International Airport, data courtesy of Environment Canada.

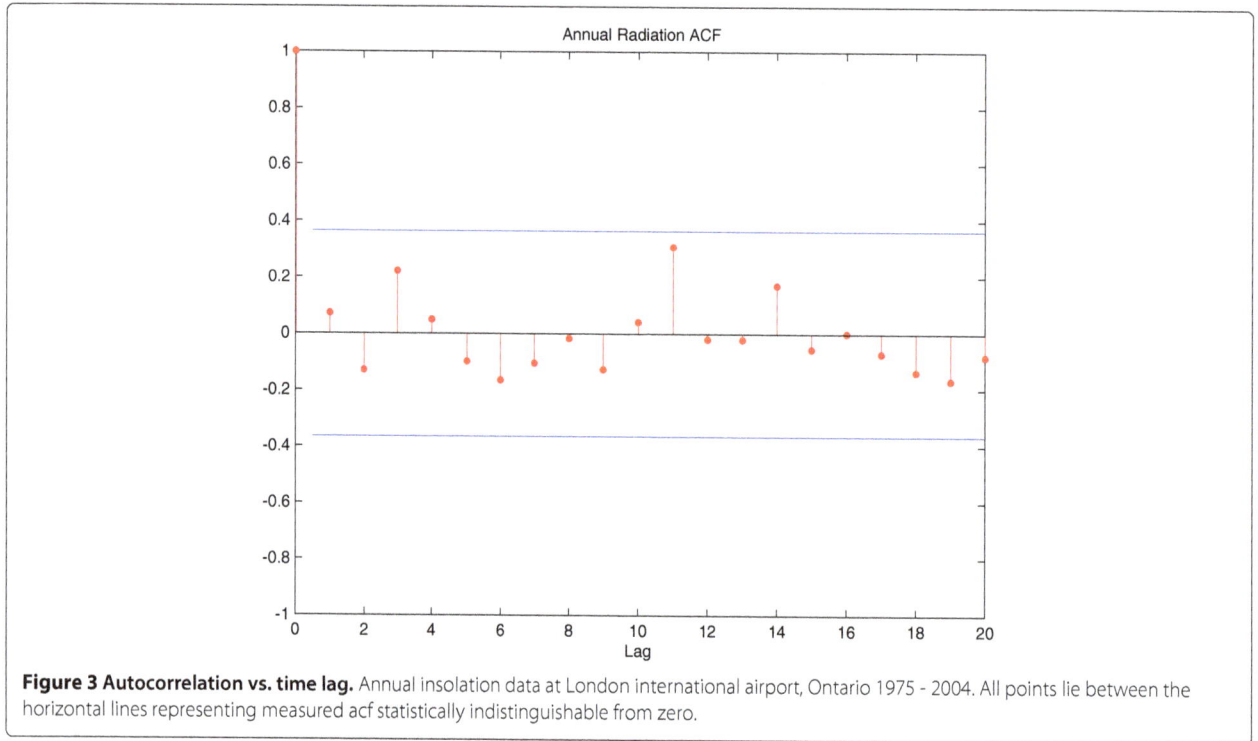

Figure 3 Autocorrelation vs. time lag. Annual insolation data at London international airport, Ontario 1975 - 2004. All points lie between the horizontal lines representing measured acf statistically indistinguishable from zero.

plant as if it were built in 1978, 1979 etc up to 1995. In this case, the sample size is small with only 8 samples in total, i.e. 1978-1997, 1979-1998 up to 1985-2004. Therefore we could work out the average repayment time under retrospective analysis.

It is impossible, given the constraints of limited data, to come up with 10,000 completely different retrospective results. In order to model more variability in solar insolation, we use the bootstrap method (Efron and Tibshirani 1994). The bootstrap is a statistical technique that strongly depends on the development of computer technology. It is extremely useful when the underlying distribution is complicated or unknown or when sample data is insufficient.

Each time, we sample with replacement from the historical annual global insolation data pool to generate a new sequence of annual insolation. We call this sequence of data a sample path. By generating N sample paths (in our simulation, N = 10,000), we are able to estimate the distribution of the repayment time. N will be chosen sufficiently large to estimate financial metrics within our desired accuracy levels. Retrospective analysis may be considered a very special case of bootstrapping in which just the sample paths which actually occurred are used.

Results and discussion

The bootstrap technique described in section 'Retrospective and bootstrap analysis' is used to generate 10,000 possible total annual insolation sequences. Using the economic model for the solar plant described in Section 'Solar PV financial basics', the time required to repay the capital costs of the plant is calculated on each of these sequences. A histogram of the resulting fraction of runs leading to a given repayment time interval is provided in Figure 5.

Figure 5 is a bell shaped curve with a slightly heavy right tail. The peak value of repayment time lies between 9.50 and 9.60 years. The average repayment time by bootstrap analysis is 9.49 years and 9.45 by retrospective analysis.

Note that the repayment time reported by (Powell et al. 2009) is about twice that reported here. This makes

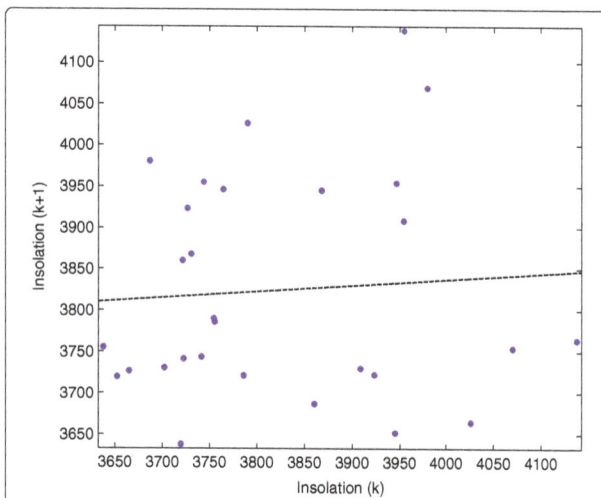

Figure 4 Insolation year (k) vs. Insolation year (k+1) from 1975 to 2004. The line of best fit through the data is $y = 0.07294x + 3545$ with $R^2 = 0.005384$.

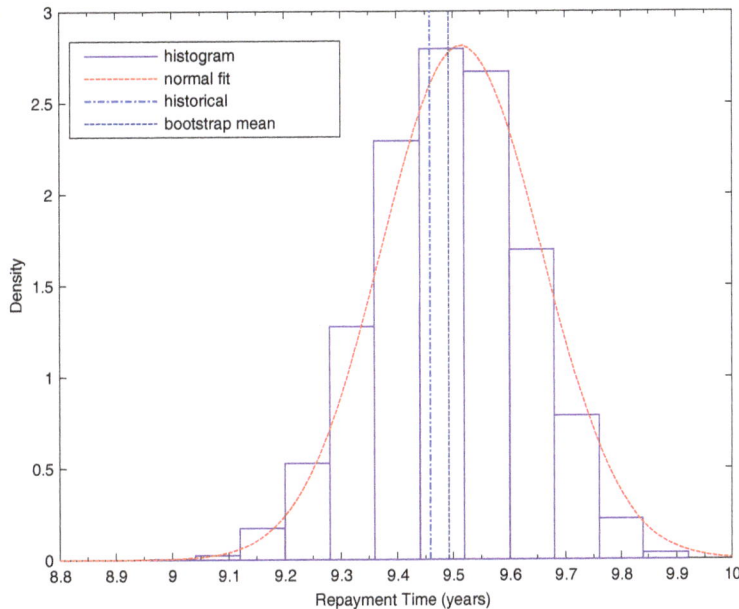

Figure 5 Histogram of repayment times by bootstrap analysis, capacity of 10MW, FIT rate $0.44/kWh, inverter efficiency 90%, system efficiency 16%, location London, ON rural area. Dashed line on the right denotes average repayment time obtained with bootstrap analysis and the dashed line on the left denotes average repayment time obtained with retrospective analysis.

sense, because in addition to modelling a completely different solar technology, (Powell et al. 2009) assume that the power was worth only the market price and not the much larger feed-in tariff price used here. According to Figure 5, the worst repayment time is 9.91 years and the best is 9.0 years. The difference between the two is less than 1 year, and the corresponding standard error of the simulated results is 0.16 years. These provide convincing evidence that FIT is an effective tool to eliminate the volume risk in financing solar PV plants. These results are helpful for those investing in solar PV plants, after which investors earn pure profits for approximately half of the contract life (10 years). Their investment are repaid after, at worst, 9.91 years. Second, the standard error of the repayment time is small. The repayment time falls in a narrow window, so even the worst case is not very much worse than the best case.

If, despite the evidence presented in section 'Data autocorrelation', there was a small auto-correlation in annual insolation, we could no longer use the naive bootstrap method presented here but would have to use the more sophisticated block bootstrap method (Hall et al. 1995). The result of such a study would likely show a positive auto-correlation would increase the probability of abnormally long repayment time while a negative auto-correlation would decrease this risk. This is most easily seen by consider the (unrealistic) limiting case of a 100% auto-correlated signal in which either insolation would always be very high (short repayment time) or very low

(long repayment time). The current random setting favors more intermediate repayment times.

Sensitivity analysis

The methodology used here is broadly consistent with that used in other broadly similar studies such as (Kirby and Davison 2010; Drury et al. 2011; Davison et al. 2012); no exact comparison study to this has been published. We also test the robustness of these results via the following parameter sensitivity study.

As stated in section 'Introduction', solar PV projects are sensitive to government subsidies and economic policies. In this section, we investigate the repayment time under different schemes: varied interest rates and varied FIT rates. We evaluate the repayment time of solar PV plant at different interest rates and FIT rates respectively such that we could see how sensitive repayment times are with respect to certain economic inputs.

Interest rate

We vary interest rates from 5% to 10% with an 0.5% increment. We compare average repayment time of bootstrap analysis, retrospective analysis and constant insolation scenario. We plot average repayment time of these three scenarios under different interest rate settings below.

Figure 6 shows the repayment time calculated by constant insolation from section 'Solar PV financial basics', retrospective analysis and bootstrap analysis. The repayment times in all three cases increase as the interest rate

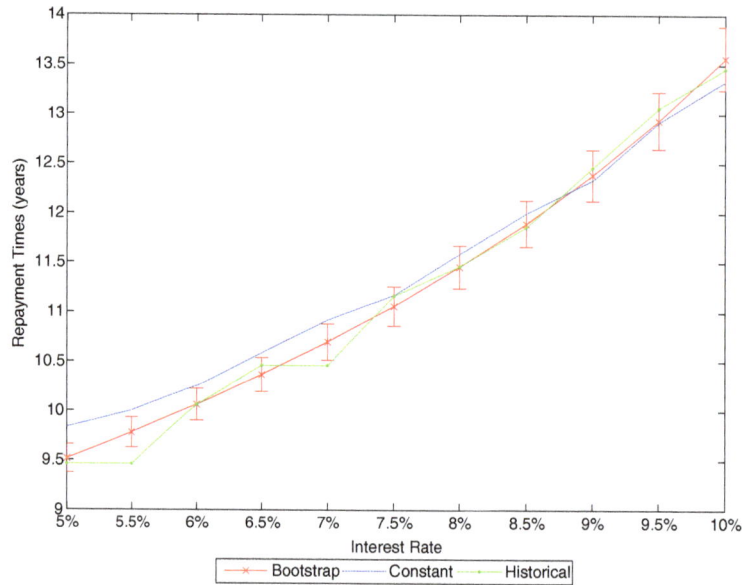

Figure 6 Repayment time, in years, capacity 10MW, FIT rate $0.44/kWh, inverter efficiency 90%, system efficiency 16%, location London, ON rural area. Solid line is obtained via bootstrap analysis, dashed line is obtained via constant insolation, and dashed line connected by circles is obtained via retrospective analysis. Bootstrap analysis resutls include ± standard error width.

increases. The repayment time computed using bootstrap analysis fluctuates around the line obtained by using retrospective analysis. The repayment time obtained from the constant insolation assumption increases less rapidly than those curves obtained from bootstrap and retrospective analyses.

FIT rates

We vary FIT rates from $0.30/$kWh$ to $0.44/$kWh$ with an increment of $0.02/$kWh$.

Figure 7 is the plot of repayment time under different FIT rates. The average value of repayment time decreases as the FIT rate increases, as does the variance.

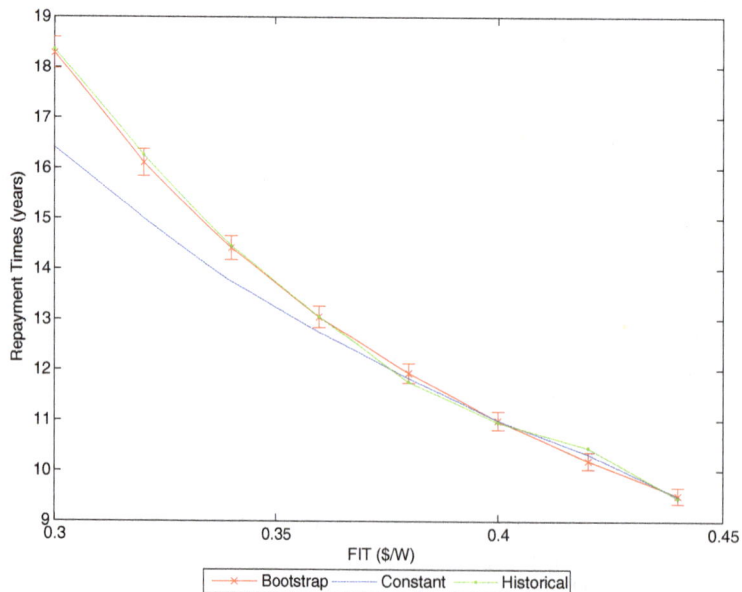

Figure 7 Repayment time varied by FIT rates, capacity 10MW, interest rate 5%, inverter efficiency 90%, system efficiency 16%, location London, ON rural area. Solid line is obtained with bootstrap analysis, dashed line via constant scenario and dashed line connected by dots is by retrospective analysis. Bootstrap analysis resutls include ± standard error width.

Moreover, comparing sensitivity to FIT rates and interest rates, suggests that repayment time is more sensitive to FIT rates than to the impact of interest rates.

Cash flow at risk

Cash Flow at Risk (CFaR) is another risk metric that can be used to quantify the risks associated with investment in solar PV plants. Cash Flow at Risk (CFaR) was first developed in 1999 by the RiskMetrics group to quantify, within some probability tolerance, the worst cash flow a company might experience over a given period (RiskMetrics Group 1999). A description of its use for non-financial firms or projects such as described in this paper is given in (Stein et al. 2001). The CFaR metric is in many ways analogous to the Value of Risk (VaR) metric used in financial risk measurement. The uses of VaR, and the way in which it may be calculated, is described in the book (Jorion 2000). The cash flow at risk measures the level below which the cash flows, normally of a non-financial firm, will not fall with some fixed probability in any given period. In mathematical terms,

$$\text{Prob}(E(C) > CFaR) = p\%. \qquad (2)$$

Here:

E(C) denotes firm/project's expected cashflow in a given duration,
CFaR denotes cashflow at risk (in dollar amount), p is some fixed probability.
Using this measurement, we can investigate the level of the cash flows at some fixed probability and time duration.

We can also investigate the probability that the cash flows do not fall below 0 in any given year using CFaR measurement. For example, Figure 8 shows the distribution of cash flows for PV plant at 9.5 years. It shows that the probability of our model PV plant being debt free in 9.5 years is more than 50%. This is, of course, fully consistent with the repayment times estimated in section 'Sensitivity analysis'. Figure 9 shows the 99% cash flow at risk for year 10 is estimated to be $(6.1294 \pm 5.3244) * 10^5$, a positive number. This shows that even under a worst case scenario, the plant load has been repaid with 10 years of FIT contract remaining.

This suggests the strong conclusion the average repayment times discussed in 2 are extremely robust to variation in insolation scenarios.

Conclusions

In this paper, we explored volume risk in a large scale solar PV plant under FIT based on a case study in Ontario,

Canada. We use two different risk metrics, repayment time and CFaR, to investigate the impact of volume risk under FIT.

It can be shown that under the current FIT rate and other economic parameters, volume risk has little impact on financing solar PV plants. The repayment time of a solar PV farm is much more rapid than the 20 years over which the current FIT agreements guarantee pricing, even allowing for the occurrence of extreme events. Examined from another angle, the worst case cash flow at risk metric, fully confirm this result. Sensitivity analyses do not change this qualitative result. We conclude that the Ontario Canada FIT 2012 program was an effective tool to encourage investment in solar PV plants. We also find that repayment time is strongly sensitive to FIT rates. So FIT is a very efficient tool to impact/control the volume risk.

The method shown here can be applied to the engineering economic analysis of other solar power projects both in other areas of Ontario and outside that province. The idea of using a bootstrap approach to simulate on environmental data to financial outcomes, including the repayment time and CFaR metrics chosen here, is generally applicable. However, it works best in a FIT style financial environment in which electricity prices are constant. If this approach were to be used in a deregulated price environment, prices would also have to be simulated and their correlation with weather also considered. The quantitative results of this study suggest that Ontario's 2012 FIT levels would all be sufficient in sunnier areas of the world.

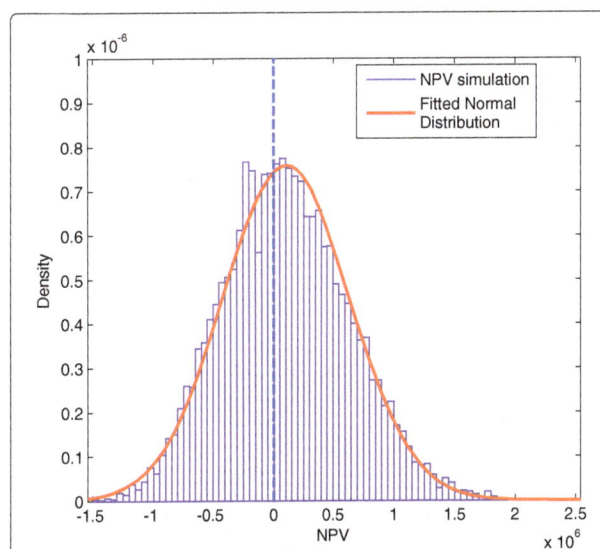

Figure 8 Cash Flow at Risk (CFaR) for solar PV plants in 9.5 years, capacity 10MW, interest rate 5%, inverter efficiency 90%, system efficiency 16%, location London, ON rural area.

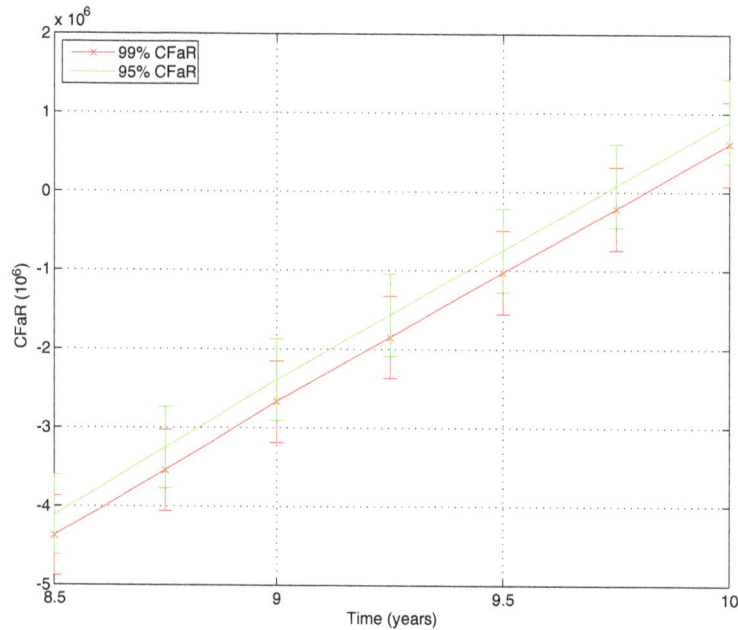

Figure 9 99% and 95% CFaR for solar PV plants in various years, from 8.5 to 10 years increased by 0.25 years, capacity 10MW, interest rate 5%, inverter efficiency 90%, system efficiency 16%, location rural area near London, ON.

Appendix A

Cloud cover model

When the Earth is 1 Astronomical Unit from the sun, the solar irradiance at the top of the earth's atmosphere on a unit area perpendicular to the beam is approximately $1367 W/m^2$ (Lorenzo 2003). This number is called the solar constant.

In general, a reasonable fit to observed clear day global radiation data is given by (Kitchin 1987):

$$G = G_0 * \epsilon_0 * 0.7^{AM^{0.678}} \qquad (3)$$

where G_0 is the solar constant, ϵ_0 is the eccentricity correction factor and AM is air mass. The air mass (AM) is defined as the relative length of the direct-beam path through the atmosphere compared with a vertical path

directly from the top of the atmosphere to sea level. For an ideal homogeneous atmosphere, AM can be expressed in terms of the zenith angle θ_{zs}, depicted in Figure 10:

$$AM = \frac{1}{\cos \theta_{zs}} \qquad (4)$$

As a sun beam passes through the earth's atmosphere, the solar radiation is modified by the interaction with components such as water droplets. In atmospheric sciences,

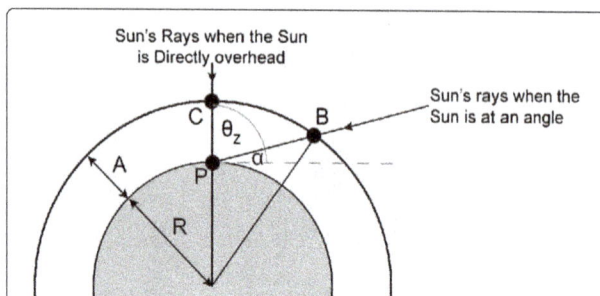

Figure 10 PB is the distance traveled through the atmosphere by the Sun's rays observed at point P when zenith angle is θ_s.
Source: figure 2.7 in (TACA 2012).

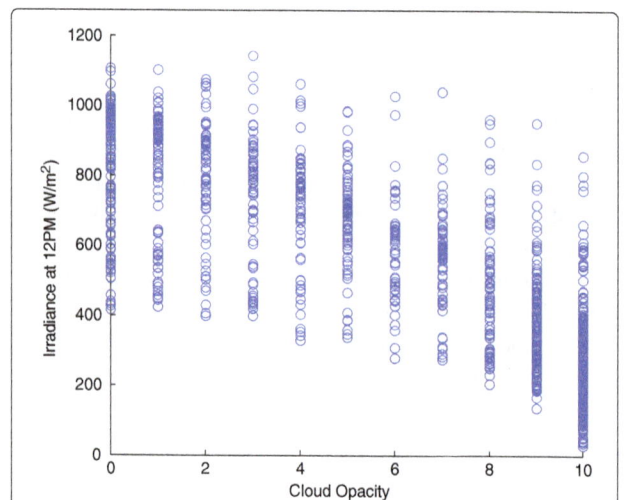

Figure 11 Daily global horizontal irradiance vs. cloud opacity at 12PM, year 2000 to 2004, data courtesy of Environment Canada.

optical depth is a measure of the proportion of radiation absorbed or scattered along a path through a partially transparent medium (See Figure 10 for a conceptual illustration). The insolation is thus defined by the following equation:

$$I = I_0 * \exp{-AM * \tau} \tag{5}$$

where τ is optical depth, I is observed intensity after a given path and I_0 is the intensity of radiation at the source.

Consider that we could abstract the main component interacting with sun beam to be cloud, we scale cloud cover data to serve as τ in above equation and substitute G in equation (3) for I_0.

Thus, the model of capturing global radiation at the ground level is:

$$I = G_0 * \epsilon_0 * 0.7^{AM^{0.678}} * \exp^{-AM * \tau} \tag{6}$$

where G_0 is the solar constant, ϵ_0 is eccentricity correction factor, AM is air mass and τ is optical depth.

Historical cloud cover data are obtained through Environment of Canada. But Figure 11 shows that the correlation between solar radiation and cloud cover is small. One explanation is that optical depth relates to cloud type as well as cloud extent.

Apparently, radiation collected on a day when the cloud cover is 100%, but made up of high white clouds is nowhere close to that collected on a similar day when the clouds were lower and dark. As a result, we dropped the idea of modelling cloud cover and instead, used insolation data directly.

Competing interests

The authors declared that they have no competing interest.

Authors' contributions

BL and MD explored volume risk in a large scale solar PV plant under FIT based on a case study in Ontario, Canada, used two different risk metrics to investigate the impact of volume risk under FIT. MD suggested the problem setup. BL carried out all computations and drafted the paper, which MD helped edit and polish. Both authors read and approved the final manuscript.

Acknowledgements

MD thanks the Canadian Natural Science and Engineering Research Council for research funding which enabled this project to be carried out.

Author details

[1]Department of Applied Mathematics, University of Western Ontario, London, ON N6A 5B7, Canada. [2]Department of Applied Mathematics, Department of Statistical and Actuarial Science, Richard Ivey School of Business, University of Western Ontario, London, ON N6A 5B7, Canada.

References

Azzopardi B, Emmott CJM, Urbina A, Krebs FC, Mutale J, Nelson J (2011) Economic assessment of solar electricity production from organic-based photovoltaic modules in a domestic environment. Energy Environ Sci 4(10): 3741–3753

Bellemare B (2003) What is a megawatt? http://www.utilipoint.com/2003/06/what-is-a-megawatt/#.TrXPH3KA-go. Accessed March 8, 2012
Borenstein S (2008) The market value and cost of solar photovoltaic electricity production. http://escholarship.org/uc/item/3ws6r3j4. Accessed March 11, 2012
Davison M, Gurtuna O, Masse C, Mills B (2012) Factors affecting the value of environmental predictions to the energy sector. Environ Syst Res 1: 4
Drury E, Denholm P, Margolis R (2011) The impact of different economic performance metrics on the perceived value of solar photovoltaics. Tech. rep., NREL, http://www.nrel.gov/docs/fy12osti/52197.pdf. Accessed March 11, 2012
Efron B, Tibshirani R (1994) An Introduction to the Bootstrap. Chapman & Hall
Environment Canada (2012) Canadian Weather Energy and Engineering Data Sets (cweeds). ftp://arcdm20.tor.ec.gc.ca/pub/dist/climate/CWEEDS_2005/ZIPPED%20FILES/ENGLISH/. Accessed March 12, 2012
Germany (2000) Germany Renewable Energy Sources Act. http://www.windworks.org/FeedLaws/Germany/GermanEEG2000.pdf. Accessed October 10, 2012
Hall P, Horowitz JL, Jing BY (1995) On blocking rules for the bootstrap with dependent data. Biometrika 82(3): 561–574
IESO (2012) Hourly Ontario Energy Price. http://www.ieso.ca/imoweb/marketdata/marketSummary.asp. Accessed March 8, 2012
International Energy Agency (2009) National survey report of PV power applications in Canada 2008. http://198.103.48.154/fichier.php/codectec/En/2009-128/2009-128_e.pdf. Accessed March 2012
Jorion P (2000) Value at risk: the new benchmark for managing financial risk, 2nd edn. McGraw-Hill
Kirby N, Davison M (2010) Using a spark-spread valuation to investigate the impact of corn-gasoline correlation on ethanol plant valuation. Energy Economics 32(6): 1221–1227
Kitchin C (1987) Stars, Nebulae and the Interstellar Medium: Observational Physics and Astrophysics. Taylor & Francis
Lorenzo E (2003) Energy collected and delivered by PV modules. In: Luque A, Hegedus S (eds) Handbook of Photovoltaic Science and Engineering. John Wiley & Sons, West Sussex, England, pp 905–970
Ontario Power Authority (2011) A Progress Report on Electricity Supply Third Quarter. http://www.powerauthority.on.ca/sites/default/files/OPA%20-%20A%20Progress%20Report%20on%20Electricity%20Supply%20-%202011%20Q3.pdf. Accessed March 12, 2012
Ontario Power Authority (2012) OPA FIT quick facts table for Solar PV. http://fit.powerauthority.on.ca/quick-facts-table-0. Accessed March 12, 2012
Pappas C, Karakosta C, Marinakis V, Psarras J (2012) A comparison of electricity production technologies in terms of sustainable development. Energy Convers and Management 64: 626–632
Perez R, Letendre S, Herig C (2012) PV and grid reliability availability of PV power during capacity shortfalls. http://www.asrc.cestm.albany.edu/perez/publications/PV%20Power%20outage/PV&Grid%20reliability-00.pdf. Accessed March 14, 2012
Powell C, Bender T, Lawryshyn Y (2009) A model to determine financial indicators for organic solar cells. Sol Energy 83(11): 1977–1984
PVResources (2012) Photovoltaic economics. http://www.pvresources.com/Economics.aspx. Accessed March 12, 2012
Rehman S, Bader MA, Al-Moallem SA (2007) Cost of solar energy generated using PV panels. Renew and Sustainable Energy Rev 11(8): 1843–1857
REN21 (2010) Renewables 2010 Global Status Report. http://www.ren21.net/Portals/97/documents/GSR/REN21_GSR_2010_full_revised%20Sept2010.pdf. Accessed March 12, 2012
RiskMetrics Group (1999) Corporatemetrics Technical Document
Rowlands IH (2005) Solar PV electricity and market characteristics: two Canadian case-studies. Renew Energy 30(6): 815–834
Sidira KD, Koukios GE (2005) The effect of payback time on solar hot water systems diffusion: the case of Greece. Energy Convers and Management 46(2): 269–280
Solar Server (2010) Solar electricity: Grid-connected photovoltaic systems. http://www.solarserver.com/knowledge/basic-knowledge/grid-connected-photovoltaic-systems.html. Accessed March 11, 2012
Stein JC, Usher SE, LaGattuta D, Youngen J (2001) A comparables approach to measuring cashflow-at-risk for non-financial firms. J Corporate Finance 13(4): 100–109

TACA (2012) Part 2: Solar energy reaching the Earths surface. http://www.
    itacanet.org/the-sun-as-a-source-of-energy/part-2-solar-energy-
    reaching-the-earths-surface/. Accessed March 8, 2012
US Treasury (2012) Clean Renewable Energy Bond Rates. https://www.
    treasurydirect.gov/GA-SL/SLGS/selectCREBDate.htm. Accessed March 12,
    2012
Weersink A, Deaton BJ, Bryan J, Meilke K (2011) Farmland prices. http://www.
    uoguelph.ca/catprn/PDF-TPB/TPB-11-01-Weersink-Deaton-Bryan-Meilke.
    pdf. Accessed March 16, 2012
White JA, Case KE, Pratt DB (2010) Principles of Engineering Economic Analysis,
    5th edn. Wiley, Hoboken, NJ

9

Spatial distribution of non-point source nitrogen in urban area of Beijing City, China

Xiaowen Ding[1,2,3*], Yongwei Gong[1], Chunjiang An[3] and Ming Lin[2]

Abstract

Background: Non-point source (NPS) pollution, has been dominant in many urban areas, causing nutrient loss, water body pollution and aqueous systems damaging. Among the pollutants, nitrogen is a key nutrient which can cause eutrophication of rivers, lakes and reservoirs. Beijing, the capital of China, has a booming economy and a huge population. Various human activities have affected nitrogen accumulation deeply and caused considerable NPS pollution. This research calculated the annual load of NPS nitrogen in urban area of Beijing City (UABC) in 2005 and simulated its spatial distribution, which may be useful for environmental planning and pollution control of the UABC and cities with intensive human activities.

Results: The total NPS nitrogen load of the UABC was is 1083.09 t in 2005. The load of agricultural land, construction land and unused land were 92.04 t, 969.29 t and 21.77 t. As far as spatial distribution is concerned, construction land was with heavy pollution. Chaoyang district, Haidian district and Fengtai district were the most major export regions of NPS nitrogen loads. The loads of them were 376.88 t, 286.87 t and 249.92 t. As for load intensity of NPS nitrogen, the high-load sources were distributed in the Xicheng district and Dongcheng district, of which the load intensities were 1.10 km^2/t and 1.11 km^2/t respectively. Among agricultural land, construction land and unused one, the high-load source was distributed in construction land.

Conclusions: Therefore, exhaust gas emission should be reduced, and roof greening as well as road sweep should be improved.

Keywords: Spatial distribution; Non-point source; Nitrogen; Urban area of Beijing city

Background

With the effective control of point source pollution, Non-point source (NPS) has become the major cause of water pollution in many countries, such as the UK (Whitehead et al. 2007), China (Ding et al., 2010) and the USA (Emili and Greene 2013). For rivers and lakes in urban areas, NPS pollution in the form of runoff from urban areas has contributed greatly to the degradation of flow in the receiving water bodies (Sartor et al. 1972; Kang et al., 2013). NPS pollution in urban areas has become an important issue for aquatic environments and

has therefore received increased attention in recent years (Rossi et al., 2006; Kang et al., 2013). NPS pollution in urban areas is not easily identified and characterized which adds significantly to the issues of source analysis, load calculation, and distribution simulation. Spatial distribution simulation in the typical urban area has its great importance of revealing the impact method of human activities on water pollution quantificationally.

Therefore, researches on NPS pollution in urban areas have been done in the fields of the runoff monitoring (Wada et al., 2006; Krometis et al., 2007), load simulation (Ward & Dudding, 2004; Brown & Peake, 2006; Duzgoren et al., 2006), and pollution control (Chimney & Pietro, 2006; Kang et al., 2006). Beijing City is the capital of China, in which water pollution caused by NPS pollution is increasingly outstanding (Jing, 2008; Wang, 2011). The Urban area of Beijing City (UABC) is the core zone of Beijing and with very intensive human activities. Intensive human activities together with torrential

* Correspondence: binger2000dxw@163.com
[1]Key Laboratory of Urban Stormwater System and Water Environment, (Beijing University of Civil Engineering and Architecture), Ministry of Education, No.1 Zhanlanguan Road, Beijing 100044, People's Republic of China
[2]Key Laboratory of Regional Energy and Environmental Systems Optimization, Ministry of Education, North China Electric Power University, No. 2 Beinong Road, Beijing 102206, People's Republic of China
Full list of author information is available at the end of the article

Table 1 Area and population of the UABC in 2010 (Leading group office of Beijing sixth national population census, Beijing Municipal Bureau of Statistics, Beijing survey office of National Bureau of Statistics of China 2011)

No.	Administrative region	Area (km^2)	Percent of the area (%)	Population (million person)	Percent of the population (%)
1	Dongcheng district	50.53	0.30	0.92	4.69
2	Xicheng district	41.86	0.25	1.24	6.34
3	Chaoyang district	455.08	2.71	3.55	18.08
4	Haidian district	430.73	2.56	3.28	16.73
5	Shijingshan district	84.32	0.50	0.62	3.14
6	Fengtai district	305.80	1.82	2.11	10.77
7	UABC	1368.32	8.14	11.72	59.74
8	Beijing city	16807.80	100.00	19.62	100.00

rains cause NPS become a major pollution source of receiving water bodies in the UABC. Studies on NPS pollution in the UABC have been also done on pollutant monitoring, component analysis and water quality assessment of runoff, as well as strategies on stormwater use and pollution control. The result indicated that NPS has become an important factor affecting urban water environment due to heavy rains and increasing impervious surface areas during the process of urbanization (Jin, 2008; Wang, 2011). Main NPS pollutants in the UABC were organic pollutants, nutrients and suspended solids (Che et al., 2002). Pollutant concentrations in rainfall-runoff were high in the early stage, especially in the first 40 minutes. Then the pollutant concentrations decreased, and reached stable values finally (Wang, 2004). Water qualities in rainfall-runoffs of various land use types were in different levels. That of grassland was better than that of land for roofed buildings, which was also superior to that of road (Liu et al., 2008). Pollution control of initial stormwater was significant for water environmental protection and stormwater utilization in the UABC (Liao et al., 2007).

It can be found that researches have been done mainly focus on field monitoring, establishing the relation between rainfall-runoff and pollution, and putting forward countermeasures and suggestions. Temporal scales of the researches focused on single or multiple storms, and spatial scales were only several monitoring sites. Researches on numerical calculation of annual load, its spatial simulation and its characteristic analysis of the UABC have not been reported. However, numerical simulation and distribution analysis in macro spatial-temporal scale is significant, which may provide more efficient and feasible decision support for pollution control and stormwater use.

Methods
Study area
Urban area of Beijing City, covering an area of 1368.32 km^2, is located in the northwest of the Huabei Plain, and accounts for 8.14% of the entire area of Beijing City (See Table 1 and Figure 1). The administrative region of UABA contains Xicheng district, Dongcheng district, Chaoyang district, Haidian district, Shijingshan district and Fengtai district (See Figure 1). It is the most developed area in the city and has a very high population density (8562 per square kilometer). About 59.74 percent of

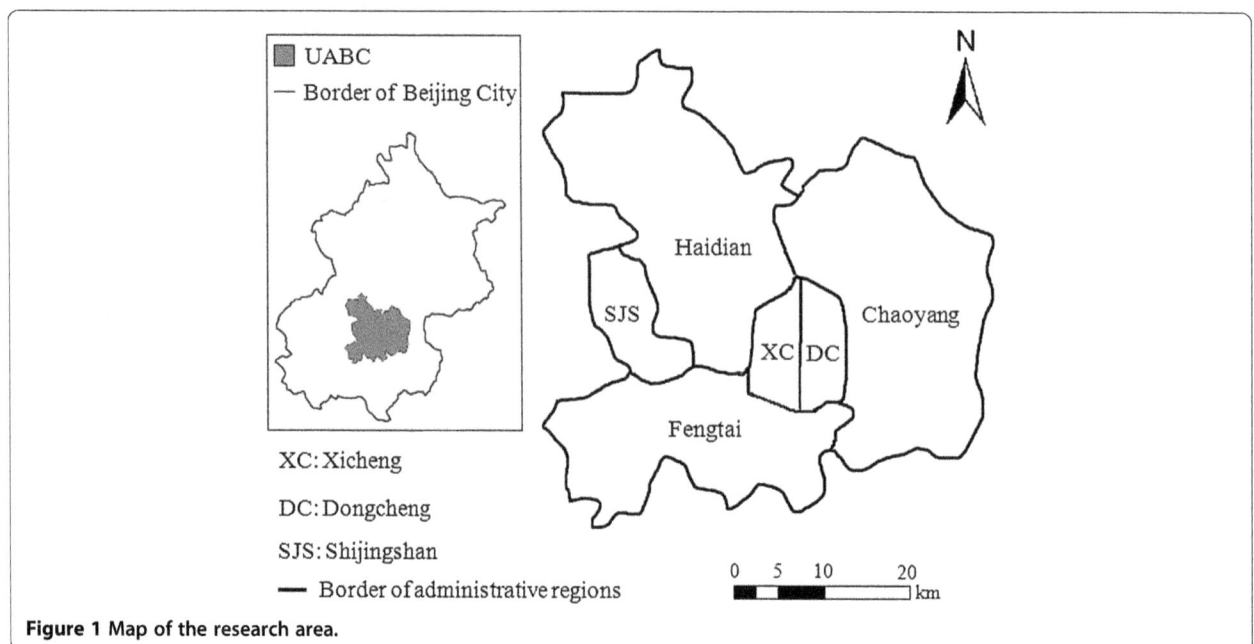

Figure 1 Map of the research area.

the city's population lives in UABC even though it accounts for much less percent of the city's area. Due to its special geographic location and atmospheric circulation pattern, torrential rain occurs frequently in summer over the UABC. Moreover, rainfall is the main driving force of NPS contamination, which leads to serious NPS pollution occurrence in the study area.

Model description

The export coefficient model (ECM), which is based on the idea that the nutrient loads exported from a watershed equals the sum of the losses from individual sources, has been used in many studies. The model allows accurate estimation and analysis of nutrient pollutants (Johns, 1996). The ECM is outlined as:

$$L = \sum_{i=1}^{n} E_i[A_i(I_i)] + p \qquad (1)$$

where L is loss of nutrients (kg), E_i is export coefficient for nutrient source i (t/km^{-2} year^{-1}), A_i is area of the catchment occupied by land use type i (km^2), I_i is the input of nutrients to source i (kg), and p is the input of nutrients from precipitation (kg). The ECM model has been applied in several watersheds, but no research is reported for a total urban area.

Data collection and preparation

Taking into account the research needs and data availability, the database for this research is as presented in Table 2. The land use map of 2005 was interpreted from the Thematic Mapper image provided by Beijing municipal people's government, China. The interpreted land use of the UABC was classified into agricultural land (plough, garden plot, woodland, grassland, and other farmland), construction land (land for roofed buildings, road, independent industrial and mining area, land for transport, and land for water conservation facilities), as well as unused land. The area of the UABC occupied

Table 2 Available data for the UABC

Type	Scale	Source
Land use	1:1 500 000	RS image digitalizing and interpreting, Beijing municipal people's government, China
Landform	1:250 000	National geomatics center of China
Administrative division	1:1 500 000	Beijing municipal bureau of statistics
Social economics	Each district	Beijing municipal bureau of statistics
Export coefficient	Each nitrogen source	Literature, statistical data, and field monitoring

Table 3 Areas of various land use types in the UABC

No.	District	Areas of various land use (km^2)		
		Agricultural land	Construction land	Unused land
1	Xicheng	0.00	50.53	0.00
2	Dongcheng	0.00	41.86	0.00
3	Chaoyang	146.60	299.27	9.21
4	Haidian	207.93	215.46	7.34
5	Shijingshan	32.84	48.37	3.12
6	Fengtai	88.67	193.25	23.87
7	UABC	476.03	848.75	43.54

by various land use types were obtained from the interpreted land use map which was digitized and evaluated in each grid (see Table 3). The input of nitrogen from precipitation was determined by monitoring data which was considered as an average value in the different land use types. The layer of export coefficient was related to each spatial unit and those coefficients were determined.

Determination of the export coefficient

The export coefficients were derived from literature sources as well as calibration using the hydrology and water quality data. An approach has been developed based on the mass balance of the pollutant to determine the export coefficients of different land use types (Ding et al., 2010).

First, the export coefficient of a specific land use type was defined as E_i, (i =1, 2, 3, ..., n), with n land use types. Next, m small sub-watersheds ($m > n$) are chosen from the entire research area and the NPS pollution

Table 4 Export coefficients of NPS nitrogen sources in the UABC (t/km$^2 \cdot$ yr)

Land use	Nitrogen source	Export coefficient
Agricultural land	Plough	0.23
	Garden plot	0.08
	Woodland	0.20
	Grassland	0.30
	Other farmland	0.15
Construction land	Land for roofed buildings	1.09
	Road	1.33
	Independent industrial and mining area	1.31
	Land for transport	1.41
	Land for water conservation facilities	1.10
Unused land		0.50

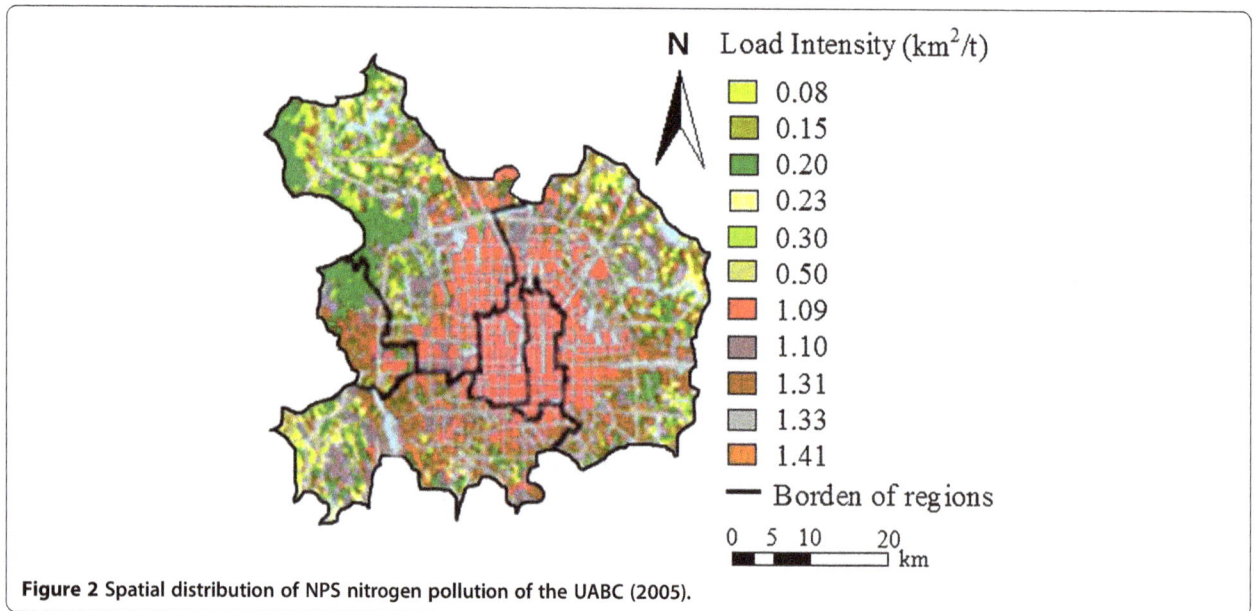

Figure 2 Spatial distribution of NPS nitrogen pollution of the UABC (2005).

yield equation for each sub-watershed is established as follows:

$$L = PS + \sum_{i-1} E[A_i(I_i)]_i^+ p \tag{2}$$

where L is the loss of nitrogen (kg), PS is the point source pollution load (kg), E_i is export coefficient of land use type i (kg/km$^2 \cdot$yr), A_i is area of land use type i (km^2), and I_i is annual input load of the NPS pollutants from land use type i (kg).

The left side of Eq. (2) can also be expressed as:

$$L = \frac{C \cdot Q}{k} \tag{3}$$

where C is the annual mean observed concentration of the NPS pollutants in the outlet of the sub-watershed (g/L), Q is the annual total stream flow measured in the outlet (m^3), and k is the loss coefficient of the NPS pollutants in the sub-watershed.

Table 5 Spatial distribution of NPS nitrogen in different administrative regions (2005)

No.	District	NPS nitrogen load (t)	Percent of the total load of UABC (%)
1	Xicheng district	55.62	5.14
2	Dongcheng district	46.35	4.28
3	Chaoyang district	376.88	34.80
4	Haidian district	286.87	26.49
5	Shijingshan district	67.46	6.23
6	Fengtai district	249.92	23.07
7	UABC	1083.09	100.00

With E_i set as unknown, the other parameters on the right side of Eq. (2) can be calculated as follows:

$$PS = \frac{C_d \cdot Q_d}{D_a \cdot k} \times 365 \tag{4}$$

where C_d is mean observed concentration of the NPS pollutants during the dry season in the outlet of the sub-watershed (g/L), Q_d is total flow volume during the dry season at the outlet (m^3), and D_d is the number of days in the dry season.

Then, Eq. (2) can be transformed to:

$$\frac{C \cdot Q}{k} = \frac{C_d \cdot Q_d}{D_d \cdot k} \times 365 + \sum_{i=1}^{n} E_i[A_i(I_i)] + p \tag{5}$$

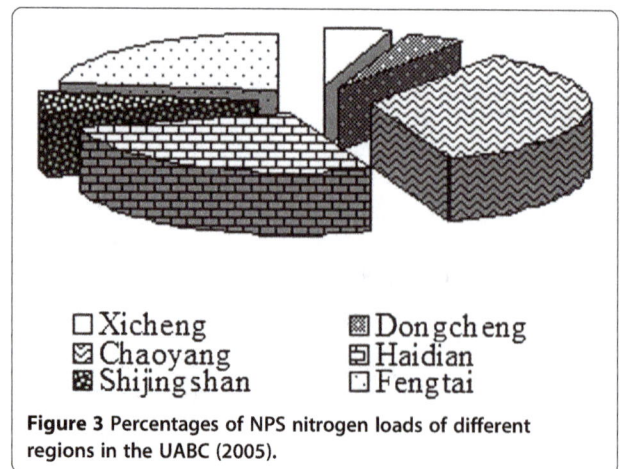

□Xicheng   ▦Dongcheng
☑Chaoyang   ⊟Haidian
▨Shijingshan   ☐Fengtai

Figure 3 Percentages of NPS nitrogen loads of different regions in the UABC (2005).

Table 6 Spatial distribution of NPS nitrogen loads in different land use types (2005)

Type of land use		NPS nitrogen load (t)	Percent of the total load of UABC (%)
Agricultural land	Plough	29.47	2.72
	Garden plot	4.47	0.41
	Woodland	43.93	4.06
	Grassland	6.62	0.61
	Other farmland	7.55	0.70
Construction land	Land for roofed buildings	743.24	68.62
	Road	40.01	3.69
	Independent industrial and mining area	68.01	6.28
	Land for transport	112.17	10.36
	Land for water conservation facilities	5.85	0.54
	Unused land	21.77	2.01
	UABC	1083.09	100.00

With the analysis described above, we can obtain m equations for the m sub-watersheds, and the following equation group is established:

$$\begin{cases} \dfrac{C_1 \cdot Q_1}{k_1} = \dfrac{C_{d1} \cdot Q_{d1}}{D_{d1} \cdot k_1} \times 365 + \sum_{i=1}^{n} E_i[A_{i1}(I_{i1})] + p_1 \\ \dfrac{C_2 \cdot Q_2}{k_2} = \dfrac{C_{d2} \cdot Q_{d2}}{D_{d2} \cdot k_2} \times 365 + \sum_{i=1}^{n} E_i[A_{i2}(I_{i2})] + p_2 \\ \cdots \\ \dfrac{C_m \cdot Q_m}{k_m} = \dfrac{C_{dm} \cdot Q_{dm}}{D_{dm} \cdot k_m} \times 365 + \sum_{i=1}^{n} E_i[A_{im}(I_{im})] + p_m \end{cases}$$

$$(6)$$

The export coefficient, E_i, for the different land use types can be deduced through optimization using the Genetic Algorithm (GA) and are listed in Table 4.

Results and discussion

Annual load of NPS nitrogen

Annual load of NPS nitrogen pollution was 1083.02 ton in 2005 and most of it occurred with urban storm. In terms of the hydrological term scale, 80 percent of the NPS nitrogen pollution concentrated in wet period (June, July and August). Therefore, it can be found that nitrogen loss is positive correlation with the rainfall, which indicates that NPS nitrogen pollution accumulates in dry seasons and occurs in wet seasons with rainfall and soil erosion.

Spatial distribution of NPS nitrogen in different regions

By geographic information system (GIS), the spatial distribution of NPS nitrogen load in the UABC in 2005 is simulated as Figure 2 shows.

The spatial distribution of NPS nitrogen load in different administrative regions is shown in Table 5. The percentages of NPS nitrogen loads of different administrative regions are shown in Figure 3. It can be found that Chaoyang district, Haidian district and Fengtai district are major regions carrying NPS nitrogen loads, while the NPS nitrogen loads in other districts are relatively low. The main reason is that the areas of the three districts are much larger than the other ones.

Spatial distribution of NPS nitrogen in various land use types

As for various land use types, spatial distributions of NPS nitrogen loads in them are shown in Table 6. The percentages of NPS nitrogen loads of various land use types are shown in Figure 4. It can be found that NPS nitrogen export varies in different land use types. The loads of agricultural land, construction land and unused land in the research area in 2005 were 92.04 t, 969.28 t, and 21.77 t. It can be seen that the region with heavy NPS nitrogen pollution is construction land which accounted for 89.49% of the total load of the UABC.

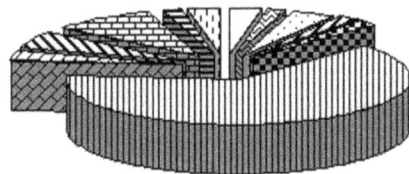

□ plough
□ woodland
▨ other farmland
▨ road
▤ land for transport
▥ unused land

▨ garden plot
▨ grassland
▥ building land (with roofs)
▨ independent industrial and mining area
▤ land for water conservation facilities

Figure 4 Percentages of NPS nitrogen loads of various land use types.

Table 7 Spatial distribution of load intensity of NPS nitrogen in various regions

No.	District	Load intensity of NPS nitrogen (km²/t)	Ratio to the average value of UABC
1	Xicheng district	1.10	1.39
2	Dongcheng district	1.11	1.41
3	Chaoyang district	0.83	1.05
4	Haidian district	0.67	0.85
5	Shijingshan district	0.80	1.01
6	Fengtai district	0.82	1.04
7	UABC	0.79	1.00

Large area and the high NPS nitrogen export are the reasons of heavy pollution in construction land. In the research, land for roofed buildings mean the land in which residential buildings, office buildings, or any other roofed buildings are constructed. Among the various types of construction lands, land for roofed buildings is the one output the most dominant pollution for its largest area and relatively high NPS nitrogen export. The pollution source of land for roofed buildings is the roof, which can't be swept routinely. NPS nitrogen accumulates in roofs during dry seasons and loses in rainfall events during wet seasons. Especially in the course of the first rainfall in a wet season, nitrogen pollutants accumulated for several months are scoured by roof runoff, which causes serious NPS nitrogen pollution in the UABC.

As for various land use types, spatial distributions of NPS nitrogen loads in them are shown in Table 6. The percentages of NPS nitrogen loads of various land use types are shown in Figure 4.

Spatial distribution of load intensity of NPS nitrogen

An important index of NPS pollution analyzing is load intensity, which means pollution load per unit area (such as t/km²) and can mitigate the effect of areas on loads). As for various administrative regions, spatial distribution of load intensity of NPS nitrogen is shown in Table 7. Load intensity of NPS nitrogen is as shown Figure 5.

The high-load sources were distributed in the Xicheng district and Dongcheng district, human activities in which are much more intensive than those of other districts. The land use of the two districts include construction land exclusively, so it can be deduced that accumulated nitrogen pollutants in roofs and those in roads caused by exhaust gas emission are the main sources of NPS nitrogen. The lowest-load source is distributed in Haidian district. This could be explained by the fact that the percent of construction land in the total district area is 49.98%, which is lower than that of any other districts. The load intensities of NPS nutrogen in other districts (i.e., Chaoyang district, Fengtai district and Shijingshan district) are in the middle level.

As for load intensity of NPS nitrogen in various land use types, the high-load source is distributed in construction land, which is the dominant one in the UABC. Large quantities of nitrogen pollutants are accumulated in building roofs and urban roads, which causes high load intensity of NPS nitrogen in construction land. The lowest-load source is agricultural land which could be explained by the fact that nitrogen export of it is relatively low due to the positive effect of soil and water conservation.

Conclusions

UABC is the most developed area in the city and has a very high population density. Sustained and cumulative human activities have serious impact on the process of nutrient accumulation and transportation. NPS pollution in urban areas has become the dominant factor affecting water quality of water bodies in Beijing City. The export coefficient model (ECM) can be used to calculate the

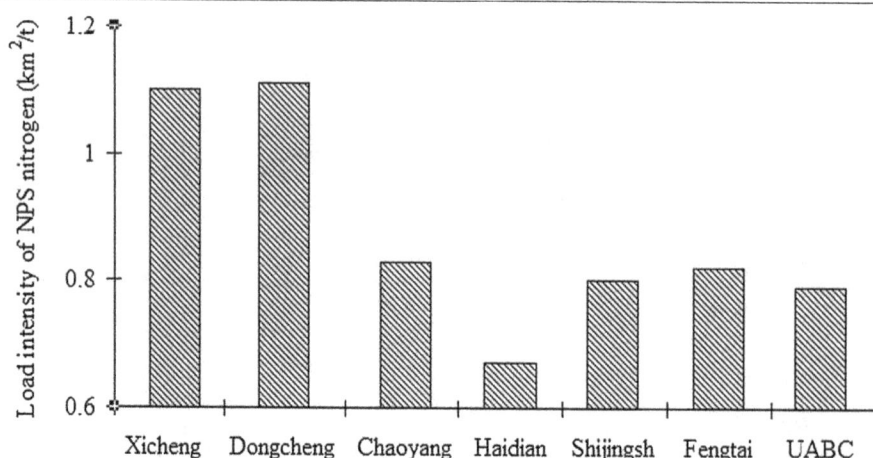

Figure 5 Load intensity of NPS nitrogen in various administrative regions.

pollution load of NPS nitrogen and the export coefficients can be calibrated using the hydrology and water quality data. Take nitrogen as a typical NPS pollutant, annual load of NPS pollution was 1083.02 ton in 2005. As for spatial distribution, Chaoyang district, Haidian district and Fengtai district are major regions carrying NPS nitrogen loads. Between various land use types, construction land is the one with heavy NPS nitrogen pollution. Among the various types of construction lands, land for roofed buildings carries the most dominant pollution. As far as various districts are concerned, high-load sources are distributed in the Xicheng district and Dongcheng district, and the lowest-load source is distributed in the Haidian district. As for load intensity of NPS nitrogen in various land use types, the high-load source was distributed in construction land, and lower percent of construction land results in lower load intensity of NPS nitrogen. So exhaust gas emission should be reduced, roof greening should be advanced, roof rainwater collection and processing should be promoted, and road sweep should be improved. In the future, analysis on pollution sources and cause factors in the study area should be advanced.

Competing interests
The authors declare that they have no competing interests.

Authors' information
[a] Key Laboratory of Urban Stormwater System and Water Environment (Beijing University of Civil Engineering and Architecture), Ministry of Education, No.1 Zhanlanguan Road, Beijing, P.R. China. 100044. [b] Key Laboratory of Regional Energy and Environmental Systems Optimization, Ministry of Education, North China Electric Power University, No. 2 Beinong Road, Beijing, P.R. China. 102206. [c] Institute for Energy, Environment and Sustainable Communities, University of Regina, 120, 2 Research Drive, Regina, Saskatchewan, Canada. S4S 7H9.

Authors' contributions
XD implemented the system architecture. She carried out the modeling and analyzed the model, results analysis and drafted the manuscript, which all the authors helped to edit and polish. YG provided data and discussed the system architecture. CA discussed and revised the manuscript. ML sorted the data and drew the figures. All the authors read and approved the final manuscript.

Acknowledgements
The work described in this study was funded by Open Research Fund Program of Key Laboratory of Urban Stormwater System and Water Environment (Beijing University of Civil Engineering and Architecture), Ministry of Education, as well as National Natural Science Foundation of China (No. 51309097).

Author details
[1]Key Laboratory of Urban Stormwater System and Water Environment, (Beijing University of Civil Engineering and Architecture), Ministry of Education, No.1 Zhanlanguan Road, Beijing 100044, People's Republic of China. [2]Key Laboratory of Regional Energy and Environmental Systems Optimization, Ministry of Education, North China Electric Power University, No. 2 Beinong Road, Beijing 102206, People's Republic of China. [3]Institute for Energy, Environment and Sustainable Communities, University of Regina, 120, 2 Research Drive, Regina, SK S4S 7H9, Canada.

References
Brown JN, Peake BM (2006) Source of heavy metals and polycyclic aromatic hydrocarbons in urban stormwater runoff. Water Sci Technol 25(8):13–19

Che W, Ou L, Wang HZ, Li JQ (2002) The quality and major influential factors of runoff in Beijing urban area. Tech Equip Environ Pollut Contr 3(1):33–37 (in Chinese)

Chimney MJ, Pietro KC (2006) Decomposition of macrophyte litter in a subtropical constructed wetland in south Florida (USA). Ecol Eng 27(4):301–321

Ding XW, Shen ZY, Hong Q, Yang ZF, Wu X, Liu RM (2010) Development and test of the export coefficient model in the upper reach of the Yangtze River. J Hydrol 383(3–4):233–244

Duzgoren ANS, Wong CSC, Songb ZG, Aydina A, Lic XD, Youb M (2006) Fate of heavy metal contaminants in road dusts and gully sediments in Guangzhou, SE China: a chemical and mineralogical assessment. Hum Ecol Risk Assess 12(2):374–389

Emili LA, Greene RP (2013) Modeling agricultural nonpoint source pollution using a geographic information system approach. Environ Manage 51(1):70–95

Jing HW (2008) The effect analysis of water eutrophication ameliorations in Liuhai lakes of Beijing. Beijing Water 1:7–10 (in Chinese)

Johnes PJ (1996) Evaluation and management of the impact of land use change on the nitrogen and phosphorus load delivered to surface waters: the export coefficient modelling approach. J Hydrol 183(3–4):323–349

Kang JH, Kayhanian M, Stensorm MK (2006) Implications of a kinematic wave model for first flush treatment design. Water Res 40(20):3820–3830

Kang OY, Lee SC, Wasewar K, Kim MJ, Liu HB, Oh TS, Janghorban E, Yoo CK (2013) Determination of key sensor locations for non-point pollutant sources management in sewer network. Korean J Chem Eng 30(1):20–26

Krometis LAH, Characklis GW, Simmons OD, Dilts MJ, Likirdopulos CA, Sobsey MD (2007) Intra-storm variability in microbial partitioning and microbial loading rates. Water Res 41(2):506–516

Leading group office of Beijing sixth national population census, Beijing Municipal Bureau of Statistics, Beijing survey office of National Bureau of Statistics of China (2011) Beijing sixth national population census data communique (in Chinese)

Liao RH, Ding YY, Hu XL, Chen JG, Zhang SH (2007) Analysis and evaluation of water quality of rainfall and runoff in urban area in Beijing. Beijing Water 1:14–16 (in Chinese)

Liu Y, Li JQ, Che W, Kuang R (2008) The urban runoff pollution control and "saving-power and decreasing-emission": Beijing as an example. Environ Pollut Contr 30(9):93–96 (in Chinese)

Rossi L, Fankhauser R, Chevre N (2006) Water quality criteria for total suspended solids in urban wet-weather discharges. Water Sci Technol 54(6–7):355–362

Sartor JD, Boyd GB, Agardy FJ (1972) Water pollution aspects of street surface contaminants. Water Pollut Contr Fed 46(3):458–467

Wada K, Yamanaka S, Yamamoto M, Toyooka K (2006) The characteristics and measuring technique of refractory dissolved organic substances in urban runoff. Water Sci Technol 53(2):193–201

Wang MM (2004) analysis and assessment of precipitation and runoff water quality in the southeast of Beijing urban area. Beijing University of Technology (in Chinese)

Wang XY (2011) Non point source pollution process mechanism and control management: taking Beijing watershed of Miyun Reservoir as an example. Sciences Press (in Chinese)

Ward NI (2004) Dudding LM (2004) Platinum emissions and levels in motorway dust samples: influence of traffic characteristics. Sci Total Environ 334–35:457–463

Whitehead PG, Heathwaite AL, Flynn NJ, Wade AJ, Quinn PF (2007) Evaluating the risk of non-point source pollution from biosolids: integrated modelling of nutrient losses at field and catchment scales. Hydrol Earth Syst Sci 11(1):601–613

10

Effects of soil properties and biosurfactant on the behavior of PAHs in soil-water systems

Hui Yu[1,2][*], Huining Xiao[1] and Dunling Wang[3]

Abstract

Background: The interactions among biosurfactant, soil components and PAHs govern the efficiency of biosurfactant enhanced remediation, which was still poorly studied. In this study, we investigated effects of biosurfactant and soil properties on sorption and desorption of phenanthrene (PHE) and pyrene (PYR) in soil – water systems. Two kinds of soil samples (ditch and under plant) from the same petroleum contaminated site in western Canada were applied.

Results: The results indicate that soil organic matter (SOM) was the predominant factor that affects PAHs sorption onto soil. The SOM content in ditch soil was half of that in under plant soil, therefore ditch soil showed less sorption affinity to PAHs than under plant soil. We also examined the combined effects of soil DOM and biosurfactant on desorption of PAHs. The results indicated that more PAHs were desorbed from ditch soil than the under plant soil under the combined conditions. The SOM was still the key factor that determined desorption of PAHs. Besides, competitions among PAHs, DOM and surfactant for sorption sites exist. In high solute concentration system, the competition for sorption site was more severely than low concentration system and more PAHs were sequenced in soil phase in high PAH concentration system. Also in low biosurfactant system, less PAHs were desorbed from soil.

Conclusions: The study results should be helpful in broadening knowledge of biosurfactant enhanced bioremediation of PAHs.

Keywords: PAHs; Biosurfactant; Soil organic matter; Desorption

Background

Polycyclic aromatic hydrocarbons (PAHs) have drawn environmental and health concern due to their known or suspected cancerogenicity and mutagenicity (Wilcke 2000; Shin et al. 2006). PAHs emitted mainly from anthropogenic activities and accumulate in environment owing to their highly hydrophobic properties. In soil/ ground water system, PAHs are largely sorbed into soil organic materials, which significantly restrict their transformation and bioavailability. The sorption behavior of PAHs in soil is greatly influenced by a number of soil characteristics, including soil organic matter (SOM), pH, clay minerals and soil texture (Cao et al. 2008). SOM is considered as the most important component affecting the behavior of hydrophobic organic contaminants (HOCs) in soil and sediment (Wen et al. 2007). The dissolved fraction of SOM, dissolved organic matter (DOM), which has the capability of mobilizing PAHs in soil water system, would also influence the mobility and bioavailability of HOCs in environment (Akkanen et al. 2005; Pan et al. 2007; Raber et al. 1998).

The use of biosurfactant to enhance the remediation is a promising approach to remove PAHs from soil system (Zhu and Aitken 2010). Surfactants are amphiphilic compounds and are known to improve the efficiency of desorption of HOCs through enhancing the solubility of HOCs in aqueous systems, thus facilitating their mobility and bioavailability (Yu et al. 2007; Abu-Zreig et al. 2003; Cheng and Wong 2006). Surfactant could greatly enhance the solubility of HOCs at concentrations higher than its critical micellar concentration (CMC). Biosurfactant is naturally produced surfactant. Compared with traditional

---

* Correspondence: bestyuhui@gmail.com
[1]Department of Chemical Engineering, University of New Brunswick, Fredericton, NB, E3B 5A3, Canada
[2]MOE Key Laboratory of Regional Energy Systems Optimization, S&C Academy of Energy and Environmental Research, North China Electric Power University, Beijing 102206, China
Full list of author information is available at the end of the article

surfactant, biosurfactant has relatively higher biodegradability, biocompatibility, and environmental resilience (Kuyukina et al. 2005; Cheng et al. 2004; Makkar and Rockne 2003). However, the efficiency of biosurfactant enhanced bioremediation can be also affected by the soil properties due to its potential to be adsorbed by soils, which would enhance contaminant retention in soil, thus their bioavailability (Rodriguez-Escales et al. 2013; Lee et al. 2005; Lu and Zhu 2012).

Therefore, the fundamental interactions among biosurfactant, soil components and PAHs jointly govern the efficiency of biosurfactant enhanced remediation and should be detailed investigated. Recently there has been increasing study on the combined effects of surfactant and SOM on HOCs partition (Cheng and Wong 2006; Wan et al. 2011; Rodriguez-Escales et al. 2013). However, these studies put more efforts on additive organic matter, such as DOM or artificial soil particles. Also, the combination of soil properties and surfactant on the partitioning of PAHs has been poorly investigated.

Given the concerns mentioned above, the objective of this study is to explore the interactive mechanisms of soil properties and biosurfactant on desorption of two PAHs: phenanthrene (PHE) and pyrene (PYR) in soil–water systems. We mainly focus on the soil DOM which has been proven to be the most important factors affecting the behavior of PAHs. Two types of soil samples collected from the same western Canadian site were compared.

Results and discussion
Soil and DOM characterization

It is known that soil properties such as organic matter, pH, cation exchange capacity (CEC), clay minerals and soil texture will affect the sorption–desorption of PAHs (Hwang 2001; Kuyukina et al. 2005; Cao et al. 2008; Chen et al. 2009a). Among them, soil SOM and clay minerals are considered as the two most chemically active components of soils (Hwang and Cutright 2003; Lu and Zhu 2012).

The characteristics of two kinds of site soils are presented in Table 1. According to the result of soil classification, the under plant soil belonged to sand loam which has a high sand mineral content (60%) and a normal clay content (28%). The ditch soil is classified as clay loam, which has relatively higher clay mineral content (37%). Of the clay minerals, the contents of smectite, illite and kaolinite in ditch soil were 7.8%, 9.4% and 2.9%, respectively, which were slightly higher than those in under plant soil.

Smectite, illites and kaolinite are the three common clay minerals which have the greatest impact on sorption/desorption because of their high surface area and CEC as well as their surface reactivities (Sheng et al.

Table 1 Properties of studied soil samples

Property		Under plant	Ditch
pH(CaCl$_2$)		6.96	7.02
Organic matter		8.65	4.3
C:N ratio		12.55	8.98
Cation exchange capacity (cmol/kg)		29.34	38.79
Coarse sand >0.25 mm (%)		14.5	2.4
Fine sand 0.25–0.02 (%)		42.5	35.6
Silt 0.02–0.002 mm (%)		15	25
Clay < 0.002 mm (%)		28	37
Classification[a]		Sand loam	Clay loam
Clay mineralogy (% wt)	Smectite	6.2	7.8
	Illites	9.3	9.6
	Kaolinite	2.4	2.9
	Quartz	54.8	42.7
	Potash feldspar	13.5	23.6
	Plagioclase	10.4	9.7
	Amphibole	1.7	1.7
	Calcite	1.7	1.2

[a]International classification system.

2001; Hwang et al. 2003). Smectite is the most expandable clay on wetting with a 2:1 ratio of Si: Al and can provide an internal surface area as high as 570–660 m^2/g (Brady and Weil 2000). Illite is nonexpandable 2:1 type minerals dioctahedral, aluminous, and contains nonexchangeable K as the major interlayer. Kaolinite is a 1:1 layered silicate with alternating silicon oxide tetrahedral sheets and aluminum hydroxide octahedral sheets, which could provide large external surface area for HOCs binding (Hwang and Cutright 2003; Woods 2004). SOM is often considered as the dominant sorption phase for organic contaminants in soil-water systems (Sheng et al. 2001). The SOM content in ditch soil is only half of that in under plant soil. For other soil characteristics given in Table 1, the pH in ditch soil was slightly higher than that in under plant soil and the CEC of ditch soil (38.79 cmol/kg) was also a little higher than under plant soil (29.34 cmol/kg).

For the DOM characterization, the concentration of DOM extracted from the under plant soil and ditch soil showed great difference, with the values of 610.2 and 207.8 mg/kg, respectively. The DOM content in under plant sample was much higher than that in ditch soil sample, which was in proportion to their total organic matter content. The E4/E6 ratio of under plant soil and ditch soil were 8.3 and 8, respectively, which means the DOM derived from ditch soil had relatively higher molecular weight. When the DOM content in all samples was adjusted to 20 ppm, the specific UV absorptions at 254 nm were 0.761 and 0.792 L mg^{-1} m^{-1} for DOM derived from under plant soil and ditch soil, respectively. The aromatic carbon content in ditch DOM sample was also higher than that in under plant DOM sample.

The structural difference of two soil DOMs was analyzed through ^{1}H NMR. The two DOM samples exhibited similar peak shapes from the ^{1}H NMR spectra (Figure 1), both showed simple and well-defined peaks, which indicated that similar compositions in the functional groups of DOM. The identified peaks in the ^{1}H NMR spectra were assigned to aliphatic H in methyl protons and main-chain methylene (0.8-1.5 ppm), carbonyl group in an acid or ester at β-C (1.8 ppm), protons to ethers or hydroxyl group (3.6 ppm), esters (4.0 ppm), water (4.8 ppm) and carboxyl group (8.4 ppm). Both samples showed these identified peaks with different responses. The DOM from ditch soil showed lower ratio of aliphatic H in methyl protons to main-chain methylene. Ditch soil DOM also showed fewer amounts of olefinic compounds as well as oxygen and nitrogen. Besides, for the spectrum of ditch DOM, there were multiple peaks between 1.8 to 3.6 ppm. This indicated the component of high molecules of polymers; H may be bonded to aromatic C in methine group or bounded to O or N in aliphatic C. However, the detailed composition of the polymers needed to be further analyzed (Figure 2).

For under plant DOM, FTIR analysis was conducted to further clarify its chemical functions and following characteristic bands can be found. A broad band at 3300 and 3500 cm^{-1} was assigned to H-bonds, OH groups.

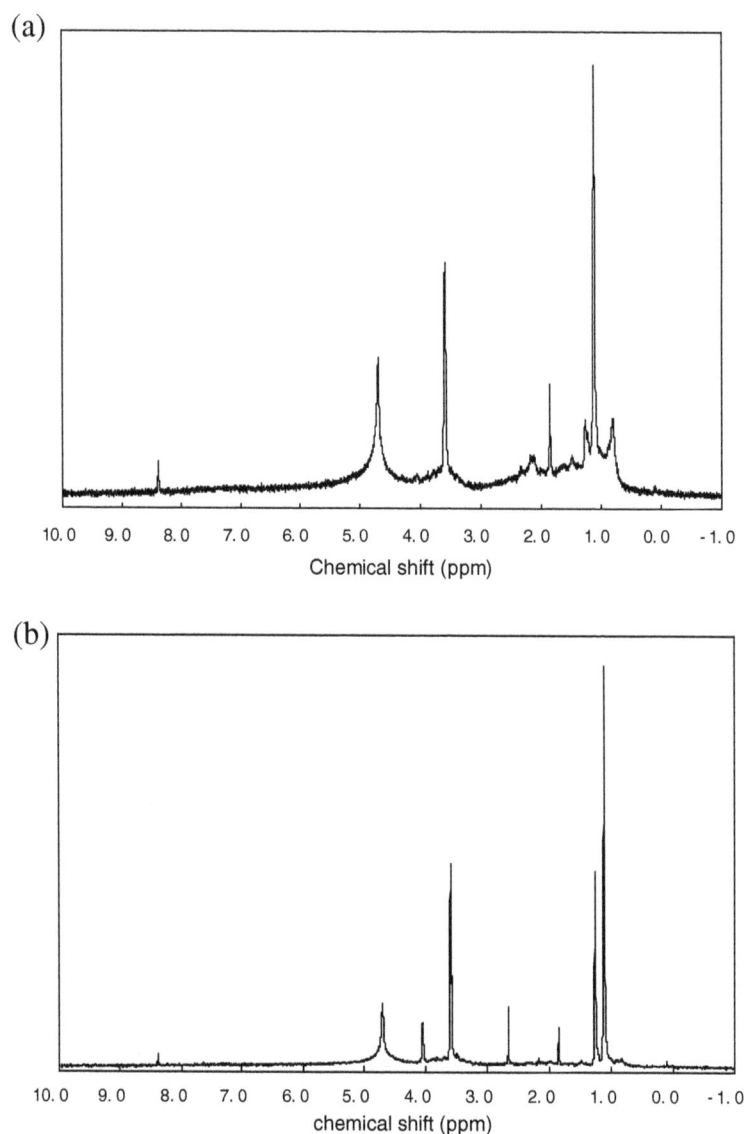

Figure 1 ^{1}H NMR spectra of DOM samples derived from (a) under plant soil, (b) ditch soil.

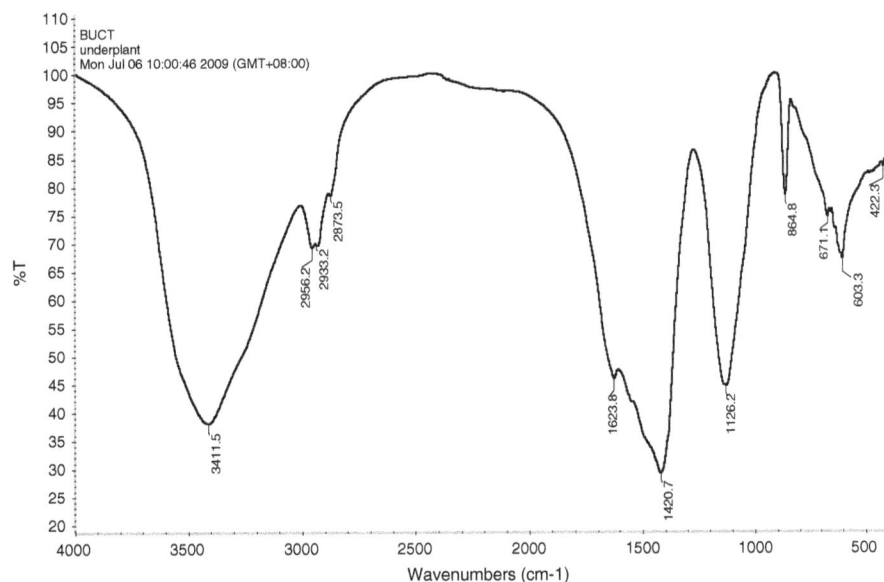

Figure 2 FTIR spectra of DOM samples derived from under plant soil.

The band at 2870 to 2970 cm^{-1} presented aliphatic carbons, of which two small peaks at 2964 and 2930 cm^{-1} were assigned to asymmetrical stretching of C-H in methyl and methylene groups, respectively. Besides, a small peak at 2873 cm^{-1} was assigned to symmetrical stretching of C-H in methyl groups; a strong absorbance at 1420 cm^{-1} exhibited paraffinic characteristics.

Effect of soil properties on the sorption of PAHs

Sorption isotherms of PHE and PYR by two soils are presented in Table 2 and Figure 3 and the data were fitted well with Freundlich sorption model. The isotherm trends for both soils were similar, which are concave-up. Whereas the binding affinity of PAHs to ditch soil was lower than that to under plant soil. The sorption coefficient (K_d) value of PHE for under plant soil can be expressed as $K_d = 1.425 \times 389 \times C_e^{0.425}$, and the values varied from 285 to 542 L/kg with the initial system PHE concentration varying from 50 to 400 mgkg – 1; K_d value of ditch soil can be expressed as $K_d = 1.517 \times 302 \times C_e^{0.517}$, and the values varied from 231 to 490 L/kg,

which was smaller than that of under plant soil. Similar results can be obtained for sorption of PYR.

The lower binding of PAHs to ditch soil may be attributed to two opposite effects. On the one hand, the relatively higher mineral content, especially the higher content of smectite and kaolinite would provide more internal surface sites, thus enhanced sorption (Sheng et al. 2001). Besides, the soil texture could also affect the sorption of PAHs. The relatively fine particle size of ditch soil may tend to sequence more PAHs. On the other hand, the SOM of ditch soil was only half of under plant soil, which would cause less the sorption of PAHs. Many previous studies have reported that the PAHs sorption was proportional to SOM content in soil-water system, that is, the higher SOM in system, the higher PAHs sorption (Gregory et al. 2005; Iorio et al. 2008; Cao et al. 2008). For the competition of these two contradicting effects, the SOM was predominant factor that determine the sorption and availability of PAHs in soil-water system (Chen et al. 2009b; Pignatello and Xing 1996).

Table 2 Isotherm parameters for the sorption of PHE and PYR from under plant and ditch soil by Freundlich simulation

Soil treatment	Under plant			Ditch		
	Log K_f	N	R^2	Log K_f	N	R^2
PHE						
DOM removed	2.712 ± 0.074	1.407 ± 0.169	0.959	2.534 ± 0.049	1.430 ± 0.145	0.969
Bulk	2.590 ± 0.051	1.425 ± 0.142	0.971	2.480 ± 0.039	1.516 ± 0.130	0.978
PYR						
DOM removed	3.899 ± 0.408	1.370 ± 0.219	0.855	3.856 ± 0.258	1.403 ± 0.216	0.933
Bulk	3.798 ± 0.1595	1.384 ± 0.136	0.963	3.811 ± 0.121	1.486 ± 0.110	0.981

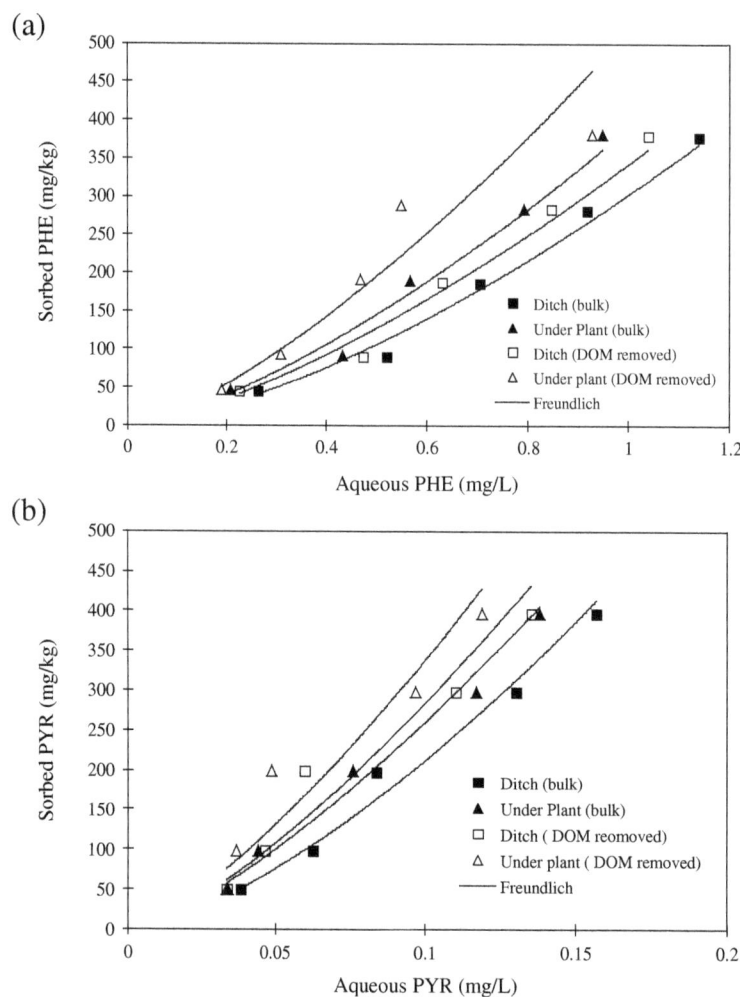

Figure 3 Sorption isotherms of (a) PHE and (b) PYR on ditch soil (bulk and DOM removed) and under plant soil (bulk and DOM removed).

The sorption isotherms of PHE and PYR to DOM removed and bulk soil in two soil samples were also compared and the results indicated that DOM had a strong effect on PAHs sorption (Figure 3). For both under plant and ditch soil, the sorption of soil treated with DOM removed was also higher than that of bulk soil. This is because the DOM in the aqueous phase will bind with PAHs, thus enhancing the PAHs concentration in aqueous phase. Also the sorption increase caused by DOM remove in under plant soil was slightly higher than ditch soil. In comparison of nuclear magnetic resonance (NMR) spectra of DOMs extracted from two soil samples, these two kinds of DOM exhibited similar compositions. While the DOM concentration in ditch soil was much lower than that in under plant soil. In addition, when comparing these two kinds of soils after removing of DOM, the sorption capacity of ditch soil was still much lower than that of under plant soil since the absolutely higher SOM content of under plant soil.

Effects of soil properties and biosurfactant on desorption of PAHs

The combined effects of DOM and biosurfactant on desorption isotherm of PHE and PYR for two kinds of soils were compared (Figure 4). For all combined systems, 50 ppm of DOM and 200 ppm of biosurfactant were added. In order to better compare the effect of DOM on desorption behavior of PAHs, under plant DOM were applied for two soils samples. All desorption isotherm data were fitted well with linear equation model, which meant that the partition was the predominant behavior (Yu et al. 2011). With the involvement of either biosurfactant or DOM, desorption extent for both PHE and PYR in the two soils were enhanced when compared with bulk soil systems. For PHE, the soil-water desorption partition coefficient (K_d) in 200 ppm biosurfactant addition system, 50 ppm DOM added system and bulk system in under plant soil were 463, 611 and 710 L/kg, respectively; the K_d values in above three systems in ditch soil were 183, 286

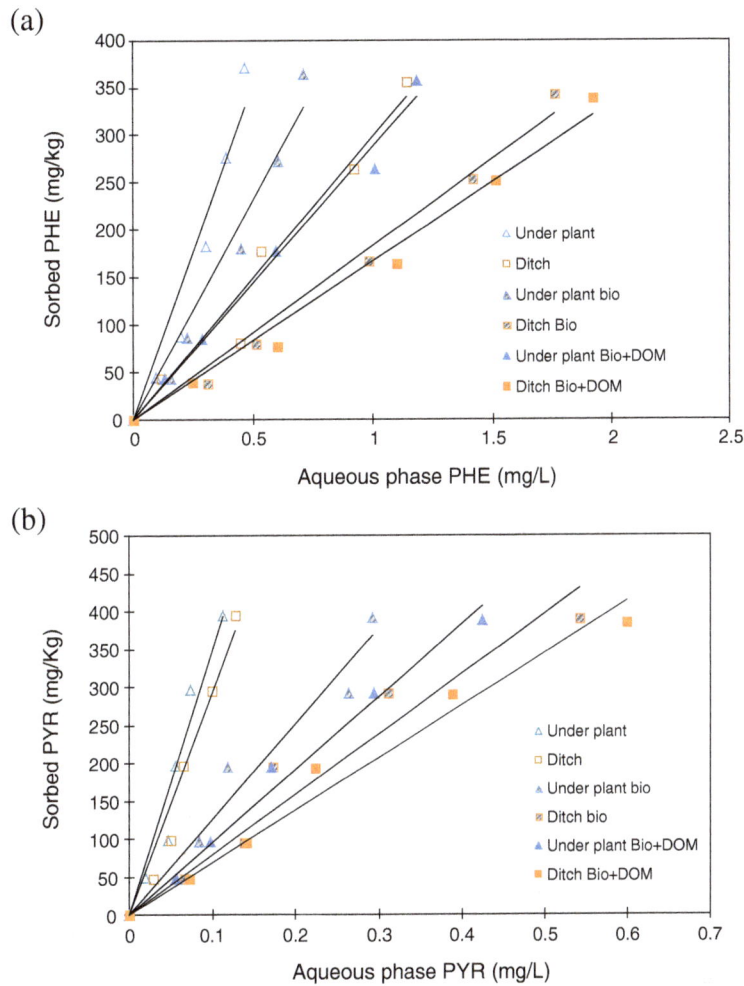

Figure 4 Combined effect of DOM and biosurfactant on the desorption of (a) PHE and (b) PYR.

and 297 L/kg, respectively. For desorption of PYR, similar results can be obtained. In addition, the desorption extent of PAHs were positively correlated with the concentrations of both DOM and biosurfactant. As discussed previous, desorption enhancement by DOM was due to the binding of PAH with DOM, which would enhance the aqueous PAHs content. Besides, biosurfactant could also enhance desorption of PAHs through increasing their solubility at concentrations greater than the CMC. According to our experimental design, biosurfactant concentrations in aqueous phase were above its CMC. With that, the desorption content of PAHs would increase with the surfactant aqueous micellar concentration.

In systems with combined DOM and biosurfactant, the desorption extent were significantly enhanced when compared with single DOM or biosurfactant system (Figures 4 and 5). That may attribute to the synergetic effects of DOM and biosurfactant in the soil water system. It can be also found that under the combined DOM and biosurfactant condition, more PAHs were

desorbed from ditch soil. The K_d value of PHE in ditch soil system and under plant soil system were 170 and 287 L/kg, which means that with the same concentration of PHE in solid phase, about 60% more of PHE were desorbed from ditch soil than under plant soil. Similar tendency can be found for PYR. Under the combined conditions, the K_d value of PYE in ditch soil system and under plant soil system were 690 and 960 L/kg, respectively. With the same concentration of PYR in solid phase, about 40% more of PYR would be desorbed from ditch soil. The desorption difference in these two soil further indicated that the soil properties would significantly influence the contaminants detached from soil during the process of surfactant remediation.

When compared with the two site soils, in systems with the same concentrations of biosurfactant and DOM, the desorption extents of ditch soil were all significantly higher than those in under plant soil, with relatively lower soil-water phase partition coefficient. That's means the binding capacity of ditch soil was weaker than under plant

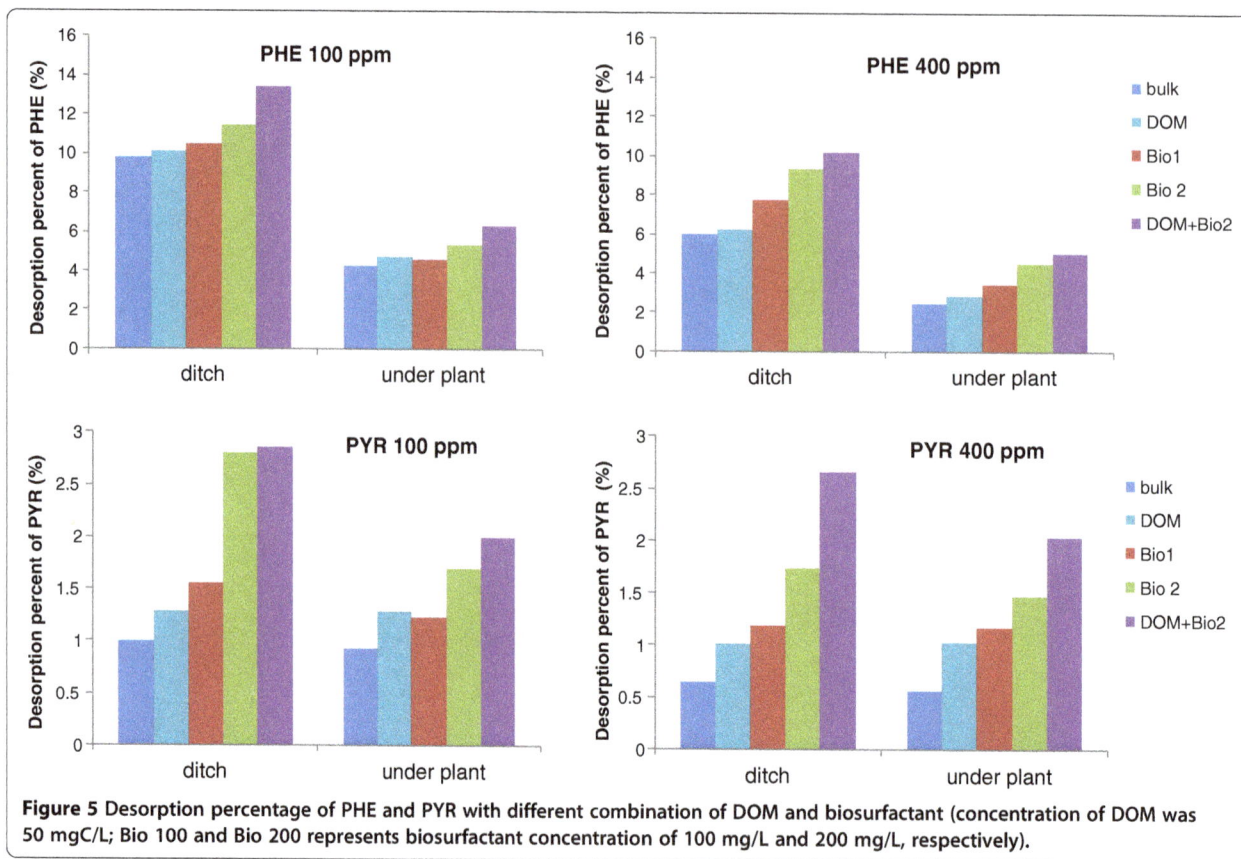

Figure 5 Desorption percentage of PHE and PYR with different combination of DOM and biosurfactant (concentration of DOM was 50 mgC/L; Bio 100 and Bio 200 represents biosurfactant concentration of 100 mg/L and 200 mg/L, respectively).

soil. The soil organic matter was still the key factor that determined the behaviors of PAHs in soil water systems.

In order to explore the interactive mechanisms of DOM and biosurfatcant with consideration of soil properties on desorption enhancement of PAHs, the mass desorption percent of PAHs in each systems were calculated (Figure 5). Generally, desorption extent in ditch soil were higher than under plant soil. And the desorption enhancement of PHE was much higher than that of PYR.

For desorption of PHE at low initial system concentration (100 ppm), the desorption enhancement in 50 mgC/L DOM addition system, 200 ppm biosurfactant addition system and the combined DOM and biosurfactant system were 0.32%, 1.65% and 3.66%, respectively when compared with the desorption in bulk system in ditch soil; the desorption enhancement in under plant soil in above three systems were 0.44%, 1.00% and 2.07%, respectively. The desorption enhancement in the combined system was larger than the sum of enhancement in two single system for both soils, which showed synergistic effects. That can be explained as follows. Both surfactant and DOM are amphiphilic compounds, when they co-existence in the system, they may interact with each other through hydrophobic surface interaction or hydrogen bonding and form mixed micelles, resulting

in synergistic effects. Similar results can be observed in previous studies (Yu et al. 2011; Wan et al. 2011; Cheng and Wong 2006). When the initial PHE concentration increase to 400 ppm, the desorption enhancement in combined system was still larger than the sum of individual system, however, the desorption extent in high concentration of PAHs system was lower than that in low concentration system in system with the same DOM/ and biosurfactant concentration. This can be explained as the competitions among PAHs, DOM and surfactant for sorption sites. In high solute concentration system, the competition for sorption site was more severely than low concentration system, with the similar aqueous concentration of DOM and biosurfactant (according to our preliminary experiment, there was no significant difference in sorption of biosurfactant and DOM in two system), Thus more PAHs were sequenced into soil phase in high PAH concentration system.

PAH desorption in 100 ppm and 200 ppm biosurfactant systems were also compared. When compared with bulk system, the desorption enhancement in 200 ppm biosurfactant system was significantly higher than that in 100 ppm biosurfactant system, however, the enhancement percentage was not in proportional with the biosurfactant concentration, which was less than double in 100 ppm biosurfactant system. This is due to the competitions among aqueous surfactant, sorbed surfactant and solute. In 100 ppm biosurfactant system, the percentage of biosurfactant sorbed by soil was higher than that in 200 ppm biosurfactant system for systems with the same sorption sites. The sorbed surfactant may also bind some of PAHs, resulting in the less PAHs desorbed from soil. Besides, the desorption of PYR showed similar tendency with that of PHE, however significantly lower desorption percentage. Competition existed between PHE and PYR for sorption site. PYR had relatively higher hydrophobic property and show preference in the competition.

Conclusions

The interactive mechanisms of soil properties, biosurfactant and pollutants in soil water system were investigated in this study and following main conclusions can be obtained: (1) SOM was predominant soil component affecting PAHs sorption/desorption and degradation; it also influenced the biosurfactant partition in soil and aqueous phase. (2) The presence of DOM in soil-water system could increase PAHs desorption from soil, however, the strong binding of DOM and PAHs would decrease their bioavailability. (3) Desorption of PAHs in the combined DOM and biosurfactant system would be significantly enhanced when compared with those in single system. The study results should be helpful in broadening knowledge of biosurfactant enhanced bioremediation of PAHs in contaminated soil.

Methods

Materials

Phenanthrene (PHE) and pyrene (PYR) were purchased from Sigma Aldrich Chemical Co. (Oakville, ON) with a purity > 98%. The biosurfactant used in this study is rhamnolipid, which is the most commonly isolated biosurfactant. A rhamnolipid solution was purchased from the Jeneil Biosurfactant Company (Saukville, Wis, USA). Specifically, Jeneil product JBR425 with a mono- to di-rhamnolipid ratio of 1:1 was used. The biosurfactant was supplied as a 10% aqueous solution.

Soil preparation and characterization

Two kinds of soil samples, under plant soil and ditch soil, respectively, were obtained from Coleville site in Saskatchewan, Canada. The under plant soil was collected at a depth of 3 to 5 feet right below the plant ground surface. The ditch soil was obtained from north garden of Coleville at a depth of 3 to 5 feet.

The soil samples were air dried at room temperature ($25 \pm 2°C$) for one week. Soils were ground and passed through a 1.0 mm stainless steel sieve in order to improve the homogeneity of the soil. The physicochemical properties of the soils are given in Table 1.

To investigate the effect of soil organic matter on the sorption and desorption of PHE and PYR, soil samples with dissolved organic matter (DOM) were used as a comparison of bulk soil. The removal of soil DOM was gone through the following procedures. The soils were extracted with 1 mM $CaCl_2$ (soil: solution, 1:20 (w/v)) for 6 h and at least 3 consecutive times and then centrifuged at 7200 g for 30 min. The residues were collected and freeze dried.

The physicochemical properties of soils were characterized in order to understand their impact on sorption, desorption and subsequent biodegradation experiments. The analytical methods listed below were applied.

(1) pH

Soil pH is a measure of the activity of ionized H (H^+) in the soil solution (Margesin and Schinner 2005). A total of 5 g air dried soil and 10 ml of 0.01 M $CaCl_2$ were mixed together. The mixture was shaken for 2 hour at 100 rpm and let stand for 10 minutes. The pH was then measured using a benchtop pH/temperature meter (410A Plus; Thermo Orion, Waltham, MA, USA).

(2) Cation exchange capacity

The cation exchange capacity (CEC) was determined using the sodium acetate-sodium chloride saturation and magnesium nitrate extraction method (Pansu and Gautheyrou 2006). The two step procedure is as follows: 1), saturation of cation exchange sites with Na^+ by "equilibration" of the soil with a 0.4 N NaOAc-0.1 N NaCl solution; 2), total Na and Cl were extracted with 0.5 N MgS04 solution so that the soluble Na from the excess saturating solution could be deducted from the total Na. This provides the exchangeable Na, which is equivalent to the CEC.

(3) SOM content

The total organic matter is routinely estimated by measuring organic carbon content. The method is described as a wet-oxidation procedure using potassium dichromate ($K_2Cr_2O_7$) with external heat and back-titration to measure the amount of unreacted dichromate (Jones 2001). The detailed procedures have been described by Mebius (1960).

(4) Soil classification

Soil classification was performed in accordance with the standard of International Soil Science Society. The International Classification system was applied to categorize the soil.

(5) Clay analysis

The soil clay mineral was determined by X-ray diffraction (XRD) (Whittig and Allardice 1982). The procedure was described by Hwang (Hwang 2001). The clay with particle size finer than 2 μm (according to the International Soil Science Society method) was applied, which was collected by elutriation. The samples were previously subjected to disintegration and dispersion processes through ultrasound. Organic residues were eliminated with a 10% H_2O_2 solution. This was done to obtain the largest amount of clay needed for the different treatments. Once the fraction smaller than 2 microns was separated, washed with demineralized water and placed on flat glass to dry. The methodology consisted of three initial pretreatments in order to identify and differentiate the principal clay group present in the sample: 1) The water suspended clay was allowed to dry in order to permit the free arrangement of the basal planes of the clays; 2) Saturated with Ethylene glycol, organic molecules occupied the interiaminar portion of the smectites in order to enlarge the basal distance. This identified the smectite group; 3) Heating of the samples to 550°C destroyed the crystal structure of the kaolinites. This identified kaolinites, chlorites and some of the interstratified groups.

Characterization of dissolved organic matter

For the DOM collection, twenty grams of clean soil sample was placed in 200 ml deionized water and agitated on a reciprocal shaker at 200 rpm and 20 ± 1°C for 24 hours. The supernatants were collected after centrifugation at 12000 g for 20 min and then filtered through a 0.45 μm sterilized membrane (PALL Corporation, Michigan, USA). The extracts could be stored at 4°C in the dark with maximum of 5 days. The filtrates were freeze-dried for further analysis.

The concentration of DOM was measured using TOC-5000A Analyzer (Shimadzu, Kyoto Japan). TOC concentration of each soil measured was the dissolved organic carbon (DOC) concentration since TOC in soil matrix was dissolved into the aqueous phase. DOM concentration was hypothesized to equal to or a fractional equivalent of the DOC concentration in this study. The ratio of absorbance of DOM at 465 and 665 nm (E4/E6) was measured to evaluate changes in the molecular weight of the DOM fractions, which is negatively correlated with molecular size (Marschner et al. 2005; Yang et al. 2007; Korshin et al. 1997).Specific UV absorbance at 254 nm was determined to estimate the aromaticity of DOM samples (Yang et al. 2007). The absorbance of water samples was measured at the selected wavelength on a 2100 spectrophotometer (Unic, Shanghai, China). Deionized water was used as reference. Functional groups of DOM were determined through ^{1}H liquid-state NMR spectroscopy analysis on a Bruker Avance 600 MHR spectrometer

(Billerica, MA) and Fourier transform infrared (FTIR) spectral analysis using aThermo Electron Nexus 8700 Spectrophotometer (Waltham, MA).

Batch sorption and desorption experiment

For both sorption and desorption isotherm experiments, triplicate tests were conducted using the standard batch equilibration method. All experiments were carried out in 25 mL Corex centrifuge tubes with Teflon-lined screw caps (Fisher Scientific, Ottawa, ON, Canada). For the sorption experiment, an appropriate volume of PHE and PYR dichloromethane stock solution was respectively added to each tube. Dichloromethane was allowed to evaporate, and then 0.5 g soil and 10 mL background solution were added to each vial. The background solution contained 0.01 M $CaCl_2$, 0.01 M NaCl, and 0.01 M NaN_3. The $CaCl_2$ and NaCl were used as electrolyte to poise the ionic strength, and the NaN_3 was used as inhibitor for bacterial growth. This approach resulted in initial soil concentrations of 50, 100, 200, 300 and 400 ppm for PHE and PYR, respectively.

The centrifuge tubes were vortexed for 20 s, and then placed on a reciprocal shaker at 20 ± 1°C and 125 rpm for 24 hours to reach the sorption equilibrium. The suspensions were then centrifuged at 5000 g for 25 min under the same temperature. PHE and PYR in the aqueous phase were extracted with dichloromethane, and their concentrations were analyzed through GC. The amount of PAHs sorbed to the soil was obtained through calculating the difference between the initial amount and remaining in the solution.

Desorption isotherm experiments were conducted immediately after the sorption experiments. The supernatants in sorption tubes were completely removed. To study the effects of DOM and biosurfactant on PAHs desorption, 10 mL fresh background solution containing different concentrations of DOM and biosurfactant was successively added into the tube, and then shaked for another 24 hours to reach desorption equilibrium. The subsequent separation of soil and aqueous phase as well as the relevant analyses were conducted as described in the sorption experiments. The sorption of PHE and PYR on the wall of centrifuge tubes was considered negligible; the amounts of PAHs blank before and after mixing (without soil) did not show significant difference between each other.

Analytical methods

The concentrations of PHE and PYR were analyzed through Gas Chromatography (Varian GC 3800-FID) system coupled with a Varian 8200 autosampler. The GC was equipped with a 25 m × 0.32 mm ID DB-5 column with 0.25 μm film thickness (J&W Scientific Inc., CA). Helium is used as the carrier gas with flow rate of

1.5 ml/min. The oven temperature was held at 40°C for 1.5 min then ramped to 175°C at a rate of 50°C/min. The temperature was held at 175°C for 1 min, then ramped to 220°C at 7°C/min, and held at this final temperature for 1 min. The injector temperature was 250°C, and the detector temperature was 230°C. Injection was made in the split mode with a ratio of 50 (since 1.75 min).

The sorption and desorption isotherms of PAHs were mathematically fitted with a Freundlich sorption and linear distribution models.

The Freundlich model has the following form:

$$C_s = K_f C_e^n \tag{6.1}$$

The linear model has the following form:

$$C_s = K_p C_e \tag{6.2}$$

where C_s is the sorbed PAH concentration (mg/kg); C_e is the solution-phase PAH concentration (mg/L); K_p (mg/kg) (mg/L) is the distribution coefficient; K_f (mg/kg) (mg/L)$^{-n}$ is the sorption constant at a given temperature; K_f represents the sorption capacity evaluated at $C_e = 1$ mg/L n is the isotherm exponent. Equation (6.1) can be linearized by a logarithmic transformation:

$$\log c_s = \log k_f + n \log c_e \tag{6.3}$$

Freundlich parameters for sorption (K_{fA}, and n_A) and desorption (K_{fD}, n_D) were respectively calculated through fitting Equation (6.3) to the observed data.

The sorption coefficient K_d can be defined as:

$$K_d = dC_s / dC_e = nK_f C^{n-1} \tag{6.4}$$

For parallel experiment, means and standard deviations are calculated for pooled results using Microsoft® Excel. To analyze the results from sorption and desorption experiments, F-tests are conducted using Minitab statistical software (Minitab Inc., State College, PA, USA). The introduced level of significance is 5% ($\alpha = 0.05$). One-way analysis of variance (F-test with $p < 0.05$) is performed to determine significant differences between different treatments.

Abbreviations
SOM: Soil organic matter; DOM: Dissolved organic matter; PAHs: Polycyclic aromatic hydrocarbons; HOCs: Hydrophobic organic contaminants; CMC: Critical micellar concentration; PHE: Phenanthrene; PYR: Pyrene; CEC: Cation exchange capacity; FTIR: Fourier transform infrared.

Competing interests
The authors declare that they have no competing interests.

Authors' contributions
HY conducted most of the experiment and drafted the manuscript. HX participated in the design of the experiments and helped to draft the manuscript. DW participated the soil characteristics analysis and helped to draft the manuscript. All authors read and approved the final manuscript.

Acknowledgements
This research was supported by National Nation Science Foundation (51109078 and 51102093) and Natural Science and Engineering Research Council of Canada.

Author details
[1]Department of Chemical Engineering, University of New Brunswick, Fredericton, NB, E3B 5A3, Canada. [2]MOE Key Laboratory of Regional Energy Systems Optimization, S&C Academy of Energy and Environmental Research, North China Electric Power University, Beijing 102206, China. [3]Ministry of Agriculture, Government of Saskatchewan, Regina, SK, S4S 0B1, Canada.

References
Abu-Zreig M, Rudra RP, Dickinson WT (2003) Effect of application of surfactants on hydraulic properties of soils. Biosyst Eng 84(3):363–372

Akkanen J, Tuikka A, Kukkonen JVK (2005) Comparative sorption and desorption of benzo [a] pyrene and 3, 4,3 ',4 '-tetrachlorobiphenyl in natural lake water containing dissolved organic matter. Environ Sci Technol 39(19):7529–7534

Brady NC, Weil RR (2000) Elements of the nature and properties of soils. Prentice-Hall, Upper Saddle River, New Jersey

Cao J, Guo H, Zhu HM, Jiang L, Yang H (2008) Effects of SOM, surfactant and pH on the sorption–desorption and mobility of prometryne in soils. Chemosphere 70(11):2127–2134

Chen JL, Wong YS, Tam NFY (2009a) Static and dynamic sorption of phenanthrene in mangrove sediment slurry. J Hazard Mater 168(2–3):1422–1429

Chen W, Hou L, Luo XL, Zhu LY (2009b) Effects of chemical oxidation on sorption and desorption of PAHs in typical Chinese soils. Environ Pollut 157(6):1894–1903

Cheng KY, Wong JWC (2006) Combined effect of nonionic surfactant Tween 80 and DOM on the behaviors of PAHs in soil-water system. Chemosphere 62(11):1907–1916

Cheng KY, Zhao ZY, Wong JWC (2004) Solubilization and desorption of PAHs in soil-aqueous system by biosurfactants produced from Pseudomonas aeruginosa P-CG3 under thermophilic condition. Environ Technol 25(10):1159–1165

Gregory ST, Shea D, Guthrie-Nichols E (2005) Impact of vegetation on sedimentary organic matter composition and polycyclic aromatic hydrocarbon attenuation. Environ Sci Technol 39(14):5285–5292

Hwang S (2001) Effect of soil properties, compound aging, and presence of cosolute on sorption, desorption, and biodegradation of polycyclic aromatic hydrocarbons in natural soils. PhD dissertation. University of Akron, Akron, Ohio, pp 65–66

Hwang S, Cutright TJ (2003) Statistical implications of pyrene and phenanthrene sorptive phenomena: effects of sorbent and solute properties. Arch Environ Con Tox 44(2):152–159

Hwang S, Ramirez N, Cutright TJ, Ju LK (2003) The role of soil properties in pyrene sorption and desorption. Water Air Soil Poll 143(1–4):65–80

Iorio M, Pan B, Capasso R, Xing BS (2008) Sorption of phenanthrene by dissolved organic matter and its complex with aluminum oxide nanoparticles. Environ Pollut 156(3):1021–1029

Jones JB (2001) Laboratory guide for conducting soil tests and plant analysis. Boca Raton, FL

Korshin GV, Li CW, Benjamin MM (1997) Monitoring the properties of natural organic matter through UV spectroscopy: a consistent theory. Water Res 31(7):1787–1795

Kuyukina MS, Ivshina IB, Makarov SO, Litvinenko LV, Cunningham CJ, Philp JC (2005) Effect of biosurfactants on crude oil desorption and mobilization in a soil system. Environ Int 31(2):155–161

Lee JF, Hsu MH, Lee CK, Chao HP, Chen BH (2005) Effects of soil properties on surfactant adsorption. J Chin Inst Eng 28(2):375–379

Lu L, Zhu LZ (2012) Effect of soil components on the surfactant-enhanced soil sorption of PAHs. J Soil Sediment 12(2):161–168

Makkar RS, Rockne KJ (2003) Comparison of synthetic surfactants and biosurfactants in enhancing biodegradation of polycyclic aromatic hydrocarbons. Environ Toxicol Chem 22(10):2280–2292

Margesin R, Schinner F (2005) Manual for soil analysis: monitoring and assessing soil bioremediation. Determination of chemical and physical soil properties, Springer Berlin Heidelberg

Marschner B, Winkler R, Jodemann D (2005) Factors controlling the partitioning of pyrene to dissolved organic matter extracted from different soils. Eur J Soil Sci 56(3):299–306

Mebius LJ (1960) A rapid method for the determination of organic carbon in soil. Anal Chim Acta 22(2):120–124

Pan B, Ghosh S, Xing B (2007) Nonideal binding between dissolved humic acids and polyaromatic hydrocarbons. Environ Sci Technol 41(18):6472–6478

Pansu M, Gautheyrou J (2006) Handbook of soil analysis: mineralogical, organic and inorganic methods. cation exchange capacity. Springer, Berlin Heidelberg

Pignatello JJ, Xing BS (1996) Mechanisms of slow sorption of organic chemicals to natural particles. Environ Sci Technol 30(1):1–11

Raber B, Kogel-Knabner I, Stein C, Klem D (1998) Partitioning of polycyclic aromatic hydrocarbons to dissolved organic matter from different soils. Chemosphere 36(1):79–97

Rodriguez-Escales P, Borras E, Sarra M, Folch A (2013) Granulometry and surfactants, key factors in desorption and biodegradation (T. Versicolor) of PAHs in soil and groundwater. Water Air Soil Pollut 224(2):1422–1433

Sheng GY, Johnston CT, Teppen BJ, Boyd SA (2001) Potential contributions of smectite clays and organic matter to pesticide retention in soils. J Agr Food Chem 49(6):2899–2907

Shin KH, Kim KW, Ahn Y (2006) Use of biosurfactant to remediate phenanthrene-contaminated soil by the combined solubilization-biodegradation process. J Hazard Mater 137(3):1831–1837

Wan JZ, Wang LL, Lu XH, Lin YS, Zhang ST (2011) Partitioning of hexachlorobenzene in a kaolin/humic acid/surfactant/water system: Combined effect of surfactant and soil organic matter. J Hazard Mater 196:79–85

Wen B, Zhang JJ, Zhang SZ, Shan XQ, Khan SU, Xing BS (2007) Phenanthrene sorption to soil humic acid and different humin fractions. Environ Sci Technol 41(9):3165–3171

Whittig LD, Allardice WR (1982) X-ray diffraction techniques: in methods of soil anaysis: part 1 - physical and mineralogical methods. Americal Society of Agronomy and Soil Science of America, Madison, WI

Wilcke W (2000) Polycyclic aromatic hydrocarbons (PAHs) in soil - a review. J Plant Nutr Soil Sc 163(3):229–248

Woods CE Jr (2004) Examination of the effects of biosurfactant concentration on natural gas hydrate formation in seafloor porous media. M.S., Mississippi State University, United States – Mississippi

Yang WC, Hunter W, Spurlock F, Gan J (2007) Bioavailability of permethrin and cyfluthrin in surface waters with low levels of dissolved organic matter. J Environ Qual 36(6):1678–1685

Yu HS, Zhu LZ, Zhou WJ (2007) Enhanced desorption and biodegradation of phenanthrene in soil-water systems with the presence of anionic-nonionic mixed surfactants. J Hazard Mater 142(1–2):354–361

Yu H, Huang GH, An CJ, Wei J (2011) Combined effects of DOM extracted from site soil/compost and biosurfactant on the sorption and desorption of PAHs in a soil-water system. J Hazard Mater 190(1–3):883–890

Zhu HB, Aitken MD (2010) Surfactant-enhanced desorption and biodegradation of polycyclic aromatic hydrocarbons in contaminated soil. Environ Sci Technol 44(19):7260–7265

11

Thickness, porosity, and permeability prediction: comparative studies and application of the geostatistical modeling in an Oil field

Shan Zhao[*], Yang Zhou, Mengyuan Wang, Xiaying Xin and Fang Chen

Abstract

Background: In this study, we applied the geostatistical modeling to analyze an oil field. The reservoir properties, thickness, porosity and permeability, were studied. Data analysis tools, such as histogram, scatter plot, variogram and cross variogram modeling, were employed to capture the interpretable spatial structure and provide the desired input parameters for further estimation. SK (simple kriging), OK (ordinary kriging), Sgism (Sequential Gaussian Simulation), SC (simple cokriging), OC (ordinary cokriging) and MM2 (Markov model 2) methods were applied to estimate reservoir properties. Estimation difference maps were generated to compare the results of each method, providing more straightforward realizations in a visual way.

Results: For thickness, results indicated that anisotropic variogram could provide better interpretations for the spatial relationships than isotropic variogram. Both SK and OK could provide better estimates. In comparison to the conventional estimation techniques, the simulation method could well reflect the reservoir's intrinsical characteristics in terms of the associated extreme values. OOIP (Original Oil In Place) was calculated later with the parameters attained before, including thickness and porosity. Estimation difference maps showed that there was no obvious difference in SK vs. OK and SC vs. OC for the study of permeability. However, OC was slightly different from OK, and there were significant discrepancies between the estimates of OC and MM2 at the unsampled locations. In addition, OC estimates were closest to the sample data of permeability with the minimum variance.

Conclusions: Geostatistical modeling is an effective way for thickness, porosity, and permeability prediction.

Keywords: Geostatistics; Variogram; Kriging; Thickness; Porosity; Permeability

Background

Maps and mapmaking are integral parts of reservoir characterization. A map is a numerical model of an attribute's (e.g., porosity, permeability, thickness, structure) spatial distribution (Malvić and Jović 2012; Huysmans and Dassargues 2013). However, mapping an attribute is rarely the goal; rather, a map is used to make a prediction about the reservoir. To paraphrase Andre Journel of Stanford University, "A map is a poor model of reality if it does not depict characteristics of the real spatial distribution of those attributes that most affect how the reservoir responds".

The enormous up-front investments for developing heterogeneous fields and the desire to increase ultimate recovery have spurred oil companies to use innovative reservoir characterization techniques (Habibnia and Momeni 2012; Abdideh and Mahmoudi 2013). Geostatistics is one of many new technologies often incorporated into the process (Cressie and Hawkins 1980; Bueno et al. 2011). For more than a decade, geostatistical techniques, especially when incorporating 3-D seismic data, have been an accepted technology to characterize petroleum reservoirs (Qi et al. 2007; Abdideh and Bargahi 2012; Esmaeilzadeh et al. 2013; Fegh et al. 2013).

Geostatistical application necessitates and facilitates cooperation between geoscientists and reservoir engineers, allowing each discipline to contribute fully. This is quite different from the past, because the mathematical formalization was often left to the reservoir engineer.

* Correspondence: shan.zhao2010@gmail.com
Faculty of Engineering and Applied Science, University of Regina, Regina, Saskatchewan S4S 0A2, Canada

Thus, part of the geostatistical philosophy is to ensure that geologic reality does not get lost during reservoir model building (Nava et al. 2010; Chen et al. 2011).

Geostatistics attempts to improve predictions by developing a different type of quantitative model. The goal is to construct a more realistic model of reservoir heterogeneity using methods that do not average important reservoir properties. Like the traditional deterministic approach, it preserves indisputable "hard" data where they are known and interpretative "soft" data where they are informative (Wilson et al. 2011).

However, unlike the deterministic approach, geostatistics provides numerous plausible results (Zarei et al. 2011). The degree to which the various models differ is a reflection of the unknown or a measurement of the "uncertainty". Some outcomes may challenge prevailing geologic wisdom and will almost certainly provide a range of economic scenarios, from optimistic to pessimistic. Having more than one result to analyze changes the paradigm of traditional reservoir analysis and may require multiple reservoir flow simulations (Schmidt and Schröder 2011; Soleymani and Riahi 2012). However, the benefits outweigh the additional time and cost.

The Stanford Geostatistical Modeling Software (SGeMS) is an open-source computer package for solving problems involving spatially related variables. It provides geostatistics practitioners with a user-friendly interface, an interactive 3-D visualization, and a wide selection of algorithms (Kelsall and Wakefield 2002). This website serves as a companion to the book Applied Geostatistics with SGeMS that provides a step-by-step guide to using SGeMS algorithms. We recommend getting the book to get the underlying theory, demonstrations of their implementation, discussion of potential limitations, and help about the choice of one algorithm over another.

Users can perform complex tasks using the embedded Python scripting language, and new algorithms can be developed using the SGeMS plug-in mechanism. SGeMS is the first software to provide algorithms for multiple-point statistics. The SGeMS package provides a versatile toolkit for Earth Sciences graduates and researchers, as well as practitioners of environmental, mining and petroleum engineering.

SGeMS provides a (fairly) comprehensive collection of geostatistical estimation and simulation algorithms and also provides a nice 3D visualization environment (Kelkar and Perez 2002; Remy 2004; Geoff 2007; Remy et al 2009). It provides a more limited selection of options for standard statistical data analysis and essentially no facilities for data management (subsetting data sets, etc.). So, you would probably want to run S-GeMS in tandem with some other data analysis & management software, such as Excel.

In the following chapters of this study, we will use SGeMS to make estimation about an oilfiled step by step. The user selects in this panel which geostatistics tool to use and inputs the required parameters. The top part of that panel shows a list of available algorithms, e.g. kriging, sequential Gaussian simulation. When an algorithm from that list is selected, a form containing the corresponding input parameters appears below the tools list. One or multiple objects can be displayed in this panel, e.g. a Cartesian grid and a set of points, in an interactive 3D environment. Visualization options such as color-maps are also set in the Visualization Panel.

Results and discussion

This study started by loading data from a file named Flow Unit 5, a DAT-format data file containing thick, porosity and permeability data from about 68 wells in Flow Unit 5 Oil Field.

Thickness

Simple kriging and ordinary kriging of isotropy

Simple kriging Figure 1a shows the gross thickness map generated with isotropic simple kriging. Overall, the map is not that smooth. In the northwest corner, the conditioning data is undersampled with respect to the rest of the grid. The isotropic simple kriging variance map is shown in Figure 1b. The estimation variance is pretty high through the entire map. Small variance only exists in places where there are wells distributed. The histogram, for the simple kriging estimates, is shown in figure. Most of the data are distributed close to the centre, with some extreme values more than 10 ft. The maximum estimate value is 24.61 ft, while the maximum for the data is 27 ft. There is almost no improvement compared to the histogram of the conditioning data. 92 points are compared and plotted in the scatter plot of gross thickness data and isotropic simple kriging estimation data. The correlation coefficient is 2.082, indicating that isotropic SK tends to overestimate the estimates.

Ordinary kriging Figure 2a shows the gross thickness map generated with isotropic ordinary kriging. Overall, the map is not that smooth. The continuity is not that good even if it is a bit better than the map generated from simple kriging. In the northwest corner, the conditioning data is undersampled with respect to the rest of the grid. The isotropic ordinary kriging variance map is shown in Figure 2b. The estimation variance is pretty high through the entire map. Small variance only exists in places where there are wells distributed. The histogram, for the isotropic ordinary kriging estimates, is shown in figure. Most of the data are distributed close to the centre, with some extreme values more than 10 ft. The maximum estimate value is 25.036 ft, while the maximum for the data is 27 ft. There is almost no big improvement compared to the histogram of the conditioning data. 92 points are

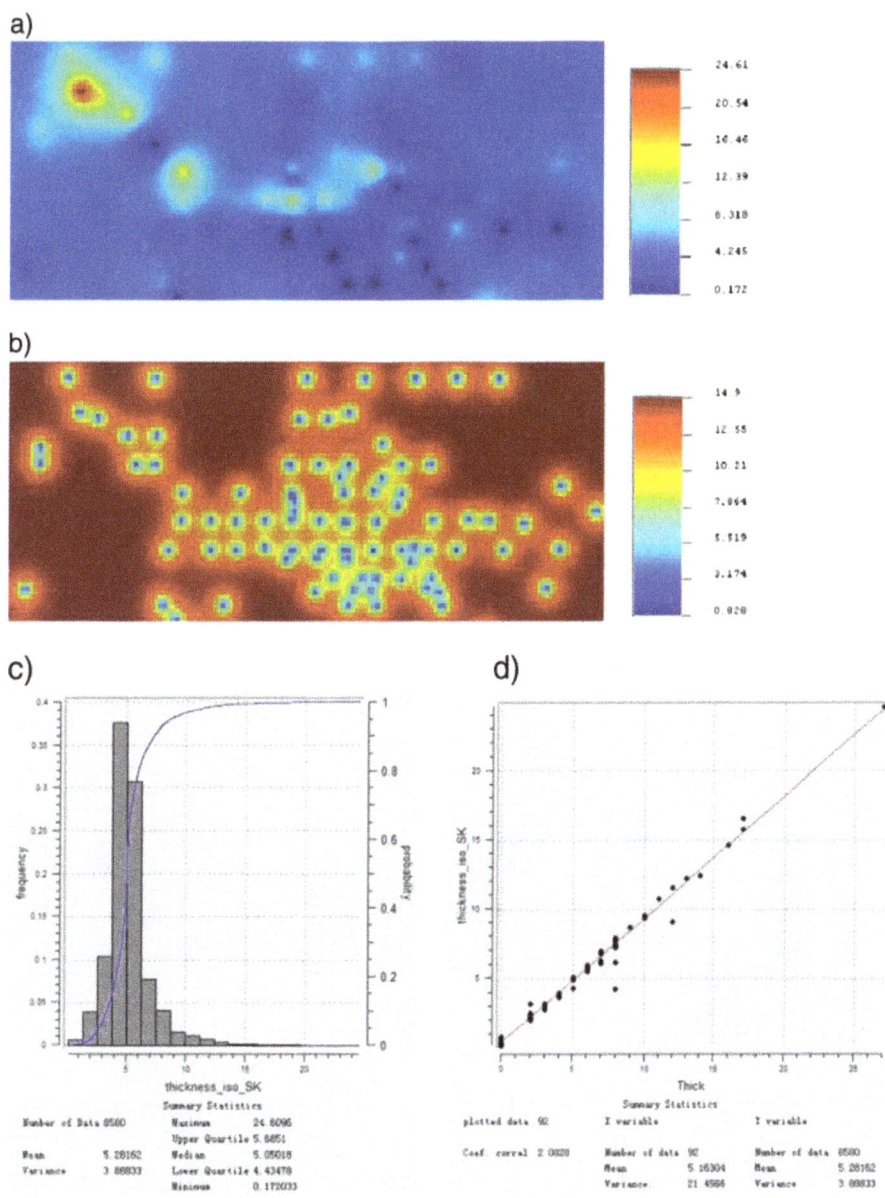

Figure 1 Thickness: a) Isotropic Simple Kriging; b) Isotropic Simple Kriging Variance; c) Histogram for Isotropic Simple Kriging; d) Compare with thickness and Isotropic simple kriging.

compared and plotted in the scatter plot of gross thickness data and isotropic ordinary kriging estimation data. The correlation coefficient is 1.384, indicating that isotropic simple kriging tends to overestimate the estimates.

Simple kriging and ordinary kriging of anisotropy
Simple kriging Figure 3a shows the gross thickness map generated with simple kriging. Overall, the map has a smooth appearance that is typical of simple kriging. The good spatial continuity from east to west corresponds to the principal direction of the variogram. The north/south trend, observed in a small area in the northwest

corner of the map, is a result from the configuration of the conditioning data and the search neighborhood. In the northwest corner, the conditioning data is under-sampled with respect to the rest of the grid. The simple kriging variance map is shown in Figure 3b. The estimation variance is small in gridblocks close to the conditioning data, and it becomes large in area far from the data. Near the conditioning data, the kriging variance becomes the nugget effect of the variogram. The histogram, for the simple kriging estimates, is shown in figure. The mean is 5.43, the median is 4.88 and the mode is around 4, all of which are pretty close to the

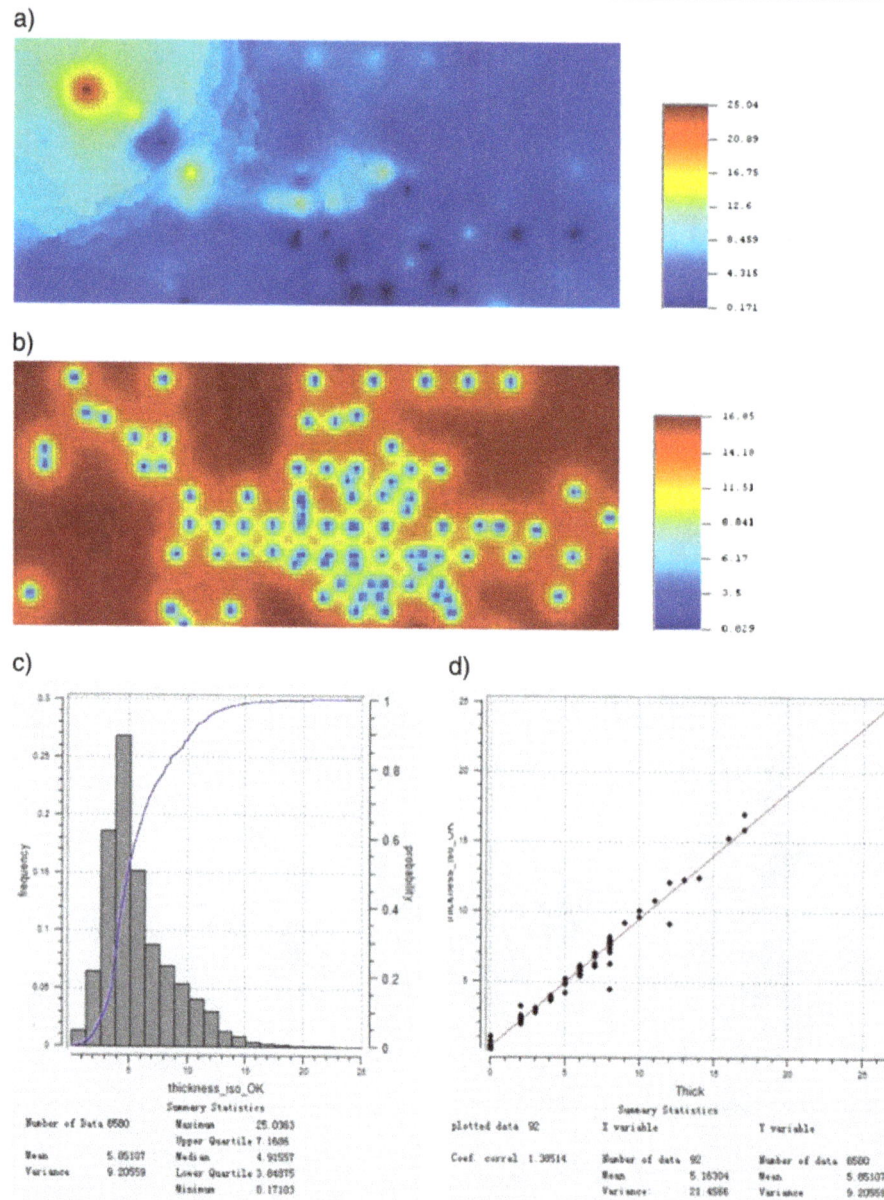

Figure 2 Thickness: a) Isotropic Ordinary Kriging; b) Isotropic Ordinary Kriging Variance; c) Histogram for Isotropic Ordinary Kriging; d) Compare with thickness and Isotropic Ordinary kriging.

conditioning data. The standard deviation of simple kriging estimates is 2.63 ft, which for conditioning data, it 4.53 ft. So it is likely that has a narrower spread than the conditioning data. In general, simple kriging does not closely reproduce the extreme values of a distribution with a gross thickness greater than 10 ft, observed in conditioning data. The maximum estimate value is 17.57 ft, while the maximum for the data is 27 ft. Simple kriging tends to make the data follow normal distribution, most of the data are distributed close to the mean value. 92 points are compared and plotted in the scatter plot of gross thickness data and simple kriging estimation data. Simple kriging provides good estimates even for values that are much larger than the mean of the data (5.05 ft). All the plotted points fall approximately along the straight line, with correlation coefficient 0.891, indicating high accuracy of the estimates.

Ordinary kriging Figure 4a shows the gross thickness map generated with ordinary kriging. Overall, the map has a more smooth appearance than that of simple kriging. Transition areas exist in ordinary kriging map, which

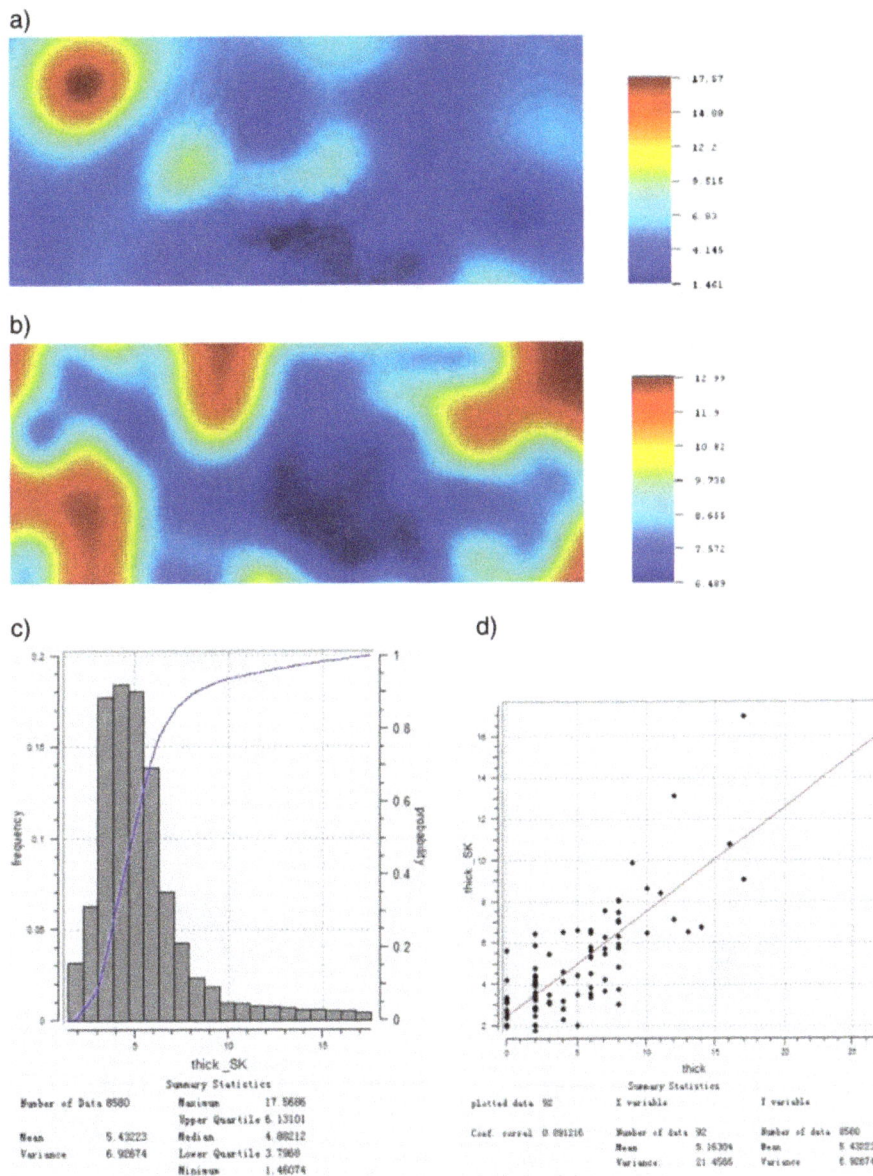

Figure 3 Thickness: a) Anisotropic Simple Kriging; b) Anisotropic Simple Kriging Variance; c) Histogram for Anisotropic Simple Kriging; d) Compare with thickness and Anisotropic simple kriging.

represent gradual change among the subareas. The good spatial continuity from east to west corresponds to the principal direction of the variogram. The north/south trend, observed in a small area in the northwest corner of the map, is a result from the configuration of the conditioning data and the search neighborhood. In the northwest corner, the conditioning data id undersampled with respect to the rest of the grid. And this area tends to be larger than that of the simple kriging. Generally, the two maps are similar with each other. The ordinary kriging variance map is shown in Figure 4b, which is smoother than that of the simple kriging variance map. The blue

area that represents small variance tends to be wider, which stands for a better estimation result compared to simple kriging. The estimation variance is small in gridblocks close to the conditioning data, and it becomes large in area far from the data. Near the conditioning data, the kriging variance becomes the nugget effect of the variogram. The histogram, for the ordinary kriging estimates, is shown in figure. The mean is 5.68, the median is 4.89 and the mode is around 4, all of which are pretty close to the conditioning data. The standard deviation of ordinary kriging estimates is 2.98 ft, while for conditioning data, it is 4.53 ft. So it is likely that has a narrower spread

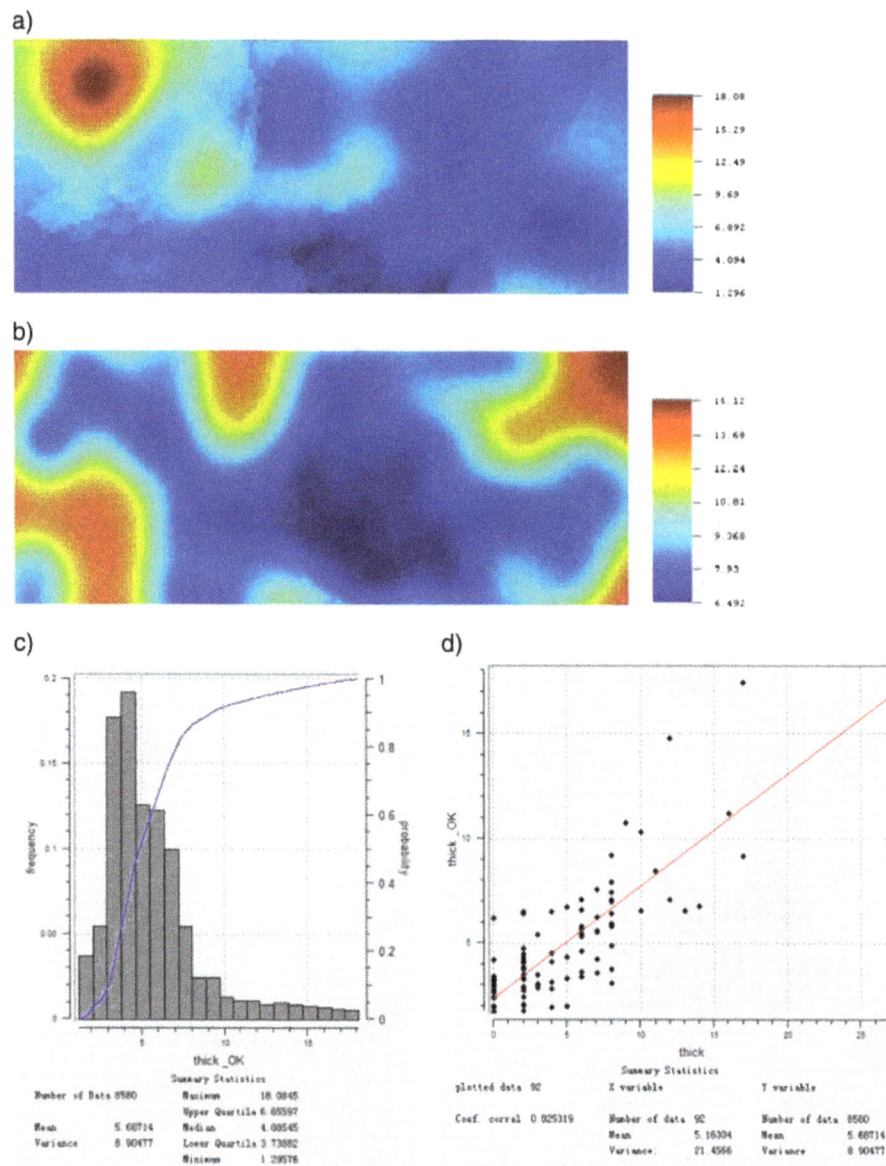

Figure 4 Thickness: a) anisotropic ordinary kriging; b) anisotropic ordinary kriging variance; c) histogram for anisotropic ordinary kriging; d) compare with thickness and anisotropic ordinary kriging.

than the conditioning data. In general, simple kriging does not closely reproduce the extreme values of a distribution with a gross thickness greater than 10 ft, observed in conditioning data. The maximum estimate value is 18.08 ft, while the maximum for the data is 27 ft. Ordinary kriging tends to make the data follow normal distribution; most of the data distribute close to the mean value. 92 points are compared and plotted in the scatter plot of gross thickness data and ordinary kriging estimation data. Simple kriging provides good estimates even for values that are much larger than the mean of the data (5.05 ft). All the plotted points fall approximately along the straight line, with correlation

coefficient 0.825,though it is slightly smaller than the simple kriging.

Comparison between isotropic and anisotropic variogram modeling

From the thickness estimate of isotropic and anisotropic modeling, as shown in Figure 1 and Figure 2, it is obvious that anisotropic variogram is much smoother than the isotropic one. From the isotropic and anisotropic variance maps, the estimation variance is much lower for the anisotropic variogram than the isotropic one. This can further be demonstrated by the scatter plot of isotropic OK and anisotropic OK variance, where the isotropic variance is much larger than the anisotropic variance. All these illustrate that anisotropic estimation is more close to the conditioning data and is more accurate.

Through the above comparison between the isotropic modeling and anisotropic modeling, we draw a conclusion that anisotropic variogram can achieve better spatial structure capture more interpretable spatial relationship. In the following analysis of porosity and permeability properties, anisotropic variogram is mainly used to generate the spatial relationship of flow unit 5.

Porosity
Simple kriging
Figure 5a shows the porosity map generated via simple kriging. Overall, the map has a smooth appearance. The

good spatial continuity from middle to west corresponds to the principal direction of the variogram. The simple kriging variance map is shown in Figure 5b. The estimation variance could be as low as 0.6369 at where gridblocks are close to the sample data. However, in the northeast corner, the estimation variance could be as large as 64.28 due to the scarcity of sample data. For porosity data, the mean and median are 0.41 and 0.97, respectively. They are very close to each other, indicating symmetry in the distribution. The coefficient of variation could be obtained via equation, and its value is 0.2735, indicating a relatively small variation within the sample. The histogram is shown in Figure 5d. The mean is 0.67, the median is 0.39 and the variance is 25.5767. Comparing with the histogram of sample data, the estimated values are generally a little smaller than the sample

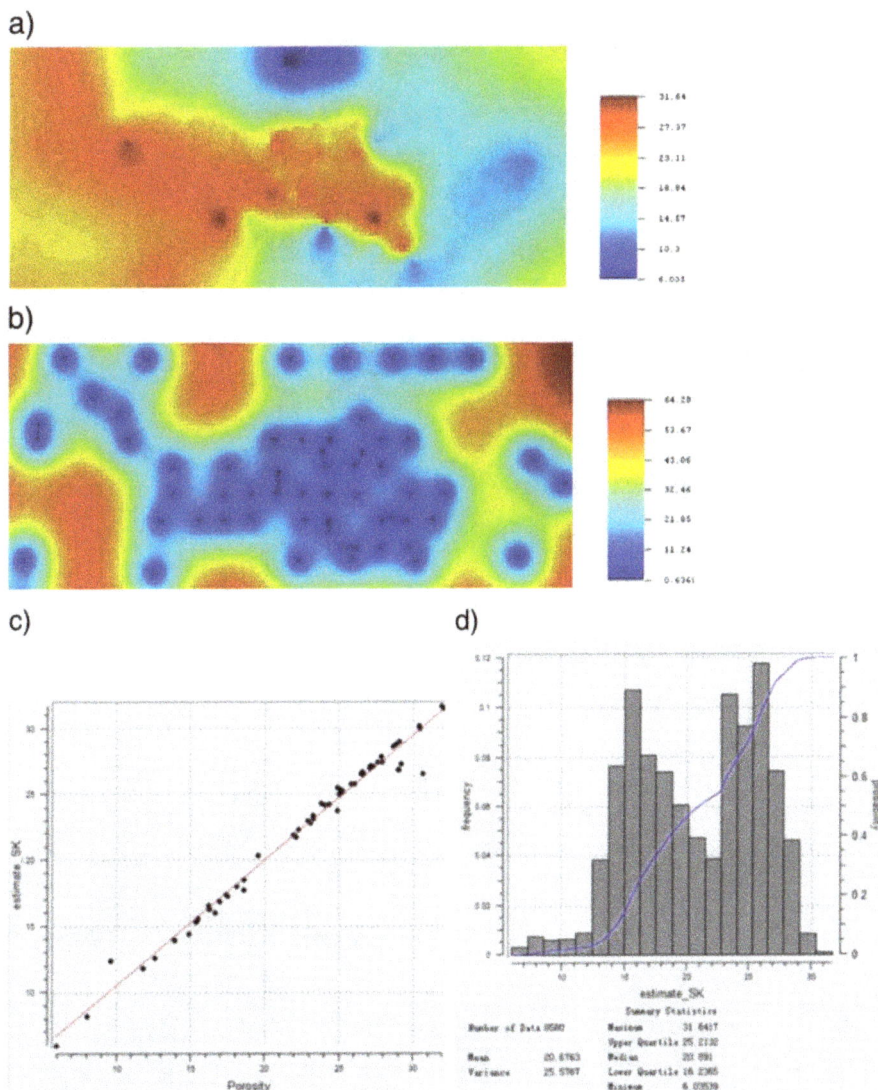

Figure 5 Porosity: a) simple kriging; b) simple kriging variance; c) coefficient of variation; d) histogram of sample data.

data. However, the maximum and minimum values are quite close to that of samples respectively, indicating a relatively fine estimation. Moreover, the estimation variance is much smaller than the sample variance, which implies a relatively low variability in the estimated values. Similar conclusion can also be obtain via the coefficient of variation, which is 0.2446.

Ordinary kriging

The porosity map from ordinary kriging is shown in Figure 6a. In general, this map is quite similar to the map generated via simple kriging. However, the appearance is much smoother. The variance map is shown in Figure 6b. The minimum variance is 0.637, which is quite close to that from simple kriging. However, the

maximum variance would be as large as 78.21, which is noticeably larger than the maximum variance from simple kriging. The true value vs. the estimated value plot is shown in Figure 6c. All the points regularly spread around the 45° line, indicating that the estimates generated via ordinary kriging can also match the sample data properly. The histogram is shown in Figure 6d. The mean is 0.7169, the median is 0.4647 and the variance is 31.3116. Comparing with the histogram of sample data, the estimated values are also slightly smaller than the sample data. However, the maximum and minimum values are also quite close to that of samples respectively, indicating a relatively fine estimation. Comparing with the histogram of simple kriging's estimation, there is no significant difference in terms of their mean

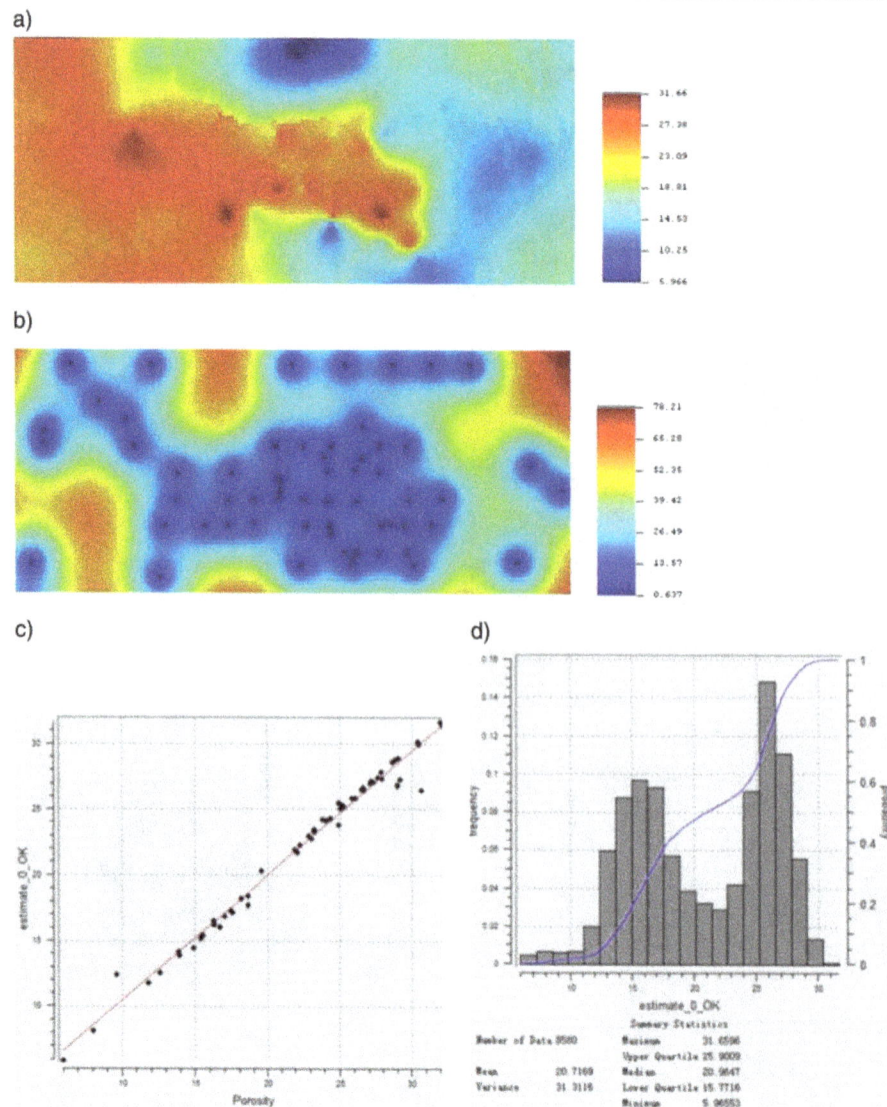

Figure 6 Porosity: a) ordianry kriging; b) ordinary kriging variance; c) coefficient of variation; d) histogram of sample data.

values. However, the estimation variance from ordinary kriging method is much larger than that from simple kriging method.

Sequential Gaussian simulation of porosity

Next, we generate five realizations of porosity using sequential Gaussian simulation. The sequential Gaussian simulation will use a normal score transform to turn the porosity values at the wells into a set of values that perfectly follow a standard normal distribution (zero mean, unit standard deviation) and will then generate grids of simulated values whose univariate distribution is also standard normal. Simple kriging is to be applied since the spatially constant mean will be assumed to be zero. In addition, we assume that the variogram of the normal-score transformed data would look very similar to the variogram of the raw data scaled to a unit sill.

The simulation map is shown in Figure 7a. As can be seen from this figure, the distribution trend of the simulated porosity is quite similar to that from the simple kriging estimation as well as the ordinary kriging estimation. High porosity locations are spreading from middle to west. Noticeably, in the south-west corner, the simulated values of porosity are higher than either of the estimation results from the two kriging estimation methods. Figure 7b shows the true value vs. the simulated value plot. All the points lie in the 45° line, indicating a perfect match between true values and simulated values at sampled locations.

Original Oil In Place (OOIP)

OOIP is calculated with gross thickness and porosity maps generated with ordinary kriging OOIP in stock

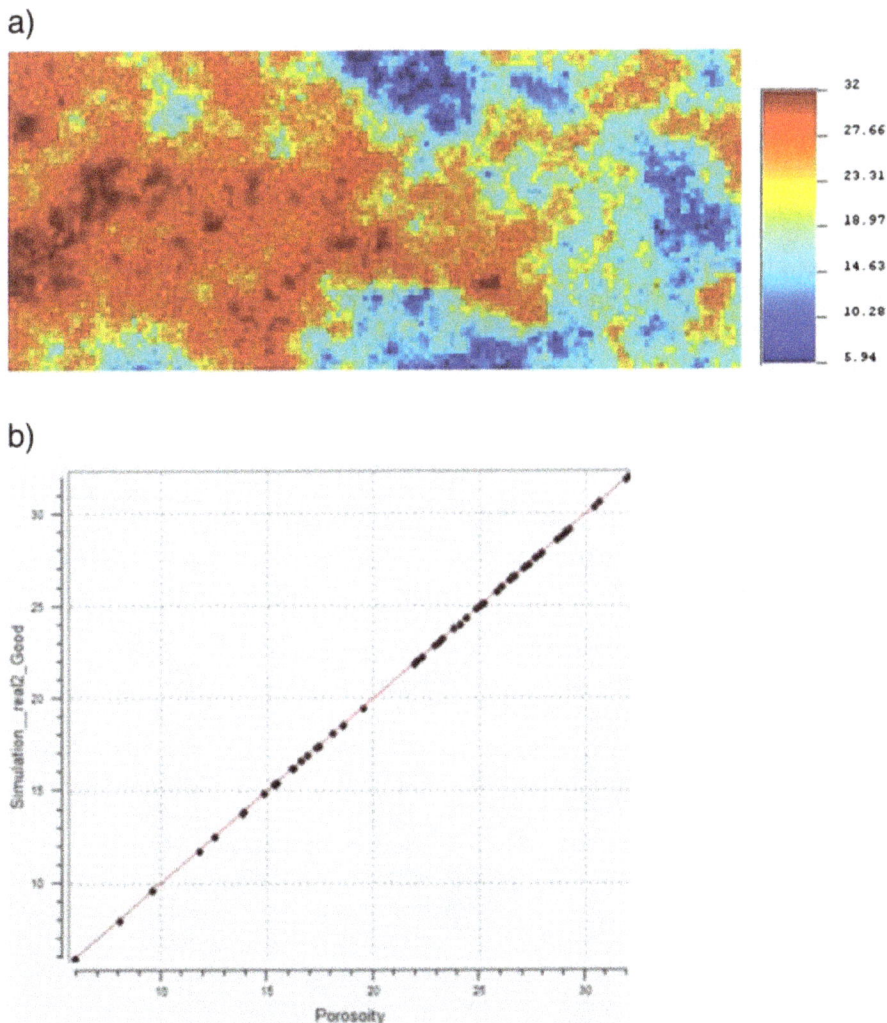

Figure 7 Porosity: a) plot of sequential gaussian simulation; b) true value vs. the simulated value plot.

tank barrel (STB) for each gridblock in the map is given by:

$$OOIP = \frac{A \times h \times ng \times (\phi/100)(1-Sw)}{5.615 \times Bo} \qquad (1)$$

A = the surface area of a block ($200 \times 200 = 40,000$ ft2), h = gross thickness(ft), ng = net to gross ratio, Φ = porosity (%), Sw = the water saturation and Bo = formation volume factor (rest bbl/STB). So:

$$OOIP = \frac{40000 \times 5.687 \times 0.7 \times (20.717/100)(1-20\%)}{5.615 \times 1.2}$$

$$\approx 3916.76$$

$$(2)$$

Permeability

When considering the permeability, the values of K range from as low as 0.01 to 750 md, the majority of the values are at the lower end of the region. This type of histogram is rarely useful for characterizing a sample because the values are clustered at one end. One way to overcome this problem is to transform the sample data in some way so that some sample characteristics are evident from the histogram plot. The most commonly used approach for permeability values is the log transform. From analysis, the log k distribution is much more symmetric than the permeability distribution. In addition, the log k and porosity histogram are remarkably similar. Both show similar trends with two peaks in the histogram plot, one of which is at the higher end of the values. Although this needs to be validated, such characteristic similarity way indicates a relationship between log k and porosity.

Simple kriging

Figure 8 shows the permeability map generated via simple kriging. Overall, the map has a smooth appearance. The good spatial continuity from middle to west corresponds to the principal direction of the variogram. The simple kriging variance map is shown in Figure 8b variance of permeability. The estimation variance could be as low as 0.6369 at where gridblocks are close to the sample data. However, in the northeast corner, the estimation variance could be as large as 64.28 due to the scarcity of sample data. Then we use the Scatter Plot to compare the value of log k (hard data) with SK value of logk. The plot shows that result of comparison is close to a 45° slant, which means the value of estimation is really match the hard data very well.

Ordinary kriging

The permeability map from ordinary kriging is shown in Figure 9a. In general, this map is quite similar to the map generated via simple kriging. However, the appearance is much smoother. The variance map is shown in Figure 9b. The minimum variance is 0.637, which is quite close to that from simple kriging. However, the maximum variance would be as large as 78.21, which is noticeably larger than the maximum variance from simple kriging.

Simple cokriging

Figure 10a shows the gross thickness map generated with simple cokriging. Overall, the map has a smooth appearance. The estimated distribution matches the original distribution well. The yellow and red areas represent the areas of which the permeability is relatively higher, while the blue areas stand for where the permeability is relatively lower. The good spatial continuity corresponds to the principal direction of the variogram. The small areas in the corner and bound of the map, is a result from the configuration of the conditioning data and the search neighborhood. In the northwest corner, the conditioning data is undersampled with respect to the rest of the grid. The simple cokriging variance map is shown in Figure 10b. The estimation variance is small in gridblocks close to the conditioning data, and it becomes large in area far from the data. Near the conditioning data, the kriging variance becomes the nugget effect of the variogram. The histogram, for the simple kriging estimates, is shown in figure. The mean is 1.90, the median is 1.95, both of which are a bit smaller than the conditioning data. The standard deviation of simple cokriging estimates is 0.78; while for conditioning data, it is 0.941. So it is likely that has a narrower spread than the conditioning data. In general, Co-simple kriging tends to gather the data to the center, making the data follow normal distribution. That is, most of the data distributes close to the mean value. 55 points are compared and plotted in the scatter plot of gross log k data and simple cokriging estimation data. Simple kriging provides good estimates even for values that are a bit smaller than the mean of the data (1.90 to 2.13). All the plotted points fall approximately along the straight line, indicating high accuracy of the estimates.

Ordinary cokriging

Figure 11a shows the gross thickness map generated with ordinary cokriging. Overall, the map has a more smooth appearance than that of simple cokriging. Transition areas exist in ordinary cokriging map, which represent gradual change among the subareas. The estimated distribution matches the original distribution well. The yellow and red areas represent the areas of which the permeability is relatively higher, while the blue areas stand for where the permeability is relatively lower. The good spatial continuity corresponds to the principal direction of the variogram. The small areas in the corner and bound of the map, is a result from the configuration of the conditioning data and the search neighborhood. In the northwest corner, the

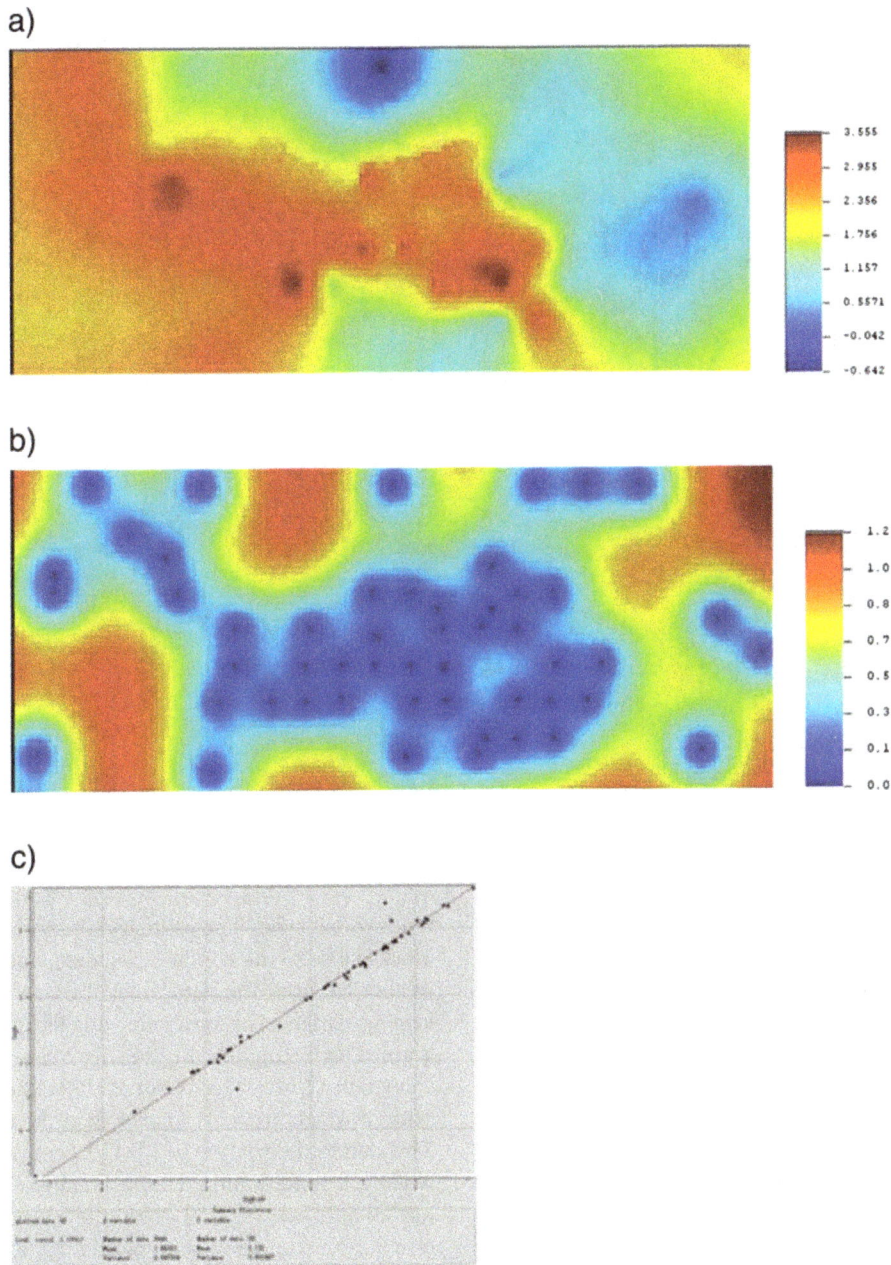

Figure 8 Permeability: a) simple kriging; b) simple kriging variance; c) coefficient of variation.

conditioning data is undersampled with respect to the rest of the grid. The ordinary cokriging variance map is shown in Figure 11b, which is more smooth than that of the simple cokriging variance map. The blue area that represents small variance tends to be wider, which stands for a better estimation result compared to simple cokriging. The estimation variance is small in gridblocks close to the conditioning data, and it becomes large in area far from the data. Near the conditioning data, the kriging variance becomes the nugget effect of the variogram.

The histogram, for the ordinary cokriging estimates, is shown in figure. The mean is 1.79, the median is 1.84, both of which are a bit smaller than the conditioning data. The standard deviation of ordinary cokriging estimates is 0.887; while for conditioning data, it is 0.941. So it is likely that has a narrower spread than the conditioning data. In general, ordinary cokriging tends to gather the data to the center, making the data follow normal distribution. That is, most of the data distributes close to the mean value. 92 points are compared and

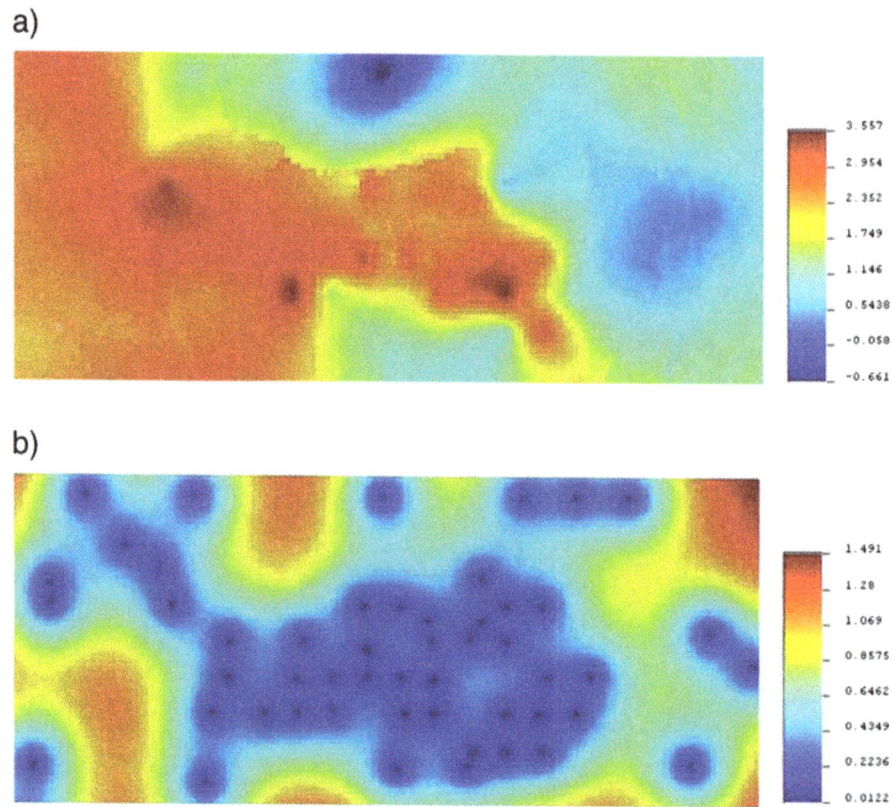

Figure 9 Permeability: a) ordinary kriging; b) ordinary kriging variance.

plotted in the scatter plot of gross log K data and ordinary cokriging estimation data. Ordinary cokriging provides good estimates even for values that are much larger than the mean of the data (1.79 to 2.13). All the plotted points fall approximately along the straight line, indicating high accuracy of the estimates.

MM2 models
Figure 12a shows the gross thickness map generated with MM2 cokriging. Overall, the map has a more smooth appearance. Transition areas exist in the estimate map, which represent gradual change among the subareas. The estimated distribution matches the original distribution well. The yellow and red areas represent the areas of which the permeability is relatively higher, while the blue areas stand for where the permeability is relatively lower. The good spatial continuity corresponds to the principal direction of the variogram. The small areas in the corner and bound of the map, is a result from the configuration of the conditioning data and the search neighborhood. In the northwest corner, the conditioning data is under-sampled with respect to the rest of the grid.

The MM2 cokriging variance map is shown in Figure 12b, which is smoother than that of the Co-simple kriging variance map. The blue area that represents small variance

tends to be wider. The estimation variance is small in grid-blocks close to the conditioning data, and it becomes large in area far from the data. Near the conditioning data, the kriging variance becomes the nugget effect of the variogram. The histogram, for the MM2 cokriging estimates, is shown in figure. The mean is 1.86, the median is 1.99, both of which are a bit smaller than the conditioning data. The standard deviation of MM2 kriging estimates is 0.879; while for conditioning data, it is 0.941. So it is likely that has a narrower spread than the conditioning data. 92 points are compared and plotted in the scatter plot of gross log K data and MM2 estimation data. All the plotted points fall approximately along the straight line, indicating high accuracy of the estimates.

Thickness SK vs OK
To have a thorough understanding of the results, the difference analysis among the three methods is required. The difference in the estimation results from the two methods is shown in Figure 13. We can see that the map is, in general, covered with light orange (less than 0.3 from color bar), indicating the slight difference between the two methods. The estimate mean of OK is a bit larger than that of SK. In the west-north region, the light blue color implies the estimate of OK mean in this

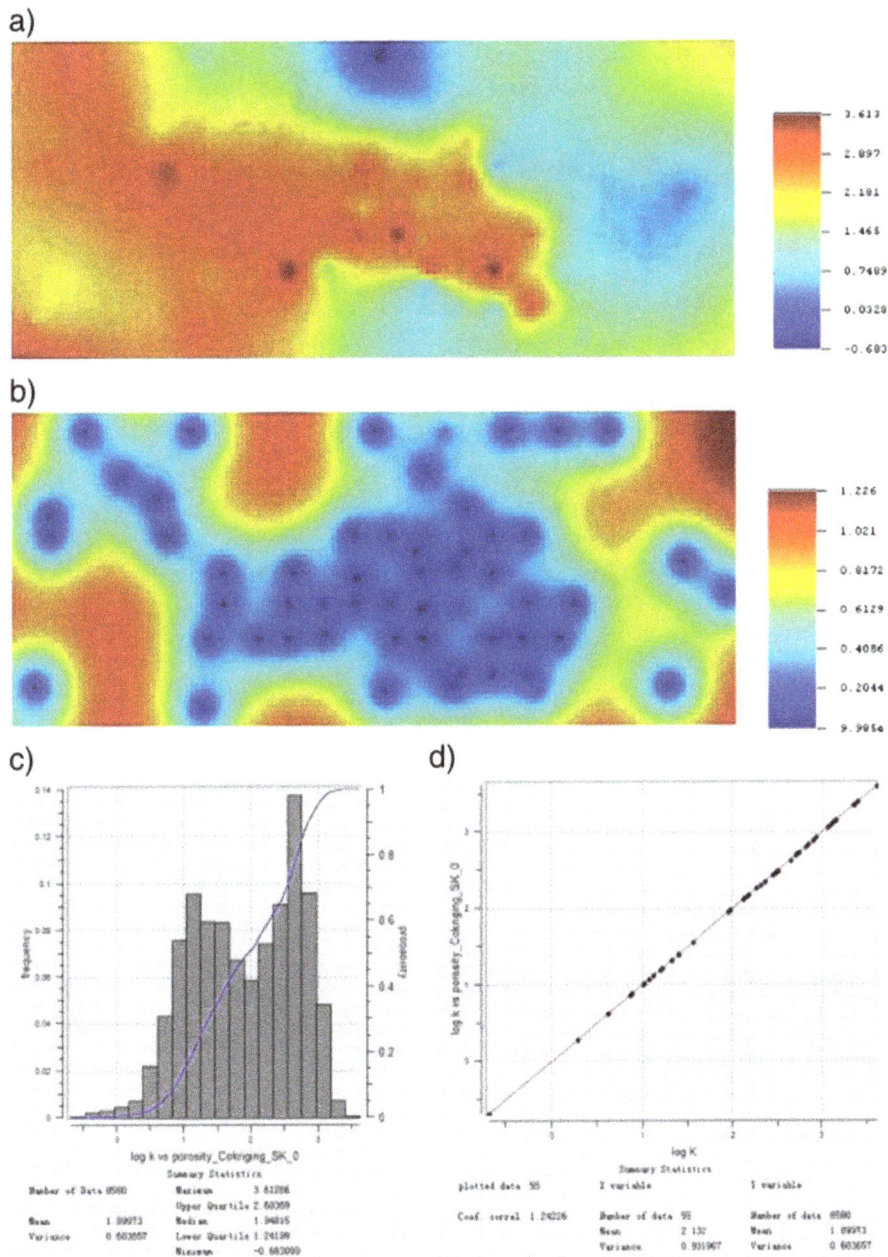

Figure 10 Permeability: a) simple cokriging; b) simple cokriging variance; c) histogram for simple cokriging; d) compare with true and simple cokriging value.

region is less than that of SK. This maybe mainly because this region is undersampled and there is less data available.

Porosity

Simple kriging estimation vs. Ordinary kriging estiamtion

The difference in the estimation results from the two methods is shown in Figure 14a. We can see that the map is, in general, covered with light blue, which implies the less difference between the two methods. After checking the sample data, we can find out that the light blue regions are where the true values exist. It thus well explains why there is less difference between the two methods. However, at where sample data are not available, the estimation difference between the two methods is remarkable. For example, in the north-east corner, the estimated values from simple kriging are significantly higher than that of ordinary kriging. In comparison, the estimated values, in the south-west region, are slightly lower than the values obtained via ordinary kriging. The

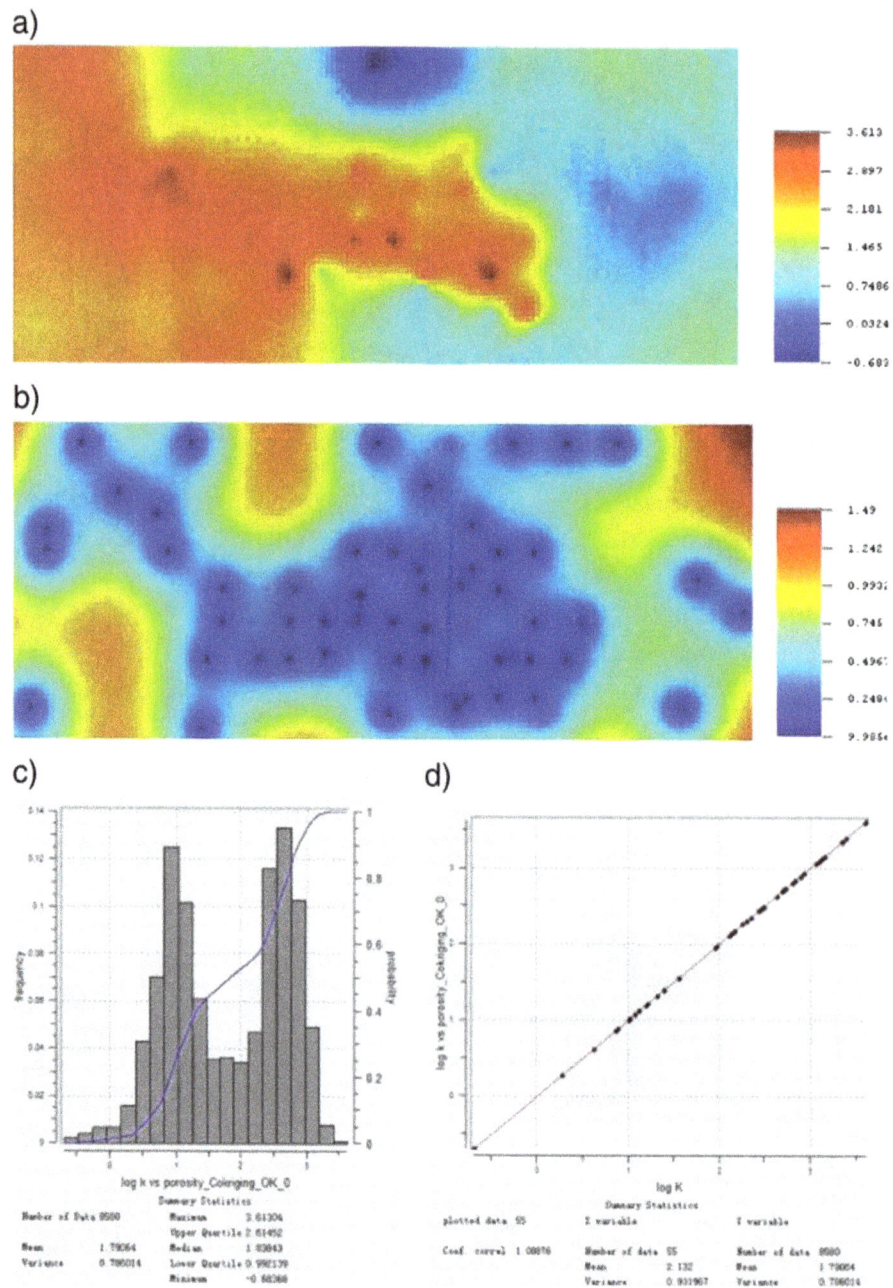

Figure 11 Permeability: a) ordinary cokriging; b) ordinary cokriging variance; c) histogram for ordinary cokriging; d) compare with logK and ordinary cokriging value.

difference in the estimation variance of the two methods is shown in Figure 14b. In general, the difference is inconsiderable. However, in the north-east corner, simple kriging estimates' variance is remarkably lower than the variance from ordinary kriging estimation.

Simple & ordinary kriging estimation vs. Simulation

The two sets of estimation results are compared with the simulation results, as shown in Figure 15. Overall, the two maps are quite similar to each other. We can

see that, in the north region, estimation results are lower than the simulation results. However, in the south region as well as the north-east corner, the estimation results are much larger than the simulation results.

Permeability
SK vs OK

The difference in the estimation results from the two methods is shown in Figure 16a. We can see that the map is, in general, covered with light blue, which implies

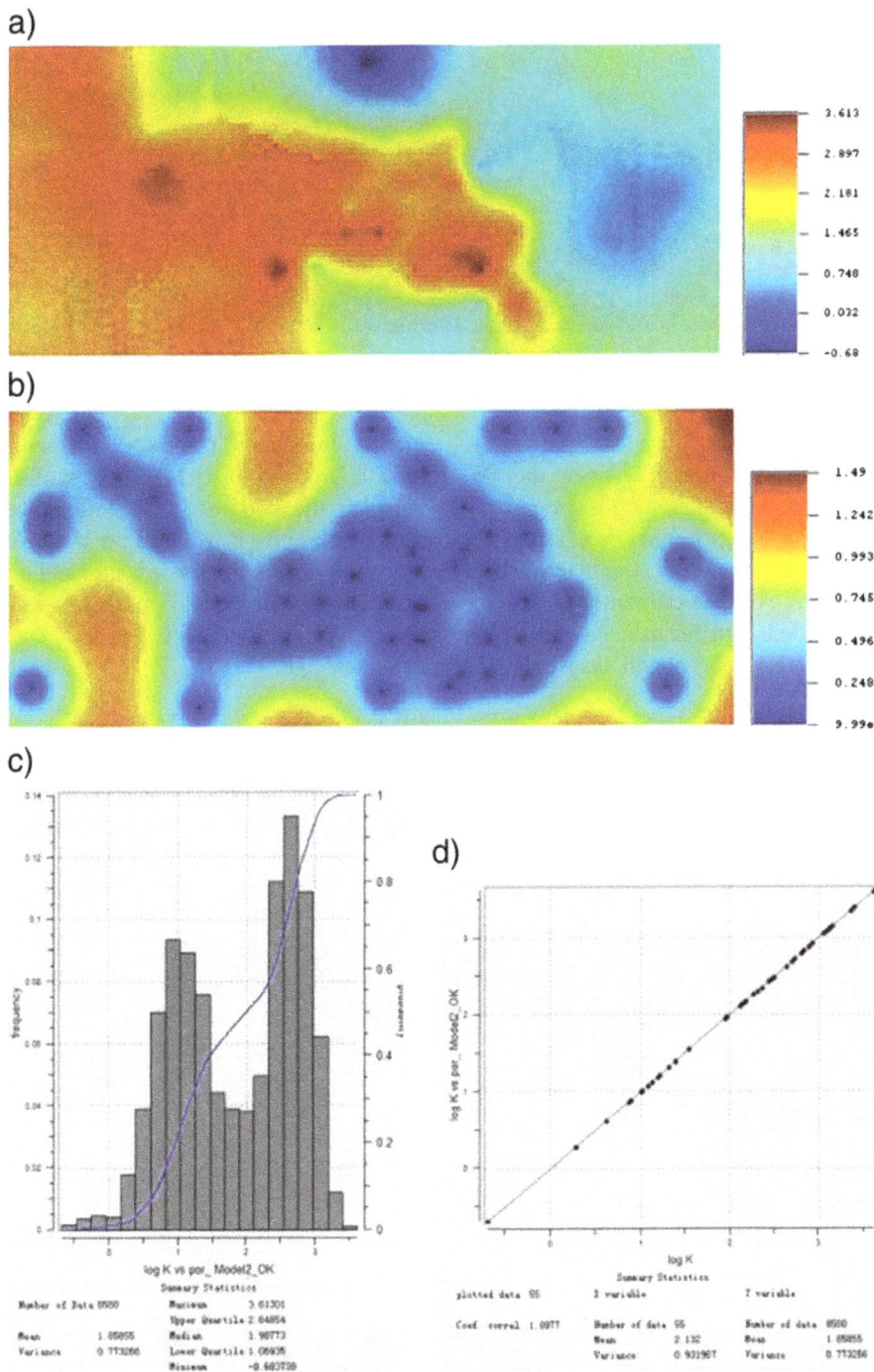

Figure 12 Permeability: a) MM2 cokriging; b) MM2 cokriging variance; c) histogram for MM2 cokriging; d) compare with logK and MM2 estimation data.

the less difference between the two methods. After checking the sample data, we can find out that the light blue regions are where the true values exist. However, at where sample data are not available, the estimation difference between the two methods is remarkable. In the northeast corner, the estimated values from simple kriging are significantly higher than that of ordinary kriging.

Simple cokriging vs ordinary cokriging
The difference in the estimation of the two methods is shown in Figure 16b. In general, the difference is inconsiderable. However, in the northeast, southwest and southeast corner, simple kriging estimate is remarkably higher than the that of ordinary kriging estimation, where sample data is undersampled.

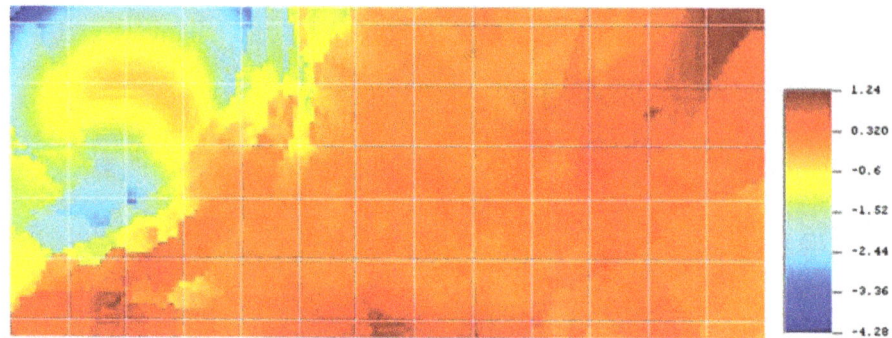

Figure 13 Difference analysis for thickness SK vs. OK.

OK vs ordinary cokriging

The difference in the estimation results from the two methods is shown in Figure 16c. The map is, in general, saturated with light blue, which implies less difference between the two methods. However, in the places where the color is dark blue and red, the estimation difference between these two methods are remarkable. This happens mainly because there is no enough data collected in these regions.

Ordinary cokriging vs MM2

The difference in the estimation results from the two methods is shown in Figure 16d. We can see that the

map is, in general, covered with orange (with the value of 0.05 from colorbar), indicating the slight difference between the two methods. And the estimate mean of ordinary kriging is a bit larger than that of MM2. Dark blue only exists in a few regions. These are where the estimate of MM2 is remarkably large than ordinary cokriging. This happens because there is not enough data collected.

Thickness comparison between SK and OK

a. The scatter plot of the simple kringing estimate and the ordinary kriging estimate is shown in Figure 17. The correlation coefficient is 0.973, indicating a high similarity

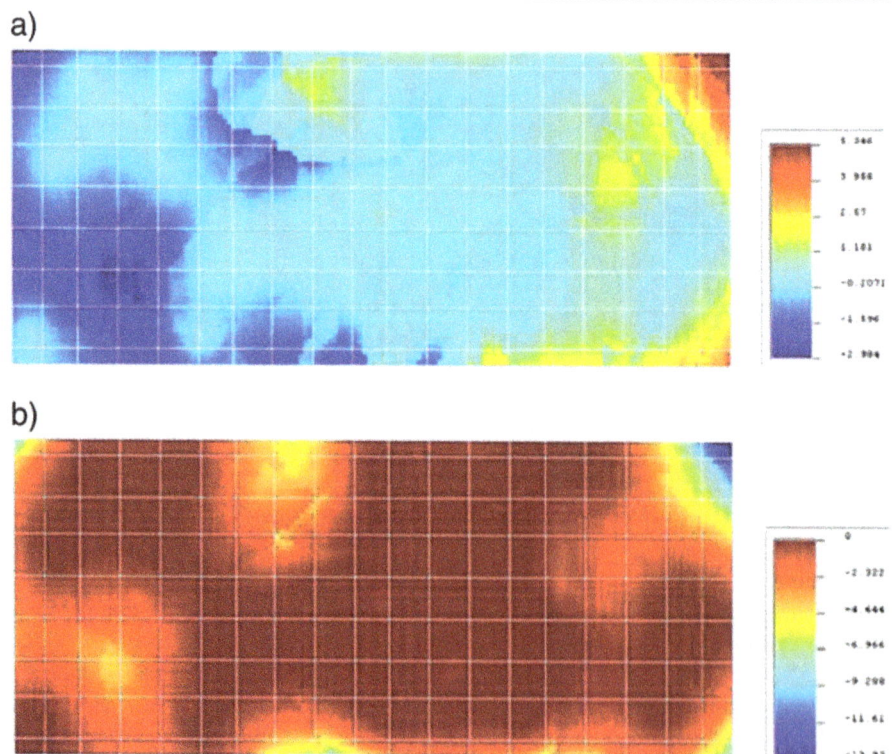

Figure 14 Porosity: a) difference in the estimation of SK and OK; b) difference in the estimation variance of SK and OK.

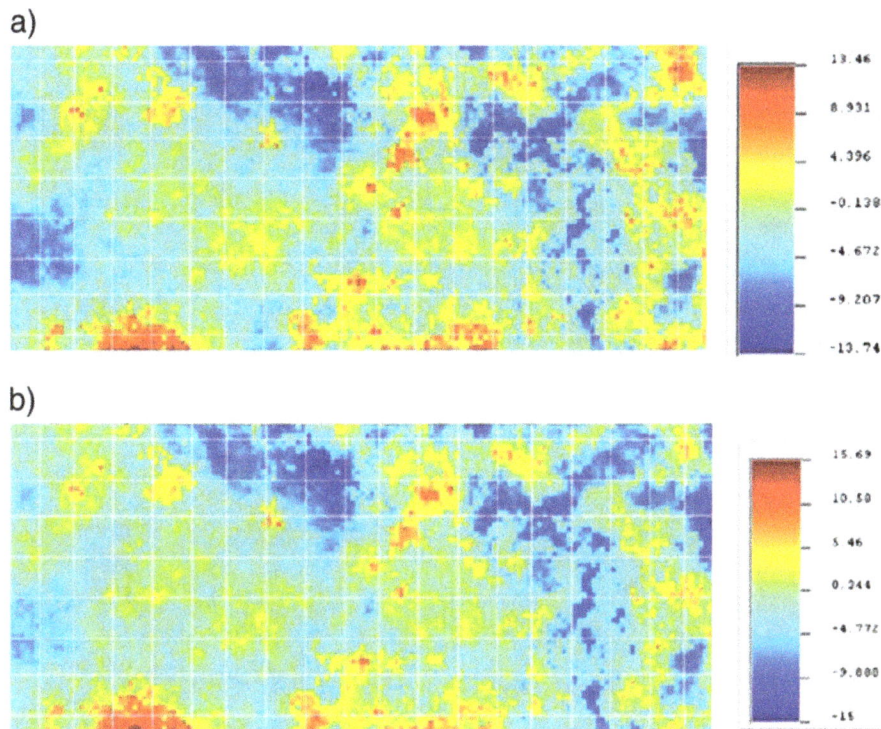

Figure 15 Porosity: simple (a) & ordinary kriging (b) estimation results are compared with the simulation results.

between these two estimates. Most differences occur for gross thickness between 4 and 7 ft. And the ordinary kriging estimate tends to be a little larger than the simple kriging estimate. b. The scatter plot of the simple kringing variance and the ordinary kriging variance is shown in Figure 17. Ordinary kriging tends overestimate thickness. Figure 17 shows that the simple kriging variance is smaller than the error variance estimated with the ordinary kriging. In view of the estimation mean and variance, although SK estimate tends to be more close to the conditioning data for this case, OK and SK both can generate good estimation result.

Porosity comparison between SK and OK
a. The scatter plot of the simple kringing estimate and the ordinary kriging estimate is shown in Figure 18. The correlation coefficient is 0.984, indicating a high similarity between these two estimates. Most differences occur for gross thickness between 12 and 19. The ordinary kriging estimate tends to be a little larger than the simple kriging estimate. b. The scatter plot of the simple kringing variance and the ordinary kriging variance is shown in Figure 18. The simple kriging variance is only slightly smaller than the error variance estimated with the ordinary kriging.

Log K comparison between SK and OK
a. The scatter plot of the simple kringing estimate and the ordinary kriging estimate is shown in Figure 19. The

correlation coefficient is 0.982, indicating a high similarity between these two estimates. Most differences occur for gross thickness between 0.5 and 1. And the ordinary kriging estimate tends to be a little larger than the simple kriging estimate. b. The scatter plot of the simple kringing variance and the ordinary kriging variance is shown in Figure 19. The correlation coefficient is 0.998, indicating there is almost no difference between the estimation variance of the two.

Log K comparison between simple cokriging and ordinary cokriging
a. The scatter plot of the simple cokringing estimate and the ordinary cokriging estimate is shown in Figure 20. The correlation coefficient is 0.972, indicating a high similarity between these two estimates. Most differences occur for log K between 1 and 3. And the ordinary cokriging estimate tends to be a little larger than the simple cokriging estimate. b. The scatter plot of the simple cokringing variance and the ordinary cokriging variance is shown in Figure 20. Figure 20 shows that the simple kriging variance is smaller than the error variance estimated with the ordinary kriging. In view of the estimation mean and variance, although simple kriging estimate tends to be more close to the conditioning data for this case, ordinary and simple kriging both can generate good estimation result.

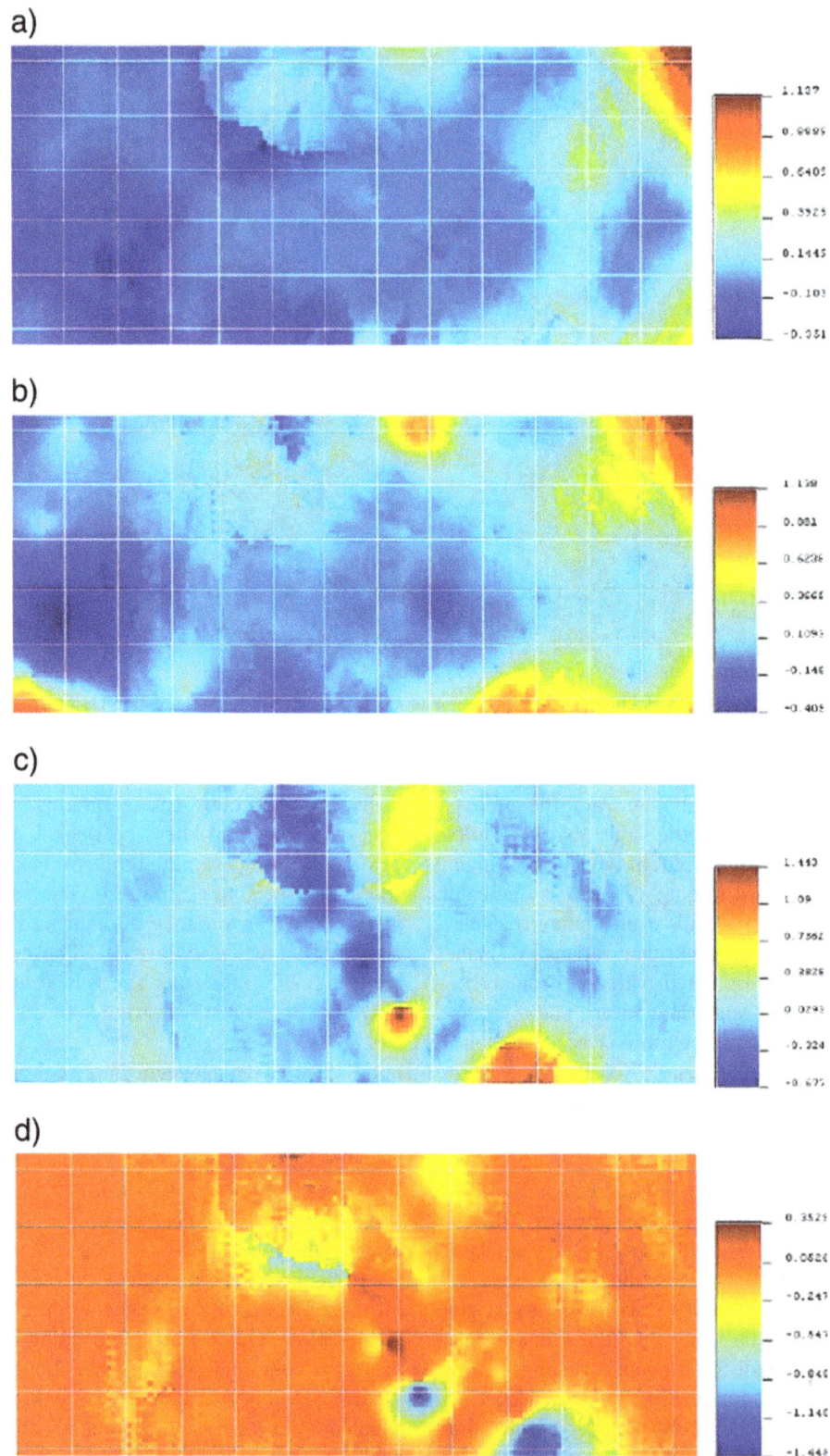

Figure 16 Permeability: a) difference in the estimation of SK and OK of log k; b) difference in the estimation of simple cokriging and ordinary cokriging of log k; c) difference in the estimation of OK and ordinary cokriging of log k; d) difference in the estimation of ordinary cokriging and MM2 of log k.

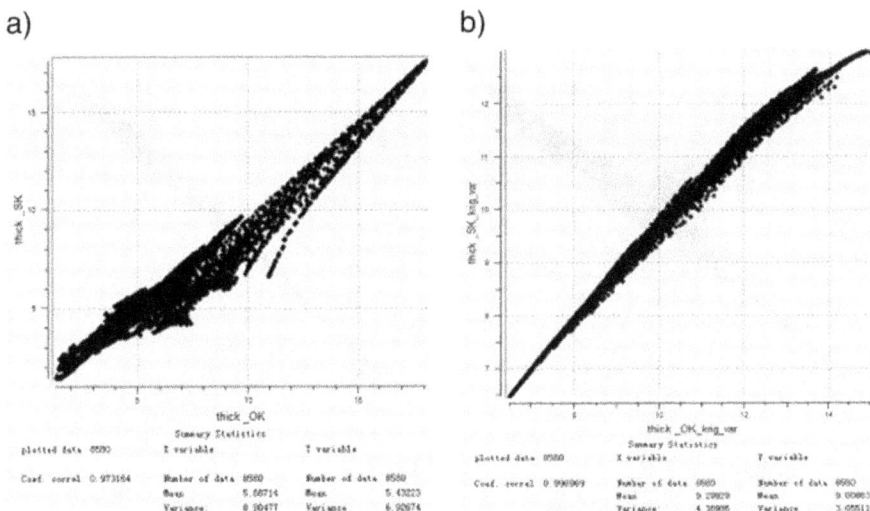

Figure 17 Thickness: a) the scatter plot of SK estimate and OK estimate for thickness; b) the scatter plot of SK variance and OK variance for thickness.

Log K comparison between ordinary kriging and ordinary cokriging

a. The spatial distributions of log K are significantly different for ordinary cokriging and ordinary kriging methods. The map, generated with ordinary cokriging is not as smooth as ordinary kriging. The influence of porosity adds some unique features to the log k data map that ordinary kriging cannot capture with a single variogram model. The ordinary cokriging and ordinary kriging estimates differ significantly, as shown in Figure 21b. Ordinary kriging tends to overestimate log K. Figure 21b shows that

the cokriging variance was smaller than the error variance estimated with the ordinary kriging. This shows that additional information used in cokriging reduces the error variance in estimates.

Log K comparison between ordinary cokriging and MM2

a. The spatial distributions of log K are significantly different for MM2 and ordinary cokriging methods. From the scatter plot of both, the MM2 estimates tend to be larger than the ordinary full cokriging estimates, as shown in Figure 22b. MM2 cokriging tends overestimate log K.

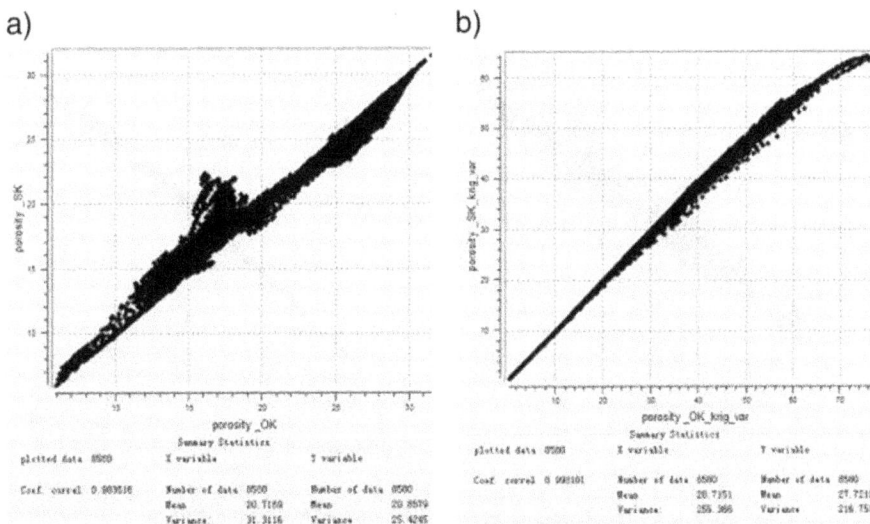

Figure 18 Porosity: a) The scatter plot of SK estimate and OK estimate for porosity; b) the scatter plot of SK variance and OK variance for porosity.

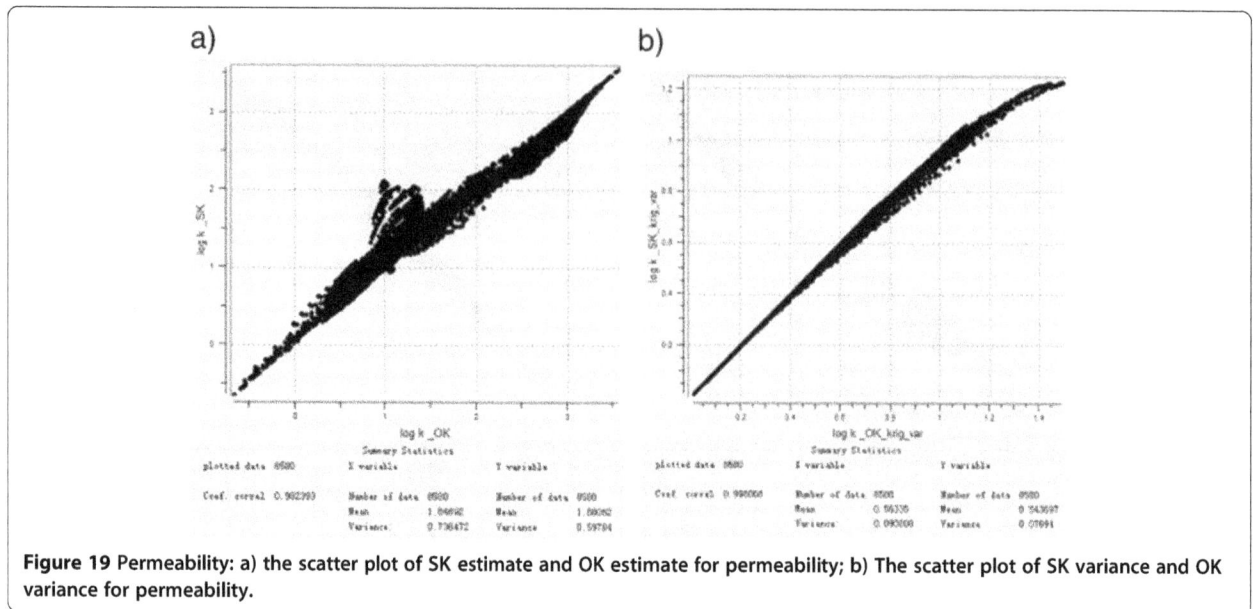

Figure 19 Permeability: a) the scatter plot of SK estimate and OK estimate for permeability; b) The scatter plot of SK variance and OK variance for permeability.

Figure 22b shows that the full cokriging variance is smaller than the error variance estimated with the MM2. This shows that additional information used in ordinary cokriging reduces the error variance in estimates.

Conclusions

Log transformation of the permeability can be applied to overcome the problem that the permeability data are clustered at the lower end of the study region. This would minimize the effect of the extreme and order-of-magnitude variations within the data points, and result a better identification of the spatial structure.

From an estimation point of view, it is always preferable to start with an isotropic variogram structure before the investigation of anisotropy. From studying the case of flow unit 5, a conclusion is made that anisotropic variogram could achieve better spatial structure and capture more interpretable spatial relationship than isotropic variogram.

Generally, the estimation results of the ordinary kriging and simple kriging method are similar. The correlation coefficient between the two estimates is quite close to 1 for thickness, porosity and log K. However, the variance maps of the estimates obtained via ordinary kriging are much smoother than that via simple kriging. Transition

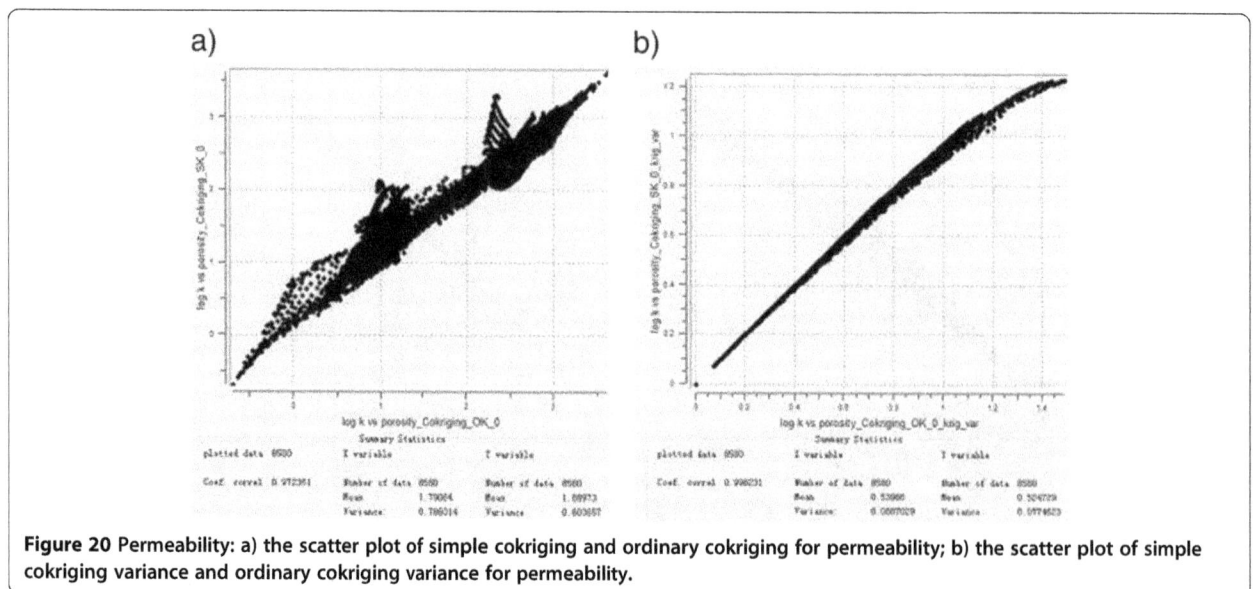

Figure 20 Permeability: a) the scatter plot of simple cokriging and ordinary cokriging for permeability; b) the scatter plot of simple cokriging variance and ordinary cokriging variance for permeability.

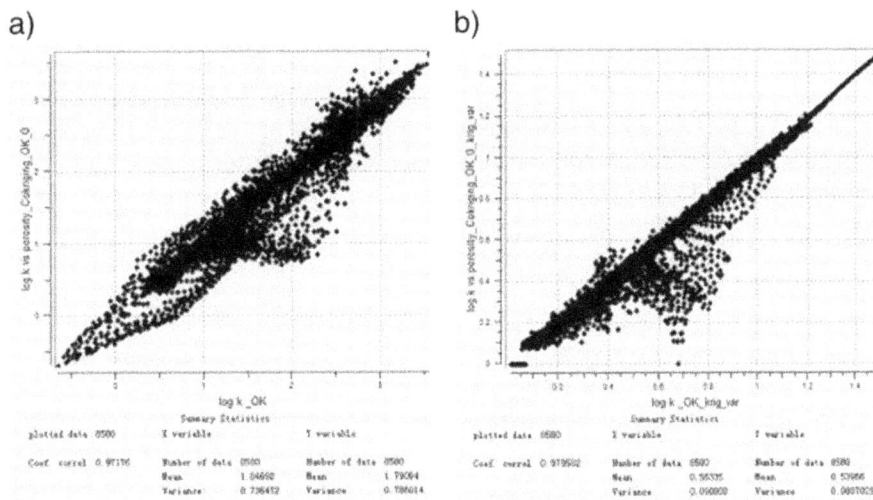

Figure 21 Permeability: a) the scatter plot of OK and ordinary cokriging for permeability; b) the scatter plot of OK variance and ordinary cokriging variance for permeability.

areas exist in ordinary kriging map, showing more gradual change among the subareas. This is because the sample mean value has been ignored in ordinary kriging estimation.

The estimation methods such as simple kriging and ordinary kriging tend to eliminate the extreme values observed in the sample data, and narrow the spread of the data distribution. Consequently, these two methods appear to gather the data to the center and make the data follow normal distribution. This is the reason why most of the data distribute close to the mean value.

Compared with conventional estimation techniques, the simulation method could well reflect the reservoir's intrinsical characteristics in terms of the associated extreme values. From the case study of the porosity, we can see that both the estimation and simulation methods could obtain the similar results. There is almost no any remarkable difference at where the sample data are sufficient. However, at the unsampled locations, the estimation results are much larger than the simulation results. For example, the south region as well as the north-east corner.

Cokriging estimation employs the secondary variable to estimate the first variable. The two variables should be linearly related and have a strong relationship with each other. Unlike ordinary kriging, in which we only use the values surrounding the sampled locations to estimate the unsampled locations, Cokriging method could

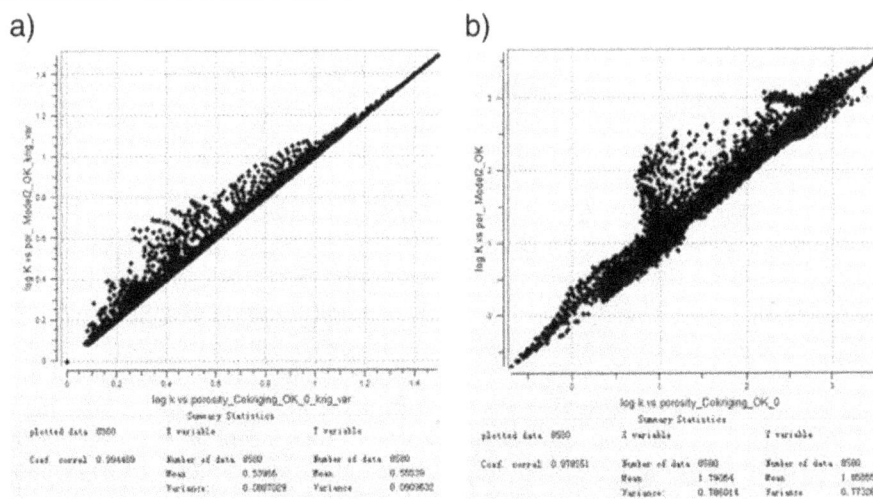

Figure 22 a) The scatter plot of ordianary cokriging and MM2 for permeability; b) the scatter plot of ordianary cokriging variance and MM2 for permeability.

be able to improve the estimation and reduce the uncertainty of the estimation with the assist of the spatial information available from the secondary variable.

Furthermore, employing porosity as the secondary variable would add some unique features to the estimation of log K, and these features cannot be captured by either of the simple kriging and ordinary kriging methods via a single variogram model. In addition, Cokriging would tend to reduce the error variance compared with the simple and ordinary kriging methods.

For thickness, both the ordinary kriging and the simple kriging methods could achieve better estimation results than others. For porosity, the simple kriging, the ordinary kriging and the Sigsm methods could generate similar estimation results. For permeability, there is no much difference between the results of simple cokriging and ordinary cokriging methods. However, ordinary cokriging method could produce better results than both of the ordinary kriging and the MM2 methods.

Cross validation can be used to select the optimum kriging parameters, which can be utilized further in the kriging methods if it is permitted. The cross-validation is verified by the existence of strong relationship between the estimate and gross data, as well as the small correlations between the estimated values and their errors. However, it is a trial and error procedure and time consuming job.

Methods

Once initial data sets are prepared, quality controlled, and loaded into the geostatistical software (like SGeMS), a typical work flow, with iterations, might be: (1) data mining; (2) spatial continuity analysis and modeling; (3) search ellipse design; (4) model crossvalidation; (5) kriging; (6) conditional simulation; (7) model uncertainty assessment.

Data mining

An early and fundamental step in any science starts at the descriptive stage. Until facts are accurately gathered and described, an analysis of their causes is premature. Because statistics generally deals with quantities of data, not with a single datum, we need some means to deal with the data in a manageable form. Thus, much of statistics deals with ways of describing the data and understanding relationships between pairs of variables. Data speak most clearly when they are organized (Isaaks and Srivastava 1989).

Because there is no one set of prescribed steps in data mining, you should follow your instincts in explaining anomalies in the data set. By using various tools, you gain clearer understanding of your data and also discover possible sources of errors. Errors are easily overlooked, especially in large data sets and when computers

are involved, because we simply become detached from our data. Thorough analysis fosters an intimate understanding of the data that can flag spurious results.

Classical statistical data analysis includes data posting, computation of means and variances, making scatterplots to investigate the relationship between two variables, and identification of subpopulations and potential outliers.

Histograms, graphical representations of the data distribution of a single variable, record how often values fall within specified intervals or classes. A bar depicts each class, and its height is proportional to the number of values within that class. The histogram shape informs us about the distribution of the data values. Ideally, we like to see a bell-shaped, symmetrical distribution around the mean value. This is referred to as a normal, or Gaussian, distribution and has a predictable shape based on the data mean and variance. Many statistical and geostatistical methods assume such a data model. If the shape is skewed to either side of the mean, then often it is necessary to adjust the shape by transforming the data into Gaussian form. Complex histograms may indicate mixing of multiple distributions. Categorization of the data (e.g., by facies) often identifies the underlying distributions.

Spatial continuity analysis and modeling

Variables of interest in the petroleum industry (e.g., porosity, permeability, saturation, sand/shale volumes, etc.) are the product of a vast number of complex physical and chemical processes. These processes superimpose a spatial pattern on reservoir rock properties, and it is important to understand the scales and directional aspects of these features for efficient hydrocarbon production. The spatial component makes these variables complicated, and we are forced to admit uncertainty about their distribution between wells. Because deterministic models do not handle uncertainties associated with these variables, a geostatistical approach is used because its foundation is probabilistic theory (covariance models) that recognizes these inevitable uncertainties.

The spatial model

Spatial continuity analysis quantifies the variability of sample properties with respect to distance and direction (geographic location is considered only if the data exhibit a trend, a property known as nonstationarity).

Quantifying spatial information involves comparing data values at one location with values of the same attribute at other locations. For example, two wells in close proximity are more likely to have similar reservoir properties than two wells farther apart. The key question—what we want to know—is what measured values tell us about reservoir properties at unsampled locations.

Geometric Anisotropy

Geometric Anisotropy typically is observed when the variograms in the directions of maximum and minimum continuity show a similar shape and sill but different range. The range in the direction of maximum continuity, u, is u a, while the range in the direction of minimum continuity is v a. We assume that the two directions are perpendicular to each other.

To model the two variograms with the same sill, we have to use the same combination of linear models in both directions expect with different ranges. For example, with a linear combination of nugget and spherical models, the model in the u direction is

$$\gamma_u(L) = C_0 + C_1 M_{Sa_u}(L). \tag{3}$$

And the model in the v

$$\gamma_v(L) = C_0 + C_1 M_{Sa_v}(L). \tag{4}$$

Modeling of Cross Variograms

Cross Variograms can be modeled with the same models used to model variograms. As before, the rule of the use of a minimum number of parameters to model the variogram applies, and the rule of the condition of positive definiteness should be satisfied.

To satisfy the condition of positive definiteness, certain additional restrictions are imposed when modeling the cross variogram. If x and y are considered as two variables, the variograms for the two variables x and y are modeled, respectively, as

$$\gamma_x(L) = C_{0x} + C_{1x} M_{Sa}(L) \tag{5}$$

$$\gamma_v(L) = C_{0v} + C_{1v} M_{Sa}(L). \tag{6}$$

Note that both variograms need to be modeled with the same linear combination of structures and that the range for both variograms must be the same. The only difference between the two structures are sill values, which can be different. To model the cross variogram between the two variables, we can write

$$\gamma_{Cxv}(L) = C_{0xv} + C_{1xv} M_{Sa}(L). \tag{7}$$

As in modeling variograms, we are restricted by the same linear combinations of models in modeling the cross variogram, and the range must also be the same for the cross-variogram structure. In addition, the coefficients of the model should be defined so that the following two conditions are satisfied.

$$C_{0x} C_{0y} > C_{0xy}^2 \tag{8}$$

$$C_{1x} C_{1y} > C_{1xy}^2. \tag{9}$$

That is, for a given model, the product of the coefficients of individual variable variogram models should be greater than the square of the coefficient of the cross variogram.

Search ellipse design

Because computers are used in mapping, we must instruct the program how to gather and use control points during interpolation. Most familiar with computer mapping know that this involves designing a search ellipse or neighborhood. We must specify the length of the search radius, the number of sectors (typically four or eight), and the number of data points per sector. Most common mapping programs allow the user to specify only one radius; thus, the search ellipse is circular (isotropic). However, during geostatistical analysis, we often find that the spatial model is anisotropic. Thus, we should design the search ellipse based on the spatial model correlation scales, aligning the search ellipse azimuth with the major axis of anisotropy.

Kriging is a geostatistical interpolation technique. It is a linear weighted-averaging method, similar to inverse weighted distance. However, kriging weights depend on a model of spatial correlation. Therefore, it is possible to create a map exhibiting strong anisotropy, resulting in a map that "looks" more geologically plausible.

Depending on the specific application, different procedures have been used for the purpose of estimation. In this study, we will introduce the Simple Kriging (CK), Ordinary Kriging (OK) and Cokriging with a case study in the following chapters. Here we will briefly recall the definition of these algorithms.

Simple kriging

Simple kriging requires a knowledge of population mean, which may not be known in practice because without prior assumptions. Therefore, this type of kriging procedure is not quite popular.

Ordinary kriging

Ordinary kriging is the most popular techbique, which eliminates the need for knowledge of mean value. It also is easier to adapt to local variations. It is, without question, is the most widely applied kriging technique.

Cokriging

The Cokriging algorithm integrates the information carried by a secondary variable related to the primary attribute being estimated. The kriging system of equations is then extended to take into account that extra information. The Markov Model 1 (MM1) and the Markov Model 2 (MM2). The Markov models (MM1 or MM2) can only be solved with simple cokriging; using ordinary cokriging would lead to ignoring the secondary variable since the sum of weights for the secondary variable must be equal to zero.

Kriging is a deterministic method that has a unique solution offering the best estimate. It does not pretend to represent the actual variability of the studied attribute. It can be used in the traditional way that other mathematical interpolation methods have been used. It has the added value of incorporating the spatial model and thus more reliably depicting the shapes of geologic features.

Abbreviations
SGeMS: Stanford Geostatistical Modeling Software; SK: Simple kriging; OK: Ordinary kriging; Sgism: Sequential Gaussian Simulation; SC: Simple cokriging; OC: Ordinary cokriging; MM2: Markov model 2; OOIP: Original oil in place.

Competing interests
• In the past five years have you received reimbursements, fees, funding, or salary from an organization that may in any way gain or lose financially from the publication of this manuscript, either now or in the future? Is such an organization financing this manuscript (including the article-processing charge)? If so, please specify.
→NO.
• Do you hold any stocks or shares in an organization that may in any way gain or lose financially from the publication of this manuscript, either now or in the future? If so, please specify.
→NO.
• Do you hold or are you currently applying for any patents relating to the content of the manuscript? Have you received reimbursements, fees, funding, or salary from an organization that holds or has applied for patents relating to the content of the manuscript? If so, please specify.
→NO.
• Do you have any other financial competing interests? If so, please specify.
→NO.
The authors declare that they have no competing interest.
• Are there any non-financial competing interests (political, personal, religious, ideological, academic, intellectual, commercial or any other) to declare in relation to this manuscript? If so, please specify.
→NO.

Authors' contributions
ZS, ZY, and WM made substantial contributions to conception and design, or acquisition of data, or analysis and interpretation of data. ZS brought up the innovation and designed the study. ZY proposed the methodology and wrote the code. WM collected the data and edited the format. XX and CF were involved in drafting the manuscript and revised it critically for important intellectual content. All authors gave final approval of the version to be published. All authors read and approved the final manuscript.

Acknowledgements
This study was part of a research project funded by Petroleum Technology Research Centre, Canada. We also feel grateful for the support from Faculty of Engineering and Applied Science, University of Regina.

References
Abdideh M, Barghi D (2012) Designing a 3D model for the prediction of the top of formation in oil fields using geostatistical methods. Geocarto Int 27(7):569–579

Abdideh M, Mahmoudi N (2013) UCS prediction: a new set of concepts for Reservoir Geomechanical Modeling. Pet Sci Technol 31(24):2629–2635

Bueno JF, Drummond RD, Vidal AC, Sancevero SS (2011) Constraining uncertainty in volumetric estimation: A case study from Namorado Field, Brazil. J Pet Sci Eng 77(2):200–208

Chen Y, Park K, Durlofsky L (2011) Statistical assignment of upscaled flow functions for an ensemble of geological models. Comput Geosci 15(1):35–51

Cressie N, Hawkins D (1980) Robust estimation of the variogram: I. Math Geol 12(2):115–125

Esmaeilzadeh S, Afshari A, Motafakkerfard R (2013) Integrating artificial neural networks technique and geostatistical approaches for 3D Geological Reservoir Porosity Modeling with an example from one of Iran's oil fields. Pet Sci Technol 31(11):1175–1187

Fegh A, Riahi M, Norouzi G (2013) Permeability prediction and construction of 3D geological model: application of neural networks and stochastic approaches in an Iranian gas reservoir. Neural Comput & Applic 23(6):1763–1770

Geoff B (2007) S-GeMS Tutorial Notes, Kansas Geological Survey

Habibnia B, Momeni A (2012) Reservoir characterization in Balal oil field by means of inversion, attribute, and geostatistical analysis methods. Pet Sci Technol 30(15):1609–1618

Huysmans M, Dassargues A (2013) The effect of heterogeneity of diffusion parameters on chloride transport in low-permeability argillites. Environ Earth Sci 68(7):1835–1848

Isaaks EH, Srivastava RM (1989) Applied Geostatistics. Oxford University Press, New York

Kelkar M, Perez G (2002) Applied Geostatistic for Reservoir Characterization

Kelsall J, Wakefield J (2002) Modeling spatial variation in disease risk. J Am Stat Assoc 97(459):692–701

Malvić T, Jović G (2012) Thickness maps of neogene and quaternary sediments in the Kloštar Field (Sava Depression, Croatia). J Maps 8(3):260–266

Nava E, Pintos S, Queipo NV (2010) A geostatistical perspective for the surrogate-based integration of variable fidelity models. J Pet Sci Eng 71(1–2):56–66

Qi L, Carr TR, Goldstein RH (2007) Geostatistical three-dimensional modeling of oolite shoals, St. Louis Limestone, southwest Kansas. AAPG Bull 91(1):69–96

Remy N, Boucher A, Wu J (2009) Applied Geostatistics with SGeMS. Cambridge University Press, UK

Remy N (2004) Geostatistical Earth Modeling Software: User's Manual

Schmidt G, Schröder W (2011) Regionalisation of climate variability used for modelling the dispersal of genetically modified oil seed rape in Northern Germany. Ecol Indic 11(4):951–963

Soleymani H, Riahi MA (2012) Velocity based pore pressure prediction—A case study at one of the Iranian southwest oil fields. J Pet Sci Eng 94–95:40–46

Wilson CE, Aydin A, Durlofsky LJ, Boucher A, Brownlow DT (2011) Use of outcrop observations, geostatistical analysis, and flow simulation to investigate structural controls on secondary hydrocarbon migration in the Anacacho Limestone, Uvalde, Texas. AAPG Bull 95(7):1181–1206

Zarei A, Masihi M, Salahshoor K (2011) Comparison of different univariate and multivariate geostatistical methods by porosity modeling of an Iranian oil field. Pet Sci Technol 29(19):2061–2076

12

A systematic approach for modelling quantitative lake ecosystem data to facilitate proactive urban lake management

Aaron N Wiegand[1*], Christopher Walker[1], Peter F Duncan[2], Anne Roiko[1] and Neil Tindale[1]

Abstract

Background: The management of the health of urban lake systems is often reactive and is instigated in response to poor aesthetic quality or physicochemical measurements, rather than from an overall assessment of ecosystem health. Interpreting physicochemical monitoring data in isolation is problematic for two main reasons: the suite of parameters that are monitored may be limited; and the contribution that any single parameter has towards water quality or health varies considerably depending on the nature of the system of interest. Extending monitoring programs to include flora and fauna results in a better dataset of ecosystem status, but also increases the complexity in interpreting whether the status is good or poor.

Results: This paper details a process by which a large set of quantitative biological, physical, chemical and social indicators may be transformed into a simple, but informative, numerical index that represents the overall ecosystem health, while also identifying the likely source and scale of pressure for remedial management action. The flexibility of the proposed approach means that it can be readily adapted to other lake systems and environments, or even to include or exclude different indicators. A case study is presented in which the model is used to assess a comprehensive longitudinal dataset that resulted from monitoring a constructed urban lake in Southeast Queensland, Australia.

Conclusions: The sensitivity analysis and case study indicate that the model identifies how changes in individual monitoring parameters result in changes in overall ecosystem health, and thus illustrates its potential as a lake management tool.

Keywords: Modelling, Urban lakes, Ecosystem health, Management, Index

Background

In Australia, design guidelines for urban lakes have been refined over the past few decades to better cater to the impacts associated with urban settings. Current design guidelines consider an urban lake to be a receiving environment for runoff, requiring that the runoff is pre-treated prior to input through various measures (e.g. retention and re-use, constructed wetlands and bioretention basins). While the contemporary design guidelines have been embraced by many levels of authority, difficulties with the subsequent management of urban lakes are still present (Lloyd et al., 2002; Bayley & Newton, 2007; Walker et al.,

2010). Management of urban lakes, and indeed many aquatic ecosystems, is often reactive and is instigated in response to poor aesthetic quality or dictated by physicochemical measurements, rather than from an overall assessment of ecosystem health (Rapport, 1998; Karr & Chu, 1999; Likens et al., 2009). Management strategies for urban lakes have generally been short term and are often reactionary to physicochemical monitoring alone, which serves to temporarily address an issue, but often fails to provide a long-term solution (e.g. macrophyte harvesting) (Walker et al. 2010). Remediation strategies that are based solely upon the limited observations provided by physicochemical monitoring, are likely to have limited effectiveness and may fail to identify and address issues with the broader ecosystem health of an urban lake.

* Correspondence: awiegand@usc.edu.au
[1]University of the Sunshine Coast, Locked Bag 4, Maroochydore DC, QLD 4558, Australia
Full list of author information is available at the end of the article

Interpreting physicochemical monitoring data as representing good or poor water quality, or even lake health as a whole, is problematic for two main reasons: the suite of parameters that are monitored may be limited; and the contribution that any single parameter has with respect to quality or health is not well identified and may vary considerably depending on the nature of the system of interest. For example, in Australia, existing guidelines, such as ANZECC (2000) and Queensland Water Quality Guidelines (DERM, 2009) present reference condition values for slightly to moderately disturbed freshwater lakes in Southeast Queensland (SEQ), but these are probably not appropriate or realistic for existing urban lakes, because they assume that urban lakes were once in an undisturbed state or that the influence of an urban catchment is minimal, as the reference condition values are based on natural lake ecosystems. Additionally, the use of reference condition values can lead to the assumption of an equilibrium, or constant state, while aquatic ecosystems are generally in a state of flux (Reeves & Duncan, 2009), driven by climatic, seasonal or external influences (e.g. stormwater runoff) and changing primary productivity. While urban lakes may demonstrate high quality and health when first constructed, this is generally considered a temporary state in the absence of effective management of the lakes (Leinster, 2006).

A further problem with interpreting a suite of physicochemical data, is how these may be used to represent the overall quality of the lake system. It is fairly common practice to express the overall quality as a single index or score, by applying a weighting to each monitored quantitative parameter that rates each parameter according to the perceived influence of that parameter on overall health (Sanchez et al., 2007; Bordalo et al., 2006; Fernández et al., 2004). Another approach is to map each measured parameter value (e.g. concentration of PO_4, or temperature) to a normalised index value by use of a graphical function (or lookup table) of parameter value (Cude, 2001).

Cude (2001) utilised eight water quality indicators (dissolved oxygen, biological oxygen demand, ammonia & nitrate nitrogen, total phosphorus, temperature, total suspended solids, pH and faecal coliforms) to establish a model which provided a scaled reference condition for streams in Oregon, USA. The model scored the geometric mean of the selected indicator scores which ranged from 0 – 100, with a score of 0 – 59 considered to be very poor, 60 – 79 poor, 80 – 84 to be fair, 85 – 89 to be good and 90 – 100 to be excellent. This modelling approach linked different indicators to present a concise summary as to the state of health in a stream and whether or not it was appropriate for recreational use. The results were easily communicable and could allow managers to prioritise at-risk areas (Cude, 2001).

Although there is debate regarding the accuracy of using an overall index, the interpretive simplicity of this approach has resulted in it being used in various forms by many agencies responsible for reporting on the quality and/or health of water systems (United Nations Environment Program, 2007; Hallock, 2002).

In order to evaluate the health status of water-based ecosystems and to pre-empt degradation, an improvement on assessing just physiochemical quality is to also assess biological indicators, which may be more representative of large-scale ecosystem health. A comprehensive study of assessing biological indicators was undertaken by Reiss and Brown (2005), who developed "a Florida Wetland Condition Index (FWCI) for isolated depressional forested wetlands in Florida". Although the detailed approach does not lend itself to be implemented as a routine monitoring program, the study concluded that the inclusion of biological indicators with physical and chemical indicators resulted in a useful index for biological integrity.

Presently, many urban lakes are degraded (Mitsch & Gooselink, 2000; Bayley & Newton, 2007) and, given that many constructed urban lakes are central features of residential developments, it is important to include social and public health indicator data in any evaluation or discussion of lake health (e.g. aesthetic satisfaction, community behaviours, microbial quality, algal and cyanobacteria risks). Social and health indicators can help assess the value the local residents place on the lake, what impacts community behaviour may have on lake health and what risks such lakes may be having upon the residents and thereby includes the local community as part of the "ecosystem".

The inclusion of a broader suite of lake-health indicators increases complexity. Evaluation of multi-parameter physicochemical data alone can be a difficult task in itself, and the inclusion of additional data types increases the complexity of the analysis, but carefully designed models can assist with the interpretation of large, complex datasets, identify problems that may yet occur and allow for pre-emptive and adaptable management.

While the approaches for reducing a large dataset to a single number are useful, they do not reflect the fact that the overall health of a system is better represented by the worst scoring parameters; individual parameter values that indicate a decline in health are typically "smoothed" and hidden when averaged against many other values. As agencies monitor more parameters, it is likely that the overall average index will become more stable and less sensitive to changes in individual parameters. Models for ecosystem health must be sensitive enough to detect when any part of the ecosystem becomes non-ideal.

The model described in this paper takes a similar approach to Cude (2001), but illustrates how a much larger, multi-disciplinary indicator set can be evaluated for overall ecosystem health, while maintaining a high degree of sensitivity to individual parameters. The model also illustrates how catchment-specific data may be used to more accurately describe the health of the constructed lake ecosystem.

Methods

The described approach aims to assist ecosystem managers in the development of a modelling tool that may be used to summarise and interpret large sets of disparate data that may result from monitoring practices. The resulting modelling tool itself is not dynamic as it does not make temporal predictions and is not based on differential equations. It is better described as a static model that may be used to summarise and simplify large volumes of disparate data for rapid interpretation and management intervention. The general approach described in this paper, or even an adaptation of the described model, is designed so it may be incorporated within explicit, temporal ecosystem models to provide a temporal "overview" of the health of the simulated ecosystem.

With respect to the constructed lake system to which this approach has been applied, three major groups of measurable environmental indicators (parameters) were identified: water quality; flora; and fauna.

A fourth group, social indicators, such as community "satisfaction" (which may perhaps be quantified through numbers of complaints to the managing authority), has not been explicitly included in the model. However, an "aesthetic index" has been derived from specific measured indicators in the water quality and floral groups. In this case, the derived aesthetic index has been used as a proxy for community satisfaction, and is presented in greater detail within the 'Model Description' section of this paper.

Water quality indicators

The water quality group of environmental health indicators include all the standard physical and chemical parameters that are typically measured in monitoring programs, such as temperature, turbidity, salinity, dissolved oxygen, pH and, in this case, additional lab-analysis values for various nutrient species.

For most water quality parameters, some measure of legislation exists which stipulates the quantitative range of values over which each water quality parameter is considered "normal" or healthy. In some cases, "normal" is defined to be based on any existing historical data for the system of interest. In the context of the lakes described in this paper, the relevant guidelines are Australian water quality guidelines, such as the Australian New Zealand Environment Conservation Council's (ANZECC) guidelines (2000) and Queensland Water Quality (QWQ) guidelines (DERM, 2009).

As some parameters aren't monitored routinely by agencies and, in some cases, the guidelines vary for different types of water bodies, the model is designed to be flexible

and easily adapted for different guidelines and different situations.

The first step in the approach is to quantify each individual parameter as being in a state that represents "ideal" or "poor" health. With this aim, the relevant water quality guidelines were used to develop functions with which each measured parameter value (e.g. concentration of PO_4, or temperature) is mapped to a Health Index (HI) for that parameter, the scale of which ranges from 0 (zero) to 1 (unity), where 0 represents the worst condition possible and 1 represents a healthy, perfect state.

$$HI_p = f_p(v_p) \qquad (1)$$

Where, for parameter p, the Health Index (HI) is related to the measured value (v) of the parameter by a function (f) that is specific to that parameter.

This approach is not new, but is often implemented through an arbitrary definition of the specific nature of the function. In many cases, the function is simply defined as a line that has been drawn on a graph of HI versus parameter value, based on nothing more than "instinct", although which is perhaps guided by water quality guidelines and experience. That is not to say that such graphically-defined functions are not effective, but they lack the rigorous and consistent approach that allows less experienced managers to generate functions suitable for their local requirements. A more mathematical approach is suggested by which, for each parameter, "regimes" of parameter values are identified that represent "terrible" (HI = 0), "perfect" (HI = 1) and "dynamic" (HI ranges between 0 and 1). Such piece-wise defined functions are readily described, tested, compared and applied in spreadsheets and models.

Two examples of functions for mapping quantitative water quality parameters to Health Indices, as used in the model described in this paper, are provided in Figure 1 (pH and Total Nitrogen).

The generation of these functions is not an arbitrary process, but follows a general procedure by which excessively low and high values are identified first, from either water quality guidelines, scientific literature or historical monitoring data. Using total nitrogen (NTOT) as an example, a complete absence of nitrogen prevents essential ecological processes from occurring, which results in a lack of vigour or productivity. Conversely, a high concentration of NTOT (1500 µg/L) infers an overly vigorous urban lake ecosystem, to the point where eutrophic conditions may be evident (i.e. very unhealthy). It follows that both a complete absence and a value exceeding 1500 µg/L are both unhealthy and are therefore mapped to a very low HI value. The "ideal" values are estimated from appropriate water quality guidelines or historical data and awarded an HI value of 1.

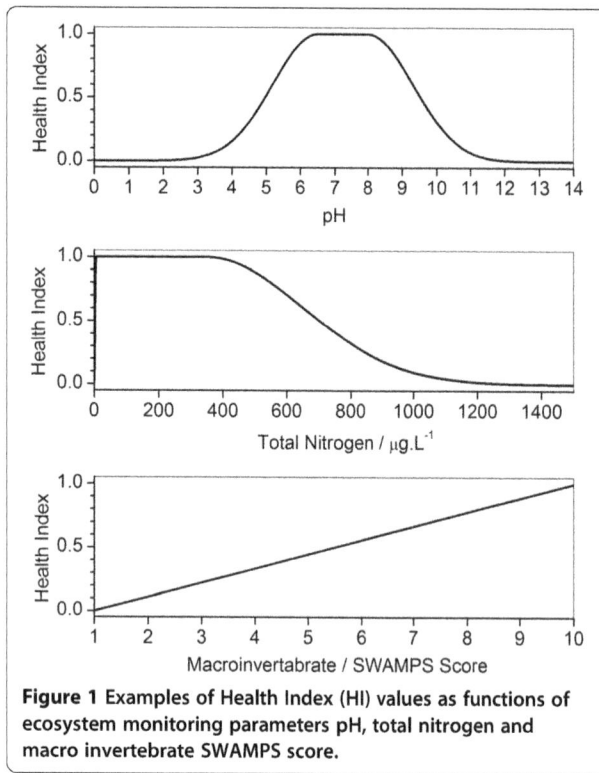

Figure 1 Examples of Health Index (HI) values as functions of ecosystem monitoring parameters pH, total nitrogen and macro invertebrate SWAMPS score.

The specific nature of the function as it changes between 0 and 1 is open to debate, but in the absence of specific empirical data, it is necessary to select a function that behaves in a predictable and reasonable manner. The simplest option for this purpose is a linear function, but this has the disadvantage of resulting in "shoulders" where the function "pieces" meet and, if the piece-wise defined function is not programmed carefully, may result in negative HI values or HI values greater than 1. This paper proposes that a simplified Gaussian function be used to describe all dynamic sections of the HI function (Equation 2).

$$HI = g(v, v_1, w) = e^{-0.5\left(\frac{v - v_1}{w}\right)^2} \qquad (2)$$

Where v is the measured value of the parameter (the variable), v_1 is the parameter value at which HI must equal 1, and w is the width coefficient, analogous to standard deviation.

The selection of this function for the dynamic regimes provides four significant advantages: i) it can be used in the same form for both cases where HI increases from zero to 1 or decreases from 1 to zero; ii) it cannot provide HI values less than zero or greater than one; iii) it provides a smooth continuum across the regimes of piece-wise defined functions and iv) the parameter-specific components (v_1 and w) are readily estimated in a systematic manner.

The width coefficient may be estimated by either of two methods. The preferred method identifies the specific

parameter value at which the health is classed as a "Fail", ie HI falls below 0.5. Having identified the boundary value of the parameter at which HI is one (v_1), identify the value at which the HI must fall to a value of 0.5 ($v_{0.5}$). The width parameter is then calculated by:

$$w = 0.85|v_1 - v_{0.5}| \qquad (3)$$

The alternative method for specifying the width parameter is to estimate the width of the domain over which the HI falls from one to near zero (Δv). The width parameter is simply estimated as $w = \Delta v / 3$. This is useful when rapid changes in HI are required. The use of this simplified Gaussian function is illustrated in Figure 2.

The complete list of HI functions used in the model described in this paper is provided in Table 1. Note that, for any given ecosystem, the list of parameters included in the model can be reduced or extended, depending on what parameters are actually measured and what data are available.

Ammonia presents a unique problem that must be discussed specifically, as it equilibrates between NH_3 and NH_4^+, the equilibrium being dependent on both pH and temperature. Although only the NH_3 form of ammonia is toxic, ammonia is typically measured as 'total ammonia nitrogen' (TAN), which is the sum concentration of NH_3 and NH_4^+. The fraction of TAN that is NH_3 can be determined as described in Körner et al. (2001), which is

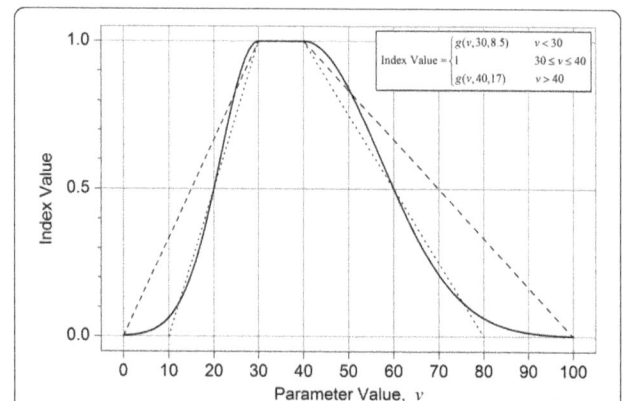

Figure 2 Illustration for the piece-wise definition of a function that maps parameter value to an index, over three distinct regimes. The solid line represents the simplified Gaussian functions as provided in the figure, in which the width parameter is defined so that the Index falls below 0.5 when the parameter value is less than 20 and greater than 60 and the "perfect" regime is between 30 and 40. The dotted lines illustrate linear functions that have equivalent "Index = 0.5" points as the Gaussian lines and the dashed lines illustrate linear functions that drop to zero at approximately the same points as the Gaussian lines. Note the lack of sharp "shoulders" in the Gaussian equations, in contrast to the linear functions.

Table 1 Model parameters that contribute to the OEHI

Parameter (p)	ANZECC guideline	Observed values	Health index mapping function HI =
Temperature (°C)	18 - 24	12 - 31	$\begin{cases} g(v, 18, 5) & v < 18 \\ 1 & 18 \leq v \leq 24 \\ g(v, 24, 2) & v > 24 \end{cases}$
pH	6.5 - 8	2 - 10	$\begin{cases} g(v, 6.5, 1.3) & v < 6.5 \\ 1 & 6.5 \leq v \leq 8 \\ g(v, 8, 1.3) & v > 8 \end{cases}$
Dissolved Oxygen (%)	90 - 110	0 -140	$\begin{cases} g(v, 100, 38) & v < 100 \\ g(v, 100, 60) & v \geq 100 \end{cases}$
NTOT: Total Nitrogen (µg/L)	0 - 350	100 - 1000	$\begin{cases} g(v, 5, 1.6) & v < 5 \\ 1 & 5 \leq v \leq 350 \\ g(v, 350, 300) & v > 350 \end{cases}$
NOx: $NO_2 + NO_3$ (µg/L)	0 - 10	0 - 150	$\begin{cases} g(v, 1, 0.3) & v < 1 \\ 1 & 1 \leq v \leq 20 \\ g(v, 20, 26) & v > 20 \end{cases}$
NH_3: Toxic Ammonia (µg/L)	0 - 10	0 - 90	$\begin{cases} 1 & v \leq 10 \\ g(v, 10, 9) & v > 10 \end{cases}$
Total Phosphorus (µg/L)	0 - 10	0 - 150	$\begin{cases} g(v, 1, 0.3) & v < 1 \\ 1 & 1 \leq v \leq 20 \\ g(v, 20, 26) & v > 20 \end{cases}$
Filtered Reactive Phosphate (PO_4) (µg/L)	0 - 5	0 - 100	$\begin{cases} g(v, 1, 0.3) & v < 1 \\ 1 & 1 \leq v \leq 15 \\ g(v, 15, 26) & v > 15 \end{cases}$
Chlorophyll-a (Chl-a) (µg/L)	0 - 5	0 - 70	$\begin{cases} 1 & v \leq 8 \\ g(v, 8, 18) & v > 8 \end{cases}$
EC: Electrical Conductivity (µS/cm)	Not Specific	0 - 200	$\begin{cases} 1 & v \leq 100 \\ g(v, 100, 300) & v > 100 \end{cases}$
TSS: Total Suspended Solids (mg/L)	Not Specific	0 - 150	$\begin{cases} 1 & v \leq 20 \\ g(v, 20, 25) & v > 20 \end{cases}$
Turbidity (NTU)	Not Specific	0 - 200	$\begin{cases} 1 & v \leq 20 \\ g(v, 20, 34) & v > 20 \end{cases}$
Macroinvertebrates (Signal2, SWAMPS)	N/A	3 - 5	$\frac{v-1}{9}$
Flora (Native, Non-weed)	N/A	0.3 – 0.5	Equal to the *Native* and *Non-weed* ratios as defined in text.

then used to determine actual NH_3 (toxic) concentration and subsequently the HI for ammonia (Table 1).

While this approach to processing the water quality indicators is similar to Cude (2001), the functions described for this specific application of the model were adjusted to relevant local and regional criteria based upon the response of a sub-tropical ecosystem to specific climatic conditions. The curves were also adjusted to suit urban lakes, which are not flowing systems, as were measured by Cude (2001).

Once all parameter values are mapped to their respective HI values, they are quantitatively summarised to create the overall water quality health index (WQHI). Cude (2001) suggests that all HI values be summarised by their geometric mean, as this is more sensitive to changes in individual variables than the arithmetic mean. It is important that the WQHI is not positively biased by large quantities of good HI values, and that very poor HI values (e.g. values of 0 which indicate a need for very urgent attention) are not "lost" through averaging with better HI values. For management purposes, it is important that the model quickly identifies when any one parameter goes bad, rather than emphasise good health. For these reasons, the WQHI is calculated from the geometric mean of the 'worst' three HI values. The use of the worst three HI ensures that the WQHI is not too sensitive to any single water quality indicator (such as if only the worst indicator was used), and is also not "buffered" or "smoothed" by using a large quantity (or all) of the indicators. The geometric mean is used instead of an arithmetic mean, as an HI of 0 for any of the three worst individual health indices results in an overall index of 0, which emphasises the fact that remedial action is required as a matter of some priority. The geometric mean also emphasises the worst case more than an arithmetic mean and is therefore more appropriate from a management perspective.

$$WQHI = \left[\prod_{i=1}^{3} \min_i \{HI_p\} \right]^{1/3} \qquad (4)$$

Where $\min_i\{HI_p\}$ is the i'th smallest value in the set of all HI values.

Floral indicators

Similarly to the WQHI, the overall floral, or vegetation health index (VHI), is calculated from individual HI values, but just two in this case. The first HI describes the proportion of all plants that are defined to be natives (as opposed to exotics), and the second HI describes the proportion of all plants that are defined to be non-weeds. Although highly structured and comprehensive survey methods have been developed for determining health of ecosystems with respect to floral indicator parameters, these are often too expansive and resource-intensive for regular surveys of small systems, especially if they are to be undertaken by local authorities with limited expertise and resources.

In this study, the method for the floral surveys was conducted through a simple visual assessment of the riparian and aquatic zones. For the case of urban lakes, a survey may be limited to riparian and visible floating and submerged aquatic vegetation. Floating aquatic macrophytes may be surveyed visually from the bank-side of an urban lake in a 10 m radius point transect (Buckland et al., 2001). At each survey point, the abundance of each observed plant species was recorded. Each species was then classified as being native or exotic (native species being preferable), and as being a weed species or non-weed species. The latter classification allows non-native species to be non-weeds; ie it simply characterises as favourable those species that do not invade and replace other species. This classification is often straight-forward as many authorities keep registers of known noxious weeds. The HI value of each classification is simply the proportion of all plants that are natives and non-weeds.

For example, if 70% of all the plants are natives, the native-vs-exotic HI value is 0.7; similarly for the non-weeds-vs-weeds HI value. Each HI for the system of interest is simply the mean value over all survey points. The VHI is calculated as the geometric mean of the two HIs

to represent the contribution of floral indicators to ecosystem health. The geometric mean is used for the same reasons as provided for the calculation of the WQHI.

$$VHI = (HI_{NATIVES} \times HI_{NONWEEDS})^{1/2} \qquad (5)$$

As floral populations do not tend to change as rapidly as, for example, water quality parameters, it is suggested that floral surveys do not need to be performed as frequently, but should be undertaken perhaps biannually or even seasonally. Exceptions to this may be following storms, weed harvesting or turnover events, after which some aquatic species may undergo exponential growth, such as *Salvinia molesta* and *Nymphaea mexicana*.

Faunal indicators

Similarly to the VHI, the overall faunal, or macroinvertebrate health index (MHI), is calculated from only two individual HI values, each being derived from separate macroinvertebrate survey approaches, as detailed by Chessman (2003) and Davis et al. (1999). The first approach is the Stream Invertebrate Grade Number – Average Level (SIGNAL2), which is typically used on Australian rivers and streams and the second approach is the Swan Wetlands Aquatic Macroinvertebrate Pollution Score (SWAMPS), which is typically applied within Australian wetland ecosystems. Each of these survey approaches are used to score the health of rivers, streams and wetlands based upon the diversity, type and abundance of macroinvertebrates (Chessman, 2003; Davis et al., 1999) and provide a numerical score from 1 to 10, 10 representing a healthy system. Because most constructed urban lakes have characteristics of both running (e.g. rivers and streams) and static (e.g. wetland) environments, it is prudent to apply both of these survey approaches. It is entirely possible to use alternative survey methods that are perhaps more suitable to a different ecosystem, or perhaps even characterise fauna in a similar manner to the approach described in this paper for floral health indices (native ratio and non-pest ratio), but discussion of this is beyond the scope of this paper. This study used SIGNAL2 and SWAMPS, because they are straight forward approaches that are recognised for Australian systems.

Each of the SIGNAL2 and SWAMPS scores are mapped linearly to HI values (Figure 1 and Table 1) and the MHI is equal to the geometric mean of the two HI values. The geometric mean is used for the same reasons as provided for the calculation of the WQHI.

$$MHI = (HI_{SIGNAL2} \times HI_{SWAMPS})^{1/2} \qquad (6)$$

Overall ecosystem health index

The geometric mean of the WQHI, VHI and MHI provides the overall ecosystem health index (OEHI), which, like the individual HI values from which it is ultimately calculated, has a range of 0 to 1.

$$OEHI = (WQHI \times VHI \times MHI)^{1/3} \qquad (7)$$

As the OEHI is a "summary" of many input parameters, it serves as a first indication that something is amiss. If, for example, the OEHI is calculated to have a value of 0.6, it suggests that the manager of the system should identify which specific aspect of the system is in need of attention. It may eventuate that the low OEHI is caused by a single high ammonia value, or a high turbidity value, which may not have been noted at the time of analysis and data entry. Alternatively, a low OEHI may be a cumulative result of several parameters slipping to low levels, which indicates a system in decline.

Community (overall aesthetic index)

In addition to characterising the overall health of the ecosystem from measurable environmental factors, the identification of specific drivers for remedial management is also a goal of this modelling approach. Pressure for remedial action may arise from adverse monitoring results, which will be flagged by the OEHI or, alternatively, it may arise from a community's attitude (for example aesthetic satisfaction) towards the lakes. It is important that any community surrounding an urban lake is considered in some way to be part of the lake ecosystem, as it is the attitudes of those communities that dictate how they interact with the lakes and this in turn can affect lake health. A community's perception of poor lake health is typically based on visual or odiferous observations, such as high turbidity or algal blooms, and the response will usually manifest itself in the form of direct complaints, either to the managing body or to media. Tracking complaints quantitatively can prove to be difficult and expensive, but if such data are available, these can be readily incorporated into this model by a similar normalisation approach as already described. In this project, no such data were available, so a proxy for community satisfaction has been developed. The overall aesthetic index (OAI) is based on the assumption that community attitudes towards a lake are primarily driven by specific, visually impacting indicators from the water quality and floral index groups. In terms of water quality, turbidity and chlorophyll-a are considered here to be the most appropriate indicators which contribute to the

OAI, as high turbidity is perceived to be "dirty" and high concentrations of chlorophyll-a are often viewed with concern, as the water is visibly green. From the floral group, the proportion of plants that are non-weeds was the parameter selected, as a dominance of weed species in Australian urban lakes may often be linked to floating macrophyte species, such as *Salvinia molesta* and *Nymphaea mexicana*. The specific functions that map these parameters to individual indices of aesthetic satisfaction (AI) are provided in Table 2.

The model assumes that community dissatisfaction (eg: formal complaints to lake council) is triggered when any one of these three indexes falls below 0.7, so to capture this, the OAI is defined to be the minimum individual AI from turbidity, chlorophyll-a and non-weed flora and is considered poor enough to generate community complaint when it falls below 0.7. While the critical value of 0.7 has been selected somewhat arbitrarily and is an assumption of community values and behaviour, this provides a conservative assessment which better allows for pre-emptive and adaptable management.

$$OAI = MIN\{AI_{Chl-a}, AI_{Turbidity}, AI_{NONWEEDS}\} \qquad (8)$$

Simulated pressure for management action

As already described, the pressure for management action (PMA) can come from either monitoring data, which describes the physical, biological and chemical health of the system, summarised as the OEHI, or from community complaint due to dissatisfaction with the appearance of the system, summarised as the OAI. Each of these indices may be used to estimate the degree of pressure for remedial action, although it is acknowledged that the specific nature of the function is open to debate. It is suggested here that lake managers must respond when OAI falls below 0.7 (to offset anticipated community complaint at an early stage) or when OEHI falls below 0.5 (which indicates a fail), so in these cases, the index value is assigned to be unity. Otherwise, the degree of pressure for remedial action is based solely on the OEHI. This is summarised by Equation 9:

$$PMAI = \begin{cases} 1 & OAI < 0.7 \text{ or } OEHI < 0.5 \\ 2 \times (1 - OEHI) & OAI \geq 0.7 \text{ and } OEHI \geq 0.5 \end{cases}$$

$$(9)$$

Table 2 Model parameters that contribute to the aesthetic index

Parameter (p)	Observed values	Aesthetic index mapping function AI =
Chlorophyll-a (Chl-a) (μg/L)	0 - 70	$\begin{cases} 1 & v \leq 30 \\ g(v, 30, 25) & v > 30 \end{cases}$ $g(v, 0, 34)$
Turbidity (NTU)	0 - 200	
Flora (Non-weed)	< 0.5	$g(v, 1, 0.34)$

An illustration of the relationships between all parts of the model is provided in Figure 3.

Assumptions and limitations

As with the development of any new modelling approach, there are several assumptions that have been made and these impose inherent limitations on the model (CRC for Catchment Hydrology, 2005).

In the model described in this paper, the specifications that define water quality parameters to be "healthy", or otherwise, are based on standard Australian guidelines (ANZECC, 2000; DERM, 2009) and on a set of monitoring data that was collected over four years. While the Australian guidelines provide reference condition values for slightly to moderately disturbed freshwater lakes, the monitoring data served to inform values more typical of the lake systems of interest. The use of monitoring data to develop specifications for individual water bodies is based on percentiles and is fully described in the above guidelines. The specific correspondence functions that map parameter values and HIs require further refinement in order for the model to be applicable to other ecosystems, particularly those with inherently different physicochemical and climatic conditions (e.g. saline lakes and temperate climates). As already described, the specific shape of each function must be defined with care

and, at the very least, undergo qualitative validation that the outcomes are reasonable.

Model sensitivity

A sensitivity analysis of the model was conducted to determine the sensitivity of the OEHI to the various health indicators and which indicators were most likely to trigger a need for remedial action. All exogenous parameters in the model (indicators) were set to optimum values (HI = 1) and the impact of each individual parameter within all groups (water quality, floral and faunal) was assessed from the lowest to the highest extremes. Examples are presented in Figure 4, which illustrate the sensitivity of the lake ecosystem to temperature, phosphate and turbidity respectively.

The sensitivity of the overall WQHI to total ammonia was considered in conjunction with temperature and pH. At higher temperatures and pH, ammonia is more persistent in its toxic, un-ionised form. As shown in Figure 5, the effect of total ammonia on ecosystem health became more pronounced as temperature and pH increased, with pH having the more substantial effect.

The indicators to which the OEHI was most sensitive to in terms of the full scale of each parameter (i.e. OEHI quickly degraded with increase/decrease over the full range) were found to be temperature, pH, turbidity, filtered

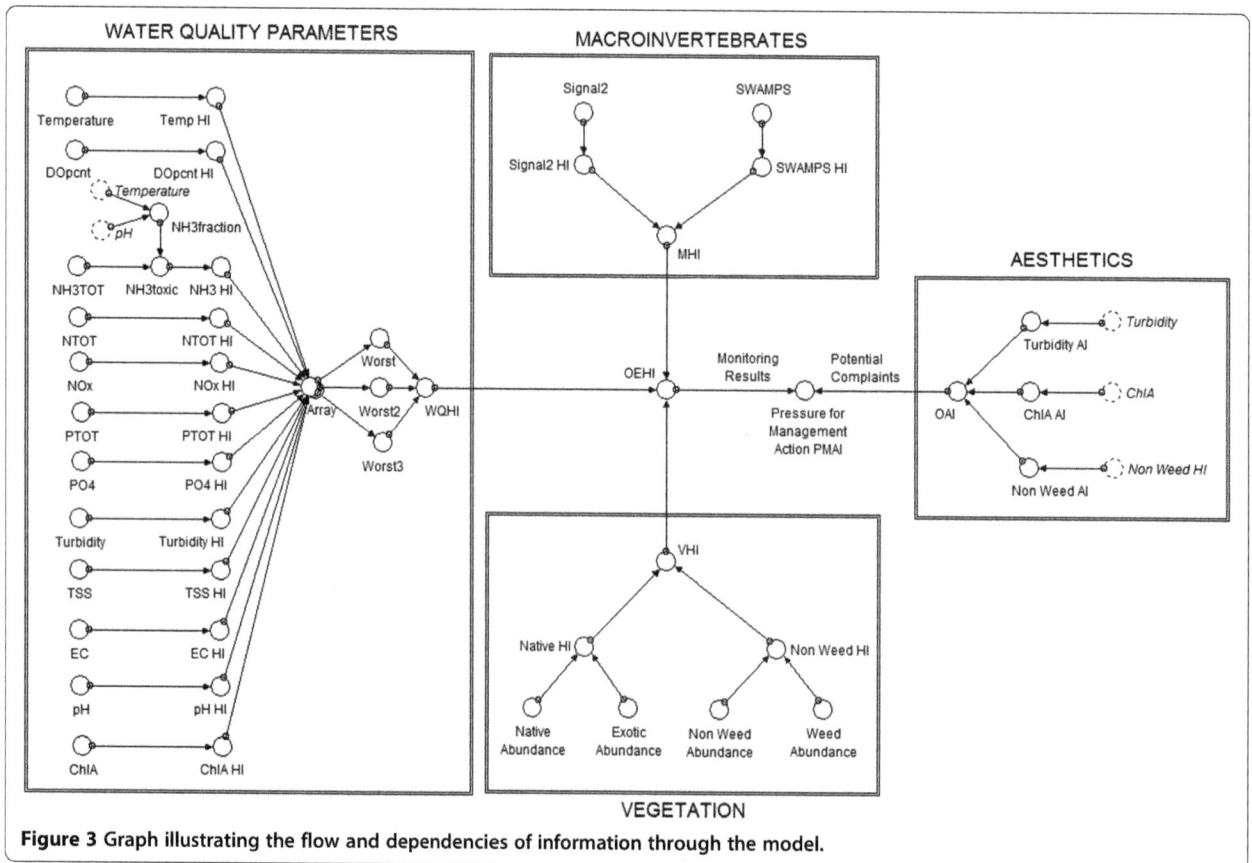

Figure 3 Graph illustrating the flow and dependencies of information through the model.

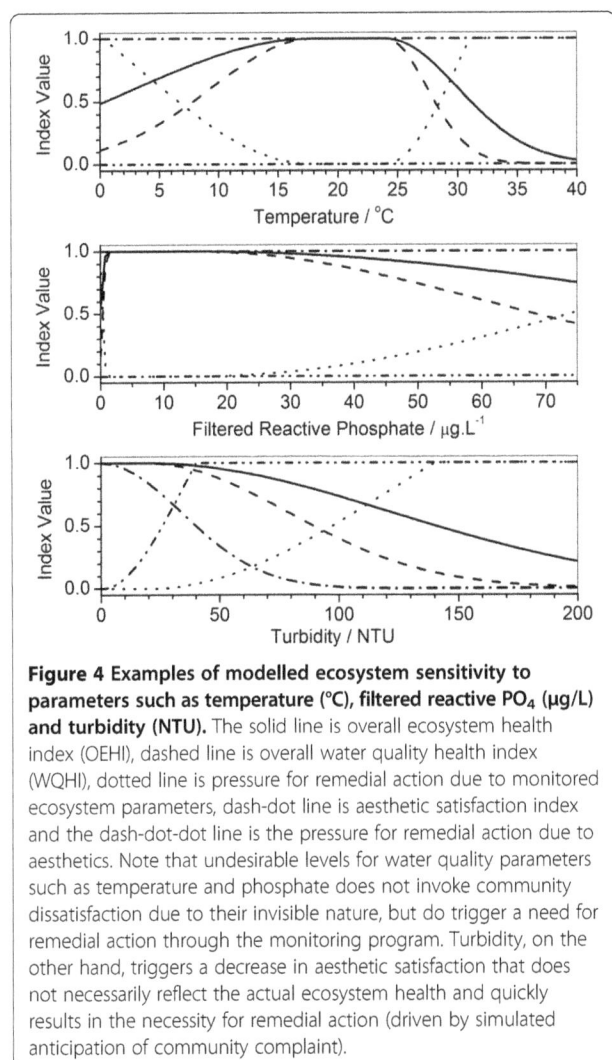

Figure 4 Examples of modelled ecosystem sensitivity to parameters such as temperature (°C), filtered reactive PO$_4$ (µg/L) and turbidity (NTU). The solid line is overall ecosystem health index (OEHI), dashed line is overall water quality health index (WQHI), dotted line is pressure for remedial action due to monitored ecosystem parameters, dash-dot line is aesthetic satisfaction index and the dash-dot-dot line is the pressure for remedial action due to aesthetics. Note that undesirable levels for water quality parameters such as temperature and phosphate does not invoke community dissatisfaction due to their invisible nature, but do trigger a need for remedial action through the monitoring program. Turbidity, on the other hand, triggers a decrease in aesthetic satisfaction that does not necessarily reflect the actual ecosystem health and quickly results in the necessity for remedial action (driven by simulated anticipation of community complaint).

reactive phosphate, total phosphorus, oxidised nitrogen, total nitrogen, total ammonia, chlorophyll-*a*, dissolved oxygen saturation and the ratio of weed to non-weed floral species. This was to be expected, as the sensitivity analysis is a reflection of the original value-to-HI functions. These indicators trigger more obvious and severe changes in the WQHI and aesthetic satisfaction, and subsequently the OEHI, and are therefore more prone to trigger a need for remedial action, which is essential if the model is to raise alerts when environmental values are not "normal".

In order to assess the sensitivity of the model to "normal" conditions, it was also necessary to determine the indicators to which the OEHI was most sensitive over the expected parameter range (20th percentile and 80th percentile from monitoring data). For example, the 20th and 80th percentile values for temperature were 18°C and 25°C respectively and the OEHI did not dramatically change in this range, indicating that it was not strongly sensitive to changes in temperature. The OEHI was most sensitive to change in dissolved oxygen saturation and to

a lesser extent, filtered reactive phosphate, over the expected range. While the remaining indicators did impact the overall ecosystem health index, such impacts were less obvious and occurred more gradually, particularly for conductivity and total suspended solids. This stable behaviour is essential if the model is not to raise alerts for fluctuations in environmental values that are "normal".

Results and discussion

In order to assess the applicability and effectiveness of the model with real-world data, a case study analysis was conducted by using the model to summarise an existing multi-disciplinary dataset that resulted from a full year of monitoring several urban lakes. This monitoring was conducted each month over 2009 (with the exception of March and April due to equipment failure) at the Chancellor Park estate in Sippy Downs, South East Queensland, Australia, which contains a series of ten linked, constructed urban lakes. Results from a specific lake in this estate, Lake 6, are presented here as a case study. Application of this model to other lakes yielded similar results to those presented here.

Site description

The Chancellor Park estate is comprised of medium to heavy residential development, with an extensive commercial development of approximately 13 ha which drains to the head of ten linearly-connected urban lakes. The lakes were constructed along the natural drainage line as a central feature of the residential development. All lakes in the system capture runoff from residential areas and the last lake in the chain discharges into a National Park. The primary functions of the lake system are twofold; the first is to capture and filter runoff from the surrounding urbanised catchment prior to discharging into a National Park. The second function is to create an aesthetic benefit for the community and provide an area of passive recreation. Of the lakes in the system, Lake 6 was selected to present as a case study as it is situated in the middle of the chained lake system and represents a typical constructed urban lake (surface area 1.67 ha, maximum depth 3.3 m, volume 26 ML, direct catchment area 67 ha).

Case study results

Water quality monitoring, for each of the water quality parameters listed in the model, was undertaken each month at the input, output and the vertical profile (i.e. the water column) at the centre. Mean values of each indicator over all measurements for each month were used to represent the value of that parameter over the entire lake. The values for the macro invertebrate and floral indicator groups were comprehensively surveyed only once during the year and therefore were represented as

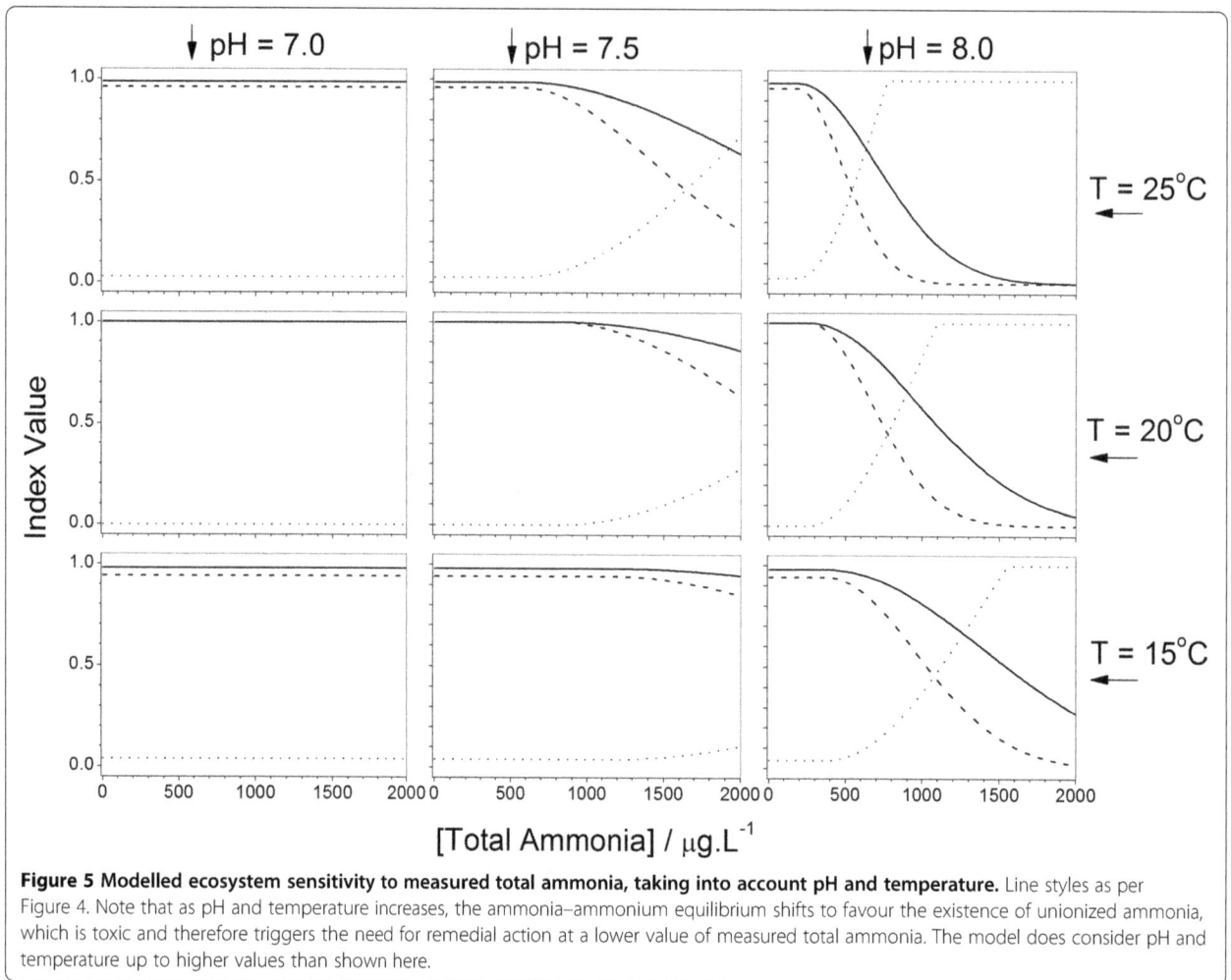

Figure 5 Modelled ecosystem sensitivity to measured total ammonia, taking into account pH and temperature. Line styles as per Figure 4. Note that as pH and temperature increases, the ammonia–ammonium equilibrium shifts to favour the existence of unionized ammonia, which is toxic and therefore triggers the need for remedial action at a lower value of measured total ammonia. The model does consider pH and temperature up to higher values than shown here.

being constant for the year (no temporal study on these was conducted). In retrospect, although potentially expensive, it is recommended that faunal and floral studies be undertaken at least twice per year to identify any seasonal changes, or more frequently if any rapid changes are observed (Figure 6).

Lake 6 demonstrated variability in the WQHI, with some distinction between the WQHI and the OEHI, the latter including the VHI and MHI. Temperatures, NOx, PO$_4$ and turbidity were the water quality parameters that had the most influence on the WQHI. In July 2009 and November 2009, the WQHI was less than 0.1 which, unto itself should raise alarm for managers, and also significantly impacted the OEHI, which is the index that managers would be watching most carefully. The low WQHI in July 2009 was a result of very high turbidity and PO$_4$ caused by construction works occurring upstream from the lake. In general of 2009, it is clear that Lake 6 had a poor OEHI, with a mean of 0.28 from January to December 2009.

The aesthetic satisfaction index was consistently low, driven by the high turbidity and a widespread presence of exotic floating weed species (*Nymphaea mexicana*, also known as the Yellow Waterlily). The lack of aesthetic satisfaction in the model anticipated that community pressure for remedial action would be forthcoming. Regardless, the very low WQHI should trigger a management response even in the case that aesthetic satisfaction had been acceptable. The low OEHI score indicates that the health of the lake should be a high management priority.

The model proved to be sensitive enough to automatically identify months in which lake health decreased or improved; closer analysis of the specific monitoring parameters for those months verifies that the model accurately reflected the health status of the lake with respect to those parameters.

Although the model indicated that there was little temporal variation between the overall ecosystem health and water quality indexes, this was likely the result of the static (non-temporal) values entered for the floral and faunal index groups. Although floral communities are less subject to observable change, bi-annual or seasonal surveys will serve not only to identify changes in floral communities, but also the effectiveness of related management strategies

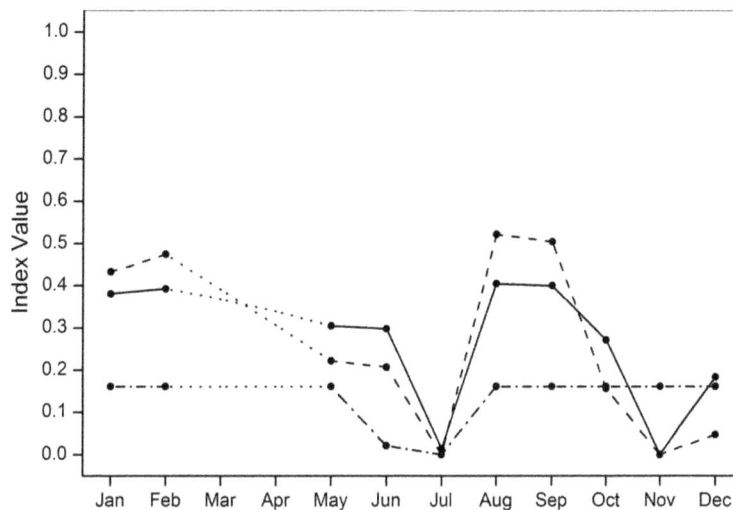

Figure 6 Temporal variability of the WQHI and OEHI in Lake 6 from January 2009 to December 2009. Line styles as per Figure 4, except the dotted lines represent the fact that water quality was not monitored in March and April. Macroinvertebrate and Vegetation indices are not included in this graph as they were modelled as being constant through the year with values of 0.33 and 0.38 respectively (based on survey data). Simulated aesthetic satisfaction was very low for the entire year, due to high population of weed species, with the June-July decrease due to increased turbidity from retro-fitting works on another lake upstream. Not included in the graph is that the model indicated an absolute need for remedial action for every month of the year, due to both low OEHI and aesthetic index.

(e.g. weed control). Faunal communities are also subject to change as a result of seasonal shifts and disturbance events, so seasonal macroinvertebrate surveys would allow stakeholders to establish patterns in faunal composition in relation to the other environmental indicators. The inclusion of more robust and variable floral and faunal data would serve to increase the sensitivity of the model to these parameters.

Model extension

Although the model provided a novel inclusion of the community in the sense that resident's attitudes were simulated by proxy, it was limited in scope and scale. A broader spectrum of community-linked satisfaction indicators may yield a more sensitive index to reflect community behaviours, interaction, satisfaction and/or attitudes towards urban lake ecosystems. For example, parameters may be included that quantify the degree and frequency of physical contact with the lakes, which could be linked to water quality parameters and expressed as a human-health risk index, but such data are not always readily available and further observational or survey type research would need to be undertaken.

Other, more physical parameters that may be included, although only limited data are often available, are microbial cyanobacteria and specific faunal hazards (e.g. mosquito larvae). These relate not only to ecosystem health, but may also be adapted in combination with quantified exposure assessments to provide an estimate of risk to the public and provide a third driver for remedial action.

Looking beyond this model itself, this simple and adaptable approach for calculating an overall index of ecosystem health directly from measured data, makes aspects of this model highly suitable for inclusion within larger, dynamic models such as those that simulate water quality parameters as a function of time.

Conclusion

This paper details a flexible approach by which a model may be developed as a means for calculating a single index value (the OEHI) to describe the overall health of a lake ecosystem, which is readily understandable to a range of stakeholders. This overall index is derived from quantifiable indicator data pertaining to physicochemical water quality parameters, floral surveys, and faunal surveys, which can be collected through a multi-disciplinary investigation of the ecosystem. The described approach is unique for several reasons namely: i) the suggested use of a simplified Gaussian function to define the dynamic regimes of mapping parameter-to-HI (makes extension of the model more accessible to non-experts); ii) the use of only the 'worst' three physicochemical water quality HI values for the calculation of the WQHI makes the model more sensitive when individual parameters become poor; iii) floral and faunal indexes are included in a simple manner as components of ecosystem health, with equal weighting as physicochemical water quality parameters; iv) simulated community satisfaction is included as a component of ecosystem health and "complaint" is generated whenever any single aesthetic indicator drops below a threshold value. The model also recognises two "drivers" for remedial action by the managing authority, specifically: monitoring data (usually undertaken by system managers), and community satisfaction. Each driver is handled differently and triggers a need for management action independently. Although the approach described in the paper is not limited to any particular indicator set and can be modified or extended to include additional indicators, it does require knowledge regarding what is considered "healthy" for any given parameter in the system and the specific correspondence functions that map parameter values and HIs require further refinement in order for the model to be applicable to other ecosystems, particularly those with inherently different physicochemical and climatic conditions.

The included case study presents a specific application of the model in which it is calibrated for urban lakes in South East Queensland, Australia, using Australian guidelines and local monitoring data. Through this, it is demonstrated that the described approach is effective for simplifying large datasets and responds to changes in individual parameters that impact the overall health of the system.

Competing interests

The authors declare that they have no competing interests.

Authors' contributions

AW assisted with collection of field data, designed the modelling approach, prepared and edited much of the manuscript. CW collected the field data, helped develop and test early drafts of the model, performed the literature review, prepared and edited the manuscript. AW, NT, PD and AR secured funding for the project, assisted with the design of the monitoring program and the interpretation of subsequent data, and helped to draft the manuscript. All authors read and approved the final manuscript.

Acknowledgements

The authors gratefully acknowledge and thank the Sunshine Coast Council for the financial support provided towards a monitoring study of lake ecosystem health, from which some data are incorporated in the case study presented in this manuscript.

Author details

[1]University of the Sunshine Coast, Locked Bag 4, Maroochydore DC, QLD 4558, Australia. [2]School of Ocean Sciences, Bangor University, Anglesey, LL59 5AB, United Kingdom.

References

ANZECC & ARMCANZ (2000) Australian and New Zealand guidelines for fresh and marine water quality. Australian and New Zealand Environment and Conservation Council & Agriculture and Resource Management Council of Australia and New Zealand, Canberra

Bayley ML, Newton D (2007) Water quality and maintenance costs of constructed waterbodies in urban areas of south east Queensland. Proceeding from 5th international water sensitive urban design conference- 'rainwater and urban design. Pub: Engineers Australia, Sydney. ISBN 1877040614

Bordalo AA, Teixeira R, Weibe WJ (2006) A water quality index applied to an international shared river basin: the case of the Colorado river. Environ Manage 38:910–920

Buckland ST, Anderson DR, Burnham KP, Laake JL, Borchers DL, Thomas L (2001) Introduction to distance sampling: estimating the abundance of biological populations. Oxford University Press, Oxford

Chessman B (2003) SIGNAL 2 – A scoring system for macro-invertebrate ('water Bugs') in Australian rivers, monitoring river heath initiative technical report No. 31. Commonwealth of Australia, Canberra

CRCfor Catchment Hydrology (2005) General approaches to modelling and practical issues on modelling choice. Series on Modelling Choice 1:1–21

Cude CG (2001) Oregon water quality index: a tool for evaluating water quality management effectiveness. J Am Water Resour Assoc 37:125–137

Davis J, Horwitz P, Norris R, Cheesman B, McGuire M, Sommer B, Trayler K (1999) Wetland bioassessment manual (macroinvertebrates). National wetlands research and development program. Environment Australia, Canberra

Department of Environment and Resource Management - DERM (2010) Monitoring and sampling manual 2009. Queensland Government, Ver. 2, Available at http://www.ehp.qld.gov.au/water/pdf/monitoring-man-2009-v2.pdf [accessed 08 Feb 2013]

Fernández N, Ramírez A, Solano F (2004) Physicochemical. Water Quality Indices – A Comparative Review Bistua: Revista de la Facultad de Ciencias Básicas 2(1):19–30

Hallock D (2002) A water quality index for Ecology's stream monitoring program, vol Publication number 02-03-052. Washington State Department of Ecology, Available at https://fortress.wa.gov/ecy/publications/summarypages/0203052.html [accessed 08 Feb 2013]

Karr JR, Chu EW (1999) Restoring life in running waters: better biological monitoring. Island Press, Washington, D.C.

Körner S, Das SK, Veenstra S, Vermaat JE (2001) The effect of pH variation at the ammonium/ammonia equilibrium in wastewater and its toxicity to Lemna gibba. Aquat Bot 71:71–78

Leinster S (2006) Delivering the final product - establishing vegetated water sensitive urban design systems. In: Ana D, Tim F (eds) Book of proceedings: 7th International Conference on Urban Drainage Modelling and the 4th International Conference on Water Sensitive Urban Design: book of proceedings /. Monash University, Melbourne., Australia, pp 1–8

Likens GE, Walker KF, Davies PE, Brookes J, Olley J, Young WJ, Thoms MC, Lake PE, Gawne B, Davis J, Arthington AH, Thompson R, Oliver RL (2009) Ecosystem science: toward a new paradigm for managing Australia's inland aquatic ecosystems. Mar Freshw Res 60:271–289

Lloyd S, Wong THF, Chesterfield CJ (2002) Water sensitive urban design – a stormwater management perspective. CRC for Catchment Hydrology 2000:1–44

Mitsch GA, Gooselink JE (2000) Wetlands, 3rd edn. John Wiley and Sons, New York

Rapport DJ (1998) Defining ecosystem health. In: Rapport DJ, Costanza R, Epstein P, Gaudet C, Levins R (eds) Ecosystem health. Blackwell Science, Oxford, UK

Reeves GH, Duncan SL (2009) Ecological history vs. social expectations: managing aquatic ecosystems. Ecol Soc 14(2):8

Reiss KC, Brown MT (2005) Pilot Study—the Florida Wetland Condition Index (FWCI): Preliminary Development of Biological Indicators for Forested Strand and Floodplain Wetlands. Howard T. Odum Center for Wetlands, University of Florida, Gainesville, FL

Sáncheza E, Colmenarejoa MF, Vicenteb J, Rubiob A, Garcíaa MG, Traviesoc L, Borja R (2007) Use of the water quality index and dissolved oxygen deficit as simple indicators of watersheds pollution. Ecol Indic 7:315–328

United Nations Environment Programme (2007) Global drinking water quality index development and sensitivity analysis report., . ISBN 92-95039-14-9

Walker C, Tindale N, Roiko A, Wiegand A, Duncan P (2010) An ecosystem health approach to assessing stormwater impacts on constructed urban lakes. Refereed proceedings from the National Conference of the Stormwater Industry Association, Sydney, Australia

13

Variation in ecosystem service values in response to land use changes in Zhifanggou watershed of Loess plateau: a comparative study

Jin Si[1], Fuzhan Nasiri[1*], Peng Han[2] and Tianhong Li[2]

Abstract

Background: Anthropological activities could lead to various land use changes. This subsequently exerts impacts on the Ecological status as the land is the carrier of Ecosystems and their services. Ecosystem Services Values (ESVs) are monetary assessment of ecosystems services. The choice of ecosystem services valuation methods highly depends on the type and conditions of the ecological environment.

Results: In this paper, Zhifangou watershed, a watershed with fragile environments in Loess Plateau, is analyzed in terms of the historical changes of land uses and their impact on Ecosystem Services Values (ESVs). The analysis presents that the watershed had seen a large cover of forest (689.8 hectare) and grassland (97.08 hectare) until 1938, which were substituted with a cropland of 663.56 hectare in the following years, and then a gradual recovery of natural land from 1978. During these years, the human activities have ranged from little disruption, to excessive cultivation, and finally to an integrated management of the watershed. The ESVs were highest at 121.77×10^4 in 1938, lowest at 43.75×10^4 in 1958, and then rebounded to 113.44×10^4 in 1999.

Conclusions: The analysis reveals that the values of soil formation and retention, biodiversity protection, and climate regulation have been associated with the largest share of the total ESVs. Among them, soil formation and retention was recognized as the most impacted service by land use changes.

Keywords: Land use change; Ecosystem services; Valuation; Anthropological activities; Zhifanggou watershed

Background

The unprecedented population growth and urbanization over the past century has changed the face of the planet as a result of many land use changes. On the other hand, the land supplies mass and energy, which are the main drivers of economic development (Darwin et al. 1996) as well as various Ecosystem Services (ES). The latter are defined as the conditions and processes through which natural ecosystems and the species that comprise them, sustain and fulfill human well-being (Daily 1997). Some examples of such services provided by land cover are biodiversity, water filtration, retention of soil, and air purification, just to name a few (Nasiri and Huang 2007). As such, there exists a close correlation between land uses and the state of ecological environment (Styers et al.

2010). An inappropriate land allocation and use may lead to significant degradation of local and regional ecological services (Collin and Melloul 2001).

Ecosystem Services Values (ESVs) are monetary assessment of ES. Costanza et al. (1997) were first to propose a list of ecosystem valuation coefficients, estimating monetary values of 17 ecosystem services generated by 16 different biomes. Although the study has faced criticisms about double counting and/or underestimating or overestimating the listed values (Fu et al. 2011), it paved the path to the science of ES valuation. Turner et al. (1998) have proposed a refined set of coefficients. On that basis, Xie et al. (2003) developed a similar set of coefficients for valuation of ecosystem services in China by increasing the weight of ecosystem services related to agriculture and decreasing that of wetlands.

As the choice of ecosystem services valuation methods highly depends on the type and conditions of the ecological environment, most of the ESV methods suggested

* Correspondence: f.nasiri@ucl.ac.uk
[1]Bartlett School of Graduate Studies, University College London (UCL), London WC1H-0NN, UK
Full list of author information is available at the end of the article

in the literature are case-specific. Kreuter et al. (2001) estimated ESVs for urban sprawl in the San Antonia area in Texas. Martínez et al. (2009) assessed the impacts of land use changes on the ES provision in tropical montage cloud forests. Li et al. (2010) investigated ES changes in response to land use changes caused mainly by climatic changes on the Zoige Plateau of Tibetan Plateau in China. Zhang et al. (2011) conducted a comparative study of ESVs based on land use changes in HaDaQi industrial corridor, in Heilongjiang Province of China.

Building on previous research by Xie et al. (2003) and Zhang et al. (2011) on the ESVs, in this paper we propose an ESV analysis of Zhifanggou watershed on the basis of historical data on land use changes. We have based our calculations on Xie et al. (2003) and Zhang et al. (2011) studies as they adjusted the ESV coefficients such that to be specific to China's ecosystems. The ESVs analysis of this area is of particular importance as it historically witnessed a period of destruction followed by a recovery as a result of employing an integrated watershed management framework. We analyze the changes in land-use and the associated ESVs, demonstrating the relationships between ecosystem changes and anthropological activities. The results may also provide insights about the impact of integrated watershed management programs, emphasizing the importance of ecosystem services conservation.

Study area

Zhifanggou watershed (109°13′46′′~109°13′45′′E, 36°42′42′′~36°46′28′′N) is located in An'sai county of Shaanxi province in China. It is part of a hilly and gully region of Loess Plateau (Zhang et al. 2008). The size of the area, as shown in Figure 1, is estimated as 8.39 km². The region is mainly affected by a semi-arid climate with an average annual rainfall of 541.2 mm. Rainfall mainly occurs from July to September and heavy storm often leads to serious soil erosion in rainy seasons (Jiang and Zheng 2004).

Prior to 1938, the ecosystem of Zhifanggou watershed was seldom disturbed by human activities. However, from the 1938 to 1960s local vegetation was destroyed heavily and soil erosion was intensified quickly due to population expansion and cultivation of large areas of slope land. Consequently, this was led to a fast deterioration of ecological environment (Wen et al. 2006). From 1973, an integrative watershed management program was started. This program was initiated to plan and manage mountains, rivers, fields and forests in the watershed under a unified framework. The main objectives were to control the soil erosion on different slopes, to aim for the recovery of the grassland, to protect the forest areas, and to promote the use of improved seeds and water-saving measures in farming. Zhifanggou watershed was selected as an experimental site for key research and development programs on integrative management of Loess Plateau (Li 1995). After 20 years of integrative watershed management, the area of the watershed that is free from soil erosion is up to 12.21 km², which is 81.4% of the whole area. As a result, the ecosystem of the watershed was gradually recovered to its previous natural state, benefiting from returning forest and grassland areas (Han et al. 2009).

The historical land use datasets of Zhifanggou watershed is divided through five milestone years of 1938, 1958, 1978, 1987 and 1999, indicating and capturing the ecological changes. The datasets used in this study were provided by the Key Laboratory for Water and Sediment

Figure 1 Location of Zhifanggou watershed.

Sciences in Peking University in China. Data from 1958, 1978 and 1987 were acquired by aerial photograph interpretation aided by ground survey and existing land use maps. Data for 1999 were obtained through monitoring, while those of 1938 were based on records of interview and surveys.

We have reclassified the above datasets into seven categories of cropland, woodland, orchard land, grassland, water body, build up and barren lands (see Table 1). Data for woodland and orchard were combined as forest category. Land use maps of (1:5000) were calibrated and coded in ARCMAP 9.2 (ESRI 2013) for the subsequent spatial analysis and ecosystem services valuations.

Methods

Xie et al. (2003) extracted the coefficients of ecosystem services per hectare of terrestrial ecosystems in China. Adopting this model, we had to make the following minor adjustments to have it customized for Zhifanggou watershed as presented in Table 2:

(1) Due to the resolution of land use datasets, classifying the land types of wetland and water is of difficulty. In this sense, this two land types are considered as one type in this study, assigning with the mean coefficients of wetland and water. This way of dealing with multiple categories of water bodies has been advocated by Zhang et al. (2011).

(2) As 1990s data is regarded as a baseline for analysis in China's statistics, the average market price of agricultural produce per hectare was calculated using the 1990s data for major crops, including rice, wheat, corns and soybeans according to National Bureau of Statistic of China (NBSC 1996–1999).

The value of food production services of agricultural produce per hectare was computed, according to Xie et al. (2003), as one-seventh of the market price of agricultural produce. The estimated monetary value of

agricultural produce per hectare was estimated as \$528.08 (i.e. an equivalent of RMB Yuan 2525.92 according to the average exchange rate at the end of 1990s). Consequently, the value of the food production service provided by agricultural lands per hectare was estimated as \$75.44 (i.e. an equivalent of RMB Yuan 528.08). Finally, the value coefficients of ecosystem services were calculated by multiplying the food production values by their service to production ratios (see Table 3), as follows:

$$VC_{kf} = R_{kf} \times V_F \tag{1}$$

where,

VC_{kf}: The ecosystem service value for ecosystem function f in land use type k (\$ per hectare per year).
R_{kf}: Ratio of ecosystem service to food production values for function f in land use type k.
V_F: Food production values of agriculture land per area per year, which is \$75.44 per hectare for 1990 base year.

Having the ecosystem service value of one unit area for each category of land use extracted, the service value for each category of land use and the service value of each function are obtained from Equations. (2) and (3):

$$ESV_k = \sum_f A_k \times VC_{kf} \tag{2}$$

$$ESV_f = \sum_k A_k \times VC_{kf} \tag{3}$$

$$ESV = \sum_k \sum_f A_k \times VC_{kf} \tag{4}$$

where

ESV_k: Ecosystem service value of land use category "k".
ESV_f: Value of ecosystem service function type "f".
ESV: Total ecosystem service value of the watershed
A_k: Area (hectare) of land use category "k".

Since the biomes we used as proxies for land use categories were not perfect matches, and as a result of the uncertainties about the correspondence of proxies for land use types and the accuracy of Xie's value coefficients embedded in the estimation, we have conducted a sensitivity analysis to determine the impact of variations in the value of coefficients on ESVs estimations. For each ESV analysis, we then calculate a coefficient of sensitivity (CS) using the following formula that resembles the standard concept of elasticity in economics (Kreuter et al. 2001):

Table 1 Land use categories in Zhifanggou watershed

Categories		Definition
Cropland		Paddy field, glebe field, irrigable land and vegetable field
Forest	Orchard	Orchard occupied by fruit, tea, mulberry and rubber trees
	Woodland	Arbor, bamboo, bush forest and protection forest along the roads, railways and costal lines
Grassland		Natural grassland and man-made grassland
Water body		Wetlands, rivers, reservoirs fishery and lakes
Buildup		Lands used for industrial, commercial, residential, transportation ends
Barren land		Lands unused or difficult for any use

Table 2 Ratio between ecosystem service value and the values of food production provided by agricultural lands (according to Xie et al. 2003)

Ecosystem service	Forest	Grassland	Cropland	Water body	Barren lands
Gas regulation	3.5	0.8	0.5	0.9	0
Climate regulation	2.7	0.9	0.89	8.78	0
Water supply	3.2	0.8	0.6	17.94	0.03
Soil formation and retention	3.9	1.95	1.46	0.86	0.02
Waste treatment	1.31	1.31	1.64	18.18	0.01
Biodiversity protection	3.26	1.09	0.71	2.495	0.34
Food	0.1	0.3	1	0.2	0.01
Raw material	2.6	0.05	0.1	0.04	0
Recreation and culture	1.28	0.04	0.01	4.945	0.01

$$CS_{kf} = \frac{\left(ESV^{(j)} - ESV^{(i)}\right)/ESV^{(i)}}{\left(VC_{kf}^{(j)} - VC_{kf}^{(i)}\right)/VC_{kf}^{(i)}} \quad (5)$$

where i and j represent the initial and adjusted values, respectively. If $CS_{kf} > 1$, then the estimated ecosystem value is elastic (i.e. highly sensitive) with respect to changes in value of VC_{kf} coefficient, but if $CS_{kf} < 1$, then the estimated ecosystem value is inelastic (i.e. not sensitive) with respect to changes in value of VC_{kf} coefficient.

Table 3 Ecosystem service value coefficients for each land use category obtained from equation (1) ($ per hectare per year[*]; revised from Xie et al. 2003)

Ecosystem service	Forest	Grassland	Cropland	Water body	Barren lands
Gas regulation	264	60	38	68	0
Climate regulation	204	68	67	662	0
Water supply	241	60	45	1353	2
Soil formation and retention	294	147	110	65	2
Waste treatment	99	99	124	1372	1
Biodiversity protection	246	82	54	188	26
Food	8	23	75	15	1
Raw material	196	4	8	3	0
Recreation and culture	97	3	1	373	1
Total	1648	546	521	4099	32

[*]The coefficients are adjusted to 1990 $ according to the exchange rate between RMB Yuan and $.

Results

Changes of land use

The data on historical land use changes (as shown in Figure 2) are summarized in Table 4. In 1938, the watershed was comprised of a large area of forest (689.8 hectare), an area of grassland (97.08 hectare) and a small area of cropland (52.19 hectare). However, this condition was interrupted in 1958, with increased farming activities, causing the area of forest to drop to 1.09 hectare. The cropland and grassland areas have increased to 663.56 hectare and 136.69 hectare, respectively. Meanwhile, the area of build up, water body and unused land had also increased to 1.51, 3.32 and 32.89 hectare, respectively. Then in 1978, as a result of introducing an integrated watershed management program, the area of forest and grassland increased to 16.98 hectare and 615.32hectare, respectively, and with decline of cropland area to 170 hectare. Until 1987, the area of forest soared to 562.11 hectare due to further transformations. Finally in 1999, the watershed was covered with a forest area of 618.1 hectare, returning to its status in 1938. Since then, the watershed consists of 93.15 hectare of cropland, 121.85 hectare of grassland and 5.97 hectare of build up (without any land use categories of water body and unused land).

A comparative analysis of the land uses over the years indicates that each land use type went through dramatic changes in four periods of 1938-1958, 1958-1978, 1978-1987 and 1987-1999. The area of forest decreased by 4.99% annually from 1938 to 1958, and increased by 72.89%, 356.71% and 0.83% annually in the next three periods. Contrastively, the area of cropland increased by 58.57% in the first period, and constantly decreased by 74.38%, 14.59% and 2.99% annually in the later periods. In case of grassland, the most significant decrease took place between 1978 and 1987 (84.27% annually), when the forest area became mature again. As for build up land uses, there has been a constant increase in all periods, with a dramatic increase of 11.81% annually in the last period. This was subject to a fast growth of urbanization during this period. The water bodies and barren lands could only be seen between 1958 and 1987.

Changes of ESVs

ESVs for each land category and the overall ESVs were obtained, according to Equations 2 and 4, for each period as reported in Table 5. The total ESVs for Zhifanggou watershed was estimated at 121.77×10^4 in 1938, 43.75×10^4 in 1958, 46.73×10^4 in 1978, 107.01×10^4 in 1987 and 113.44×10^4 in 1999. From 1938 to 1958, the total ESVs dropped by 64.07%, or 3.20% annually. From 1958 to 1978, the total ESVs increased by 6.81%, or 0.34% annually. And from 1978 to 1987, the total ESVs rose largely by 129%, or 14.33% annually. Finally from 1987 to 1999, the total ESVs retained a slow increase by 6.01% (or 0.5% annually).

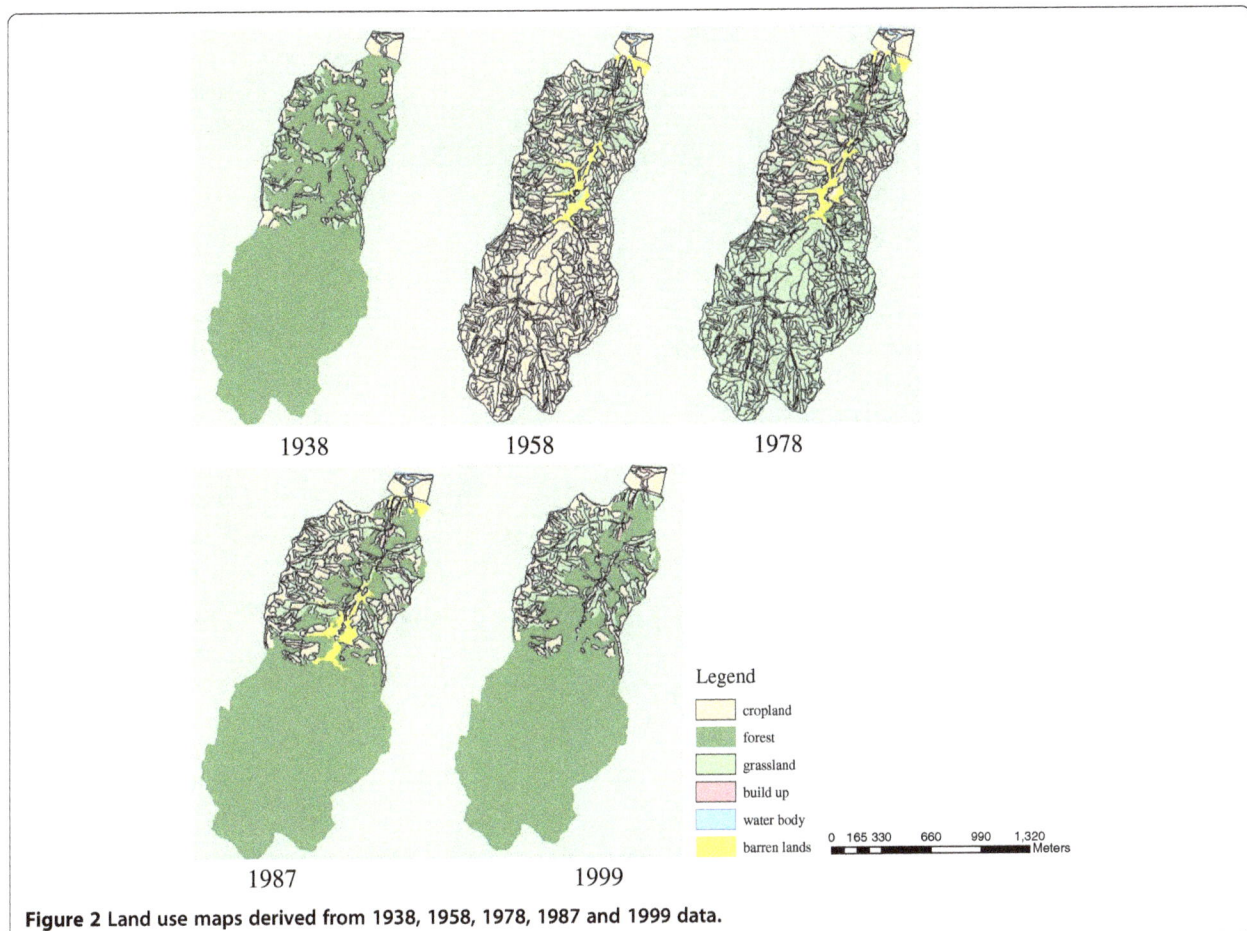

Figure 2 Land use maps derived from 1938, 1958, 1978, 1987 and 1999 data.

From Table 5, among various types of land, forest produced the largest ESVs due to its highest ecosystem services coefficients and large areas. Among other types of land use, the grassland provided the watershed with higher ESVs. Contrastively, the cropland supplied the lowest ESVs, except the period beginning in 1958 when it was associated with a large area. Although, the water body could provide high ESVs in theory, but this was insignificant in Zhifanggou watershed with small water body areas. The sum of ecosystem service values provided by forest, grassland and cropland accounted for over 90% of the total value, indicating their main role in ecosystem service provision in Zhifanggou watershed.

According to Equation 3, the ecosystem service values provided by individual ecosystem functions were also estimated (see Table 6). The ranking for all the functions were based on its contributions to the total ESVS. The impact of each ecosystem function on total value of the ecosystem services was represented by an upward arrow "↑" for increasing (or reinforcing) impacts, downward arrow "↓" for decreasing (or balancing) impacts, and a dash "–" for no or negligible impacts. The overall ranking for the studied periods were estimated based on the average impact of individual ESVs on total ESVS. These impacts, from high to low, are: soil formation and retention, biodiversity protection, climate regulation, gas regulation, water supply, climate regulation, waste treatment, raw material, recreation and culture, and food. The share of soil formation and retention service was the highest (about 20%), followed by biodiversity protection, gas regulation and climate regulation (each around 14%), all of which are major ecosystem functions mainly derived from the forest areas.

Discussion

Using Equation 5, we conduct a sensitivity analysis with an extreme 50% adjustment in the value of the ecosystem services coefficients. The results, as reported in Table 7, indicate that CS values in all studied periods were below 1, which means that the estimated overall ecosystem value (ESV) was considered to be inelastic (low sensitive) to the change in coefficient values. We can conclude that the ESV calculations are consistent and reliable with a minimum impact from variations in estimation of VC values caused by occasional data inaccuracies. The highest CS scores belong to forest in 1938, 1987 and 1999, cropland in 1958, and grassland in 1978 (i.e. the land uses with large areas and big value coefficients).

Table 4 Land-use changes from 1938 to 1999

Land Uses		Years				
		1938	1958	1978	1987	1999
Forest	Land-use (hectare)	689.8	1.09	16.98	562.11	618.1
	% of Change	-	−99.84	1457.80	3210.42	9.96
	Average Annual % of Change	-	−4.99	72.89	356.71	0.83
Grassland	Land-use (hectare)	97.08	136.69	615.32	96.79	121.85
	% of Change	-	40.80	350.16	−84.27	25.89
	Average Annual % of Change	-	2.04	17.51	−9.36	2.16
Cropland	Land-use (hectare)	52.19	663.56	170	145.19	93.15
	% of Change	-	1171.43	−74.38	−14.59	−35.84
	Average Annual % of Change	-	58.57	−3.72	−1.62	−2.99
Build up	Land-use (hectare)	0.00	1.51	2.04	2.47	5.97
	% of Change	-	-	35.10	21.08	141.70
	Average Annual % of Change	-	-	1.75	2.34	11.81
Water body	Land-use (hectare)	0.00	3.32	3.32	3.32	0.00
	% of Change	-	-	0.00	0.00	−100.00
	Average Annual % of Change	-	-	0.00	0.00	−8.33
Unused land	Land-use (hectare)	0.00	32.89	31.39	29.18	0.00
	% of Change	-	-	−4.56	−7.04	−100.00
	Average Annual % of Change	-	-	−0.23	−0.78	−8.33

Table 5 Ecosystem service values of Zhifanggou watershed from 1938 to 1999

ESVs		Years				
		1938	1958	1978	1987	1999
Forest	ESV ($10^4$$/year)	113.75	0.18	2.8	92.69	101.92
	% of Change	-	-99.84	1455.56	3210.36	9.96
	Average Annual % of Change	-	-4.99	72.78	356.71	0.83
Grassland	ESV ($10^4$$/year)	5.3	7.46	33.6	5.28	6.65
	% of Change	-	40.75	350.4	-84.29	25.95
	Average Annual % of Change	-	2.04	17.52	-9.37	2.16
Cropland	ESV ($10^4$$/year)	2.72	34.64	8.87	7.58	4.86
	% of Change	-	1173.53	-74.39	-14.54	-35.88
	Average Annual % of Change	-	58.68	-3.72	-1.62	-2.99
Water body	ESV ($10^4$$/year)	0	1.36	1.36	1.36	0.00
	% of Change	-	-	0.00	0.00	-100.00
	Average Annual % of Change	-	-	0.00	0.00	-8.33
Barren land	ESV ($10^4$$/year)	0	0.11	0.10	0.10	0.00
	% of Change	-	-	-9.09	0.00	-100.00
	Average Annual % of Change	-	-	-0.45	0.00	-8.33
Total	ESV ($10^4$$/year)	121.77	43.75	46.73	107.01	113.44
	% of Change	-	-64.07	6.81	129.00	6.01
	Average Annual % of Change	-	-3.20	0.34	14.33	0.50

Table 6 ESVs associated with ecosystem service functions (10⁴$, with 1990 as the base year)

Ecosystem services	Years										Rank
	1938		1958		1978		1987		1999		
	ESVs	% Change	ESVs	% Change	ESVs	% Change	ESVs	% Change	ESVs	% Change	
Gas regulation	18.99	15.60	3.39	7.76	4.81	10.29	15.99	14.95	17.40	15.34	↓
Climate regulation	15.08	12.39	5.62	12.84	5.89	12.60	13.32	12.45	14.06	12.40	↑
Water supply	17.44	14.32	4.29	9.80	5.32	11.39	15.24	14.24	16.05	14.15	↓
Soil formation and retention	22.28	18.30	9.37	21.41	11.44	24.48	19.57	18.29	20.99	18.50	↑
Waste treatment	8.44	6.93	10.05	22.97	8.83	18.89	8.78	8.21	8.48	7.48	↑
Biodiversity protection	18.05	14.82	4.88	11.15	6.53	13.96	15.54	14.53	16.71	14.73	↓
Food	1.17	0.96	5.30	12.11	2.71	5.80	1.77	1.65	1.47	1.30	↑
Raw material	13.60	11.17	0.61	1.39	0.72	1.53	11.17	10.44	12.24	10.79	↓
Recreation and culture	6.73	5.52	0.25	0.56	0.49	1.06	5.62	5.25	6.04	5.33	↓
Total	121.77	100	43.75	100	46.73	100	107.01	100	113.44	100	-

The historical changes in ecosystem services values of Zhifanggou watershed were triggered by several factors described as follows:

Population: There were only 94 local residents in Zhifanggou watershed in 1938. This number increased to 221 in 1958 as a result of advances in nationwide medical care and increase in farming activities. From 1970s, the policy of birth control began to implement, and the growth of local population was slowed to 476 people in 1990, with a balancing impact on build up land use.

Land Policy: Prior to 1978, there was no local ownership of the croplands. This was a major factor in low productivity of croplands as there was little motivation for locals to cultivate efficiently. In 1978, when the third Plenary Session of the Eleventh Central Committee in China took place, a directive for local management of croplands was proposed. This came into effect in 1985, triggering an increase in productivity of croplands and reducing the need to more land-uses of this type.

Technological Change: Technological developments in farming also contributed to improvements in croplands productivity. Before 1970s, farming was mainly carried out using low-tech methods, leading to unproductive practices and overexploitation and deterioration of soil. From 1970s, there were technological advances such as new cultivation approaches for intercropping, mulching technologies, and efficient pest control methods, all of which contributed to high productivity of croplands.

Water and Soil Conservation: In Zhifanggou watershed, notorious conditions such as little vegetations, loose soil and heavy rain impose pressure on local environmental protection. From 1977, locals began to cultivate on slope lands, modifying them to terraces and paying higher attention to soil conservation. In addition, Zhifanggou

watershed was chosen as a test zone for water and soil conservation projects in late 1980s, when the local government began to implement integrated water, land and economic development plans in the region, increasing the ecosystem services value of the watershed.

Conclusions

We have customized the ecosystem services valuation approach developed by Costanza et al. (1997) according to Xie et al. (2003) to account for specific characteristics of ecosystems in China. In addition, to address the overestimation or underestimation of service values in Xie et al. (2003) approach (as shown by Fu et al. (2011), we have revised some of the coefficient. A sensitivity analysis was also employed to test of the impact of variations in coefficients on estimation of ecosystem services values. The analysis revealed that the estimation of ecosystem service coefficients was robust despite the imposed fluctuations.

We have used the land use as a proxy measure of ecosystem services. It should be mentioned that the accuracy of satellites images could significantly influence the accuracy of the results, generating more detailed data, such as information on individual trees or hidden water bodies. Moreover, the biomes and land categories are not always matched. For example, the woodland and orchard uses were jointly considered as forest in this study.

In Zhifanggou watershed, with little human activities until 1938, the watershed was then covered with large areas of forest (689.8 hectare) and grassland (97.08 hectare). During 1938–1958, the growth of population mixed with cropland expansion led to a dramatic decline of forest and grassland at a rate of 2.04% each year. The subsequent decline of ecosystem services values was finally recognized by the local government, and an integrative watershed management program was introduced. With the support of government policies, the area of forest and grassland recovered at a rate of 168.58% per year

Table 7 Sensitivity of ecosystem valuation coefficients

Land use	Ecosystem services	Year				
		1938	1958	1978	1987	1999
Forest	Gas regulation	0.15	0.00	0.01	0.14	0.14
	Climate regulation	0.12	0.00	0.01	0.11	0.11
	Water supply	0.14	0.00	0.01	0.13	0.13
	Soil formation and retention	0.17	0.00	0.01	0.15	0.16
	Waste treatment	0.06	0.00	0.00	0.05	0.05
	Biodiversity protection	0.14	0.00	0.01	0.13	0.13
	Food	0.00	0.00	0.00	0.00	0.00
	Raw material	0.11	0.00	0.01	0.10	0.11
	Recreation and culture	0.05	0.00	0.00	0.05	0.05
Grassland	Gas regulation	0.00	0.02	0.08	0.01	0.01
	Climate regulation	0.01	0.02	0.09	0.01	0.01
	Water supply	0.00	0.02	0.08	0.01	0.01
	Soil formation and retention	0.01	0.05	0.19	0.01	0.02
	Waste treatment	0.01	0.03	0.13	0.01	0.01
	Biodiversity protection	0.01	0.03	0.11	0.01	0.01
	Food	0.00	0.01	0.03	0.00	0.00
	Raw material	0.00	0.00	0.01	0.00	0.00
	Recreation and culture	0.00	0.00	0.00	0.00	0.00
Cropland	Gas regulation	0.00	0.06	0.01	0.01	0.00
	Climate regulation	0.00	0.10	0.02	0.01	0.01
	Water supply	0.00	0.07	0.02	0.01	0.00
	Soil formation and retention	0.00	0.17	0.04	0.01	0.01
	Waste treatment	0.01	0.19	0.05	0.02	0.01
	Biodiversity protection	0.00	0.08	0.02	0.01	0.00
	Food	0.00	0.11	0.03	0.01	0.01
	Raw material	0.00	0.01	0.00	0.00	0.00
	Recreation and culture	0.00	0.00	0.00	0.00	0.00
Water body	Gas regulation	0.00	0.00	0.00	0.00	0.00
	Climate regulation	0.00	0.01	0.00	0.00	0.00
	Water supply	0.00	0.01	0.01	0.00	0.00
	Soil formation and retention	0.00	0.00	0.00	0.00	0.00
	Waste treatment	0.00	0.01	0.01	0.00	0.00
	Biodiversity protection	0.00	0.00	0.00	0.00	0.00
	Food	0.00	0.00	0.00	0.00	0.00
	Raw material	0.00	0.00	0.00	0.00	0.00
	Recreation and culture	0.00	0.00	0.00	0.00	0.00
Barren land	Gas regulation	0.00	0.00	0.00	0.00	0.00
	Climate regulation	0.00	0.00	0.00	0.00	0.00
	Water supply	0.00	0.00	0.00	0.00	0.00
	Soil formation and retention	0.00	0.00	0.00	0.00	0.00
	Waste treatment	0.00	0.00	0.00	0.00	0.00

Table 7 Sensitivity of ecosystem valuation coefficients *(Continued)*

Biodiversity protection	0.00	0.00	0.00	0.00	0.00
Food	0.00	0.00	0.00	0.00	0.00
Raw material	0.00	0.00	0.00	0.00	0.00
Recreation and culture	0.00	0.00	0.00	0.00	0.00

from 1978 to 1999. In 1999, the area of forest was fully recovered to its original state.

Land use changes in Zhifanggou watershed resulted in changes in ecosystem services. In 1938, the sum of ESVs was estimated at 121.77×10^4. It had dramatically decreased to 43.75×10^4 in 1958. From 1958 to 1978, the sum of ESVs increased marginally owing to the local government policies. From 1978, Zhifanggou watershed was chosen to serve as a test zone for the government's integrative watershed management program. This further increased the sum of ESVs to 107.01×10^4 in 1987. As a result of such steady improvements the sum of ESVs for Zhifanggou watershed reached to 113.44×10^4 in 1999, nearly back to its peak in 1938. Reviewing specific ecosystem services, the values of soil formation and retention, biodiversity protection, gas regulation and climate regulation were forming the largest part of the total ESVs. Among these services, soil formation and retention was recognized as the most impacted service by land use changes, corresponding to the highest rate of change. This was due to the fact that the low productivity of land forced local residents to farm more intensively, which led to much severer soil erosion.

This study can be extended in several ways. First, the impact of crop yield differences in the study area over the years can be investigated. These differences might be due to natural causes or triggered by the use of improved farming technology. In addition, due to space limitations, we have only used a 50% adjustment rate for our sensitivity analysis. It would be interesting to investigate the outcomes of various sensitivity analyses approaches with varied rates of adjustment. It should also be mentioned that the choice of ESV calculation methodology and the associated coefficients are case-specific. We have developed a calculation framework on the basis of Xie et al. (2003) coefficients for China's ecosystems. Comparing the outcomes of ESV calculations obtained for a case area using different bases for estimation of coefficients would certainly enhance our understanding of their applicability and limitations.

Competing interests
The authors declared that they have no competing interests.

Authors' contributions
The authors have co-developed the research agenda and analysis. JS and FN have drafted and revised the manuscript. All authors have read and approved the final manuscript.

Acknowledgements
This research was supported by the National Natural Science Foundation of China. The authors are very much thankful to two reviewers of this paper whose comments and suggestions were very helpful in improving the manuscript.

Author details
[1]Bartlett School of Graduate Studies, University College London (UCL), London WC1H-0NN, UK. [2]The Key Laboratory of Water and Sediment Science, Department of Environmental Engineering, Peking University, Beijing 100871, China.

References
Collin ML, Melloul AJ (2001) Combined land-use and environmental factors for sustainable groundwater management. Urban Water 3:229–237
Costanza R, Arge RD, Groot RD, Farber S, Grasso M, Hannon B, Limburg K, Naeem S, Neill RVO, Paruelo J, Raskin RG, Sutton P, Belt MVD (1997) The value of the world's ecosystem services and natural capital. Nature 386:253–260
Daily GC (1997) Nature's Services: Societal Dependence on Natural Ecosystems. Island Press Washington D.C, USA
Darwin R, Tsigas M, Lewandrowski J, Raneses A (1996) Land use and cover in ecological economics. Ecol Econ 17:157–181
ESRI (2013) ARCGIS for Desktop, ESRI ltd. http://www.esri.com/software/arcgis/arcgis-for-desktop (Last accessed on: August 10–2013)
Fu BJ, Wei YP, Willett IR, Lu YH, Liu GH (2011) Double counting in ecosystem services valuation: causes and countermeasures. Ecol Res 26:1–14
Han P, Si J, Wang YG (2009) Contrastive analysis of valuation methods of ecosystem services value: a case study of Zhifanggou watershed in a hilly and gully region of loess plateau. J Basic Sci Eng 17:102–111
Jiang ZS, Zheng FL (2004) Assessment on benefit of sediment reduction by comprehensive controls in the Zhifanggou Watershed. J Sediment Res 2:56–61
Kreuter UP, Harris HG, Matlock MD, Lacey RE (2001) Change in ecosystem service values in the San Antonio area, Texas. Ecol Econ 39:333–346
Li BC (1995) The remote sensing monitoring of soil erosion and integrated management in watersheds. Press of Science, China
Li JC, Wang WL, Hu GY, Wei ZH (2010) Changes in ecosystem service values in Zoige Plateau, China. Agric Ecosyst Environ 139:766–770
Martínez ML, Perez-Maqueo O, Vazquez G, Castillo-Campos G, García-Franco J, Mehltreter K, Equihua M, Landgrave R (2009) Effects of land use change on biodiversity and ecosystem services in tropical montane cloud forests of Mexico. For Ecol Manage 258:1856–1863
Nasiri F, Huang GH (2007) Ecological viability assessment: A fuzzy multiple-attribute analysis with respect to three classes of ordering techniques. Ecol Inform 2:128–137
NBSC (1996–1999) National Bureau of Statistic of China (1996–1999 yearly data sets). http://www.stats.gov.cn/english/statisticaldata/yearlydata (Last accessed on: August 10–2013)
Styers DM, Chappelka AH, Marzen LJ, Somers GL (2010) Developing a land-cover classification to select indicators of forest ecosystem health in a rapidly urbanizing landscape. Landsc Urban Plan 94:158–165
Turner RK, Adger N, Brouwer R (1998) Ecosystem services value, research needs and policy relevance: a commentary. Ecol Econ 25:61–65
Wen ZM, Jiao F, He X, Yang Q, Liu B (2006) Increase of land productivity and its implication for eco-environment improvement: a case study in Zhifanggou catchment in loess hilly areas. Trans Chin Soc Agric Eng 22:91–95

Xie GD, Lu C, Leng Y, Zheng D, Li S (2003) Ecological assets valuation of the
    Tibetan Plateau. J Nat Resour 18:189–196
Zhang CX, Xie GD, Yang QK (2008) Assessment of human activities on soil
    conservation value in hilly and gully region of Loess Plateau. J Nat Resour
    23:1035–1042
Zhang S, Wu C, Liu H, Na X (2011) Impact of urbanization on natural ecosystem
    service values: a comparative study. Environ Monit Assess 179:575–588

14

Goodness-of-fit testing for the inverse Gaussian distribution based on new entropy estimation using ranked set sampling and double ranked set sampling

Amer Ibrahim Al-Omari[1*] and Abdul Haq[2]

Abstract

Background: Entropy is a measure of uncertainty and dispersion associated with a random variable. Several goodness-of-fit tests based on entropy are available in literature and the entropy been widely used in many applications.

Results: Goodness-of-fit test for the inverse Gaussian distribution is studied based on new entropy estimation using simple random sampling (SRS), ranked set sampling (RSS) and double ranked set sampling (DRSS) methods. The critical values of the new tests are obtained using Monte Carlo simulations. The power values of the suggested tests based on several alternative hypotheses using SRS, RSS, and DRSS are also presented. It is observed that the proposed tests are more powerful as compared to the test under SRS. Also, it turns out that the test based on DRSS is superior to the RSS test for all of the cases considered in this study.

Conclusion: Since the suggested goodness-of-fit tests for the inverse Gaussian distribution using DRSS are more efficient than that based on RSS, one may consider them using multistage RSS.

Keywords: Entropy, Goodness-of-fit test, Inverse Gaussian, Root mean square error, Simple random sampling, Ranked set sampling, Double ranked set sampling

Background

Entropy is a measure of uncertainty and dispersion associated with a random variable. It is not uniquely defined, there exist axiom systems that justify the particular entropies. Shannon (1948) defined the entropy $H(f)$ of the random variable X as

$$H(f) = -\int_{-\infty}^{\infty} f(x)\log f(x)dx, \qquad (1)$$

where X is a continuous random variable with probability density function (pdf) $f(x)$ and cumulative

distribution function (cdf) $F(x)$. Vasicek (1976) defined $H(f)$ as

$$H(f) = \int_{0}^{1} \log\left(\frac{d}{dp}F^{-1}(p)\right)dp. \qquad (2)$$

Let $X_1, X_2, \ldots, X_n$ be a simple random sample of size n from $F(x)$ and let $X_{(1)} \le X_{(2)} \le \cdots \le X_{(n)}$ be the order statistics of the sample. Vasicek (1976) estimator of $H(f)$ is given by

$$VE_{(m,n)} = \frac{1}{n}\sum_{i=1}^{n} \log\left\{\frac{n}{2m}\left(X_{(i+m)} - X_{(i-m)}\right)\right\}, \qquad (3)$$

where m is a positive integer, known as a window size, $m < n/2$. Here $X_{(i)} = X_{(1)}$ if $i < 1$ and $X_{(i)} = X_{(1)}$ if $i > n$. It is of interest to note that $VE_{(m,n)} \to PH(f)$ as $n \to \infty$, $m \to \infty$ and $m/n \to 0$.

* Correspondence: alomari_amer@yahoo.com
[1]Department of Mathematics, Faculty of Science, Al al-Bayt University, Mafraq 25113, Jordan
Full list of author information is available at the end of the article

Van Es (1992) suggested another entropy estimator based on spacing's, given by

$$VE_{(m,n)} = \frac{1}{n-m} \sum_{i=1}^{n-m} \log\left(\frac{n+1}{m}\left(X_{(i+m)} - X_{(i)}\right)\right) + \sum_{k=m}^{n} \frac{1}{k} + \log\left(\frac{m}{n+1}\right). \quad (4)$$

They proved the consistency and asymptotic normality of this estimator under some conditions.

Ebrahimi et al. (1994) suggested a new estimator by assigning different weights in Vasicek (1976) entropy estimator, and proposed the following estimator

$$EE_{(m,n)} = \frac{1}{n} \sum_{i=1}^{n} \log\left(\frac{n}{c_i m}\left(X_{(i+m)} - X_{(i-m)}\right)\right), \quad (5)$$

where

Table 1 Monte Carlo RMSEs and bias values of the entropy estimators $VE_{(m,n)}$ and $AE_{(m,n)}$ for the uniform distribution, $H(f) = 0$

n	m	SRS				RSS			
		$VE_{(m,n)}$		$AE_{(m,n)}$		$VE_{(m,n)}$		$AE_{(m,n)}$	
		Bias	RMSE	Bias	RMSE	Bias	RMSE	Bias	RMSE
10	1	−0.519826	0.569537	−0.046482	0.521035	−0.396308	0.443439	−0.343522	0.396739
	2	−0.415135	0.452358	−0.298609	0.350332	−0.304078	0.329233	−0.189664	0.228762
	3	−0.422613	0.453818	−0.249056	0.298944	−0.327681	0.343991	−0.154894	0.186380
	4	−0.458940	0.487054	−0.229082	0.281422	−0.371538	0.383103	−0.143218	0.171767
	5	−0.502063	0.527918	−0.215077	0.270468	−0.425903	0.436521	−0.137821	0.168029
20	1	−0.393900	0.418346	−0.366867	0.392622	−0.343340	0.365754	−0.314244	0.338695
	2	−0.271880	0.290818	−0.212993	0.236696	−0.217937	0.233026	−0.162729	0.183187
	3	−0.253931	0.270200	−0.168961	0.192998	−0.205321	0.216879	−0.117939	0.136570
	4	−0.260596	0.274678	−0.144016	0.167779	−0.214042	0.222524	−0.100304	0.118284
	5	−0.276800	0.288985	−0.133179	0.157805	−0.235141	0.242179	−0.091608	0.108584
	6	−0.299321	0.310256	−0.125960	0.150733	−0.258899	0.264554	−0.085981	0.101365
	7	−0.322084	0.332301	−0.121244	0.146386	−0.285310	0.290156	−0.084733	0.099613
	8	−0.348254	0.357901	−0.118562	0.144786	−0.314138	0.318471	−0.083482	0.098588
	9	−0.374620	0.383864	−0.116399	0.143986	−0.343410	0.347711	−0.083926	0.099430
	10	−0.402840	0.411741	−0.117057	0.145063	−0.371780	0.375737	−0.848235	0.101014
30	1	−0.352853	0.368369	−0.334631	0.351096	−0.319230	0.333509	−0.300423	0.316118
	2	−0.223356	0.235685	−0.184969	0.199765	−0.190866	0.201625	−0.152577	0.165665
	3	−0.197719	0.208362	−0.141411	0.156683	−0.165182	0.173360	−0.106329	0.119047
	4	−0.196240	0.205882	−0.118803	0.133958	−0.162899	0.169841	−0.087046	0.099566
	5	−0.202003	0.210395	−0.105711	0.120861	−0.172441	0.178293	−0.078599	0.088072
	6	−0.213804	0.221385	−0.097719	0.113216	−0.185622	0.190458	−0.069898	0.081972
	7	−0.226688	0.233521	−0.092957	0.109089	−0.200036	0.204048	−0.066053	0.077716
	8	−0.242599	0.248992	−0.089259	0.105818	−0.217704	0.221309	−0.064713	0.076188
	9	−0.259471	0.265356	−0.087074	0.103535	−0.235661	0.238850	−0.062931	0.073734
	10	−0.276934	0.282548	−0.085151	0.102071	−0.254437	0.257257	−0.062044	0.072402
	11	−0.295302	0.300725	−0.841357	0.101314	−0.273700	0.276336	−0.062243	0.072977
	12	−0.313803	0.319255	−0.083206	0.102002	−0.293398	0.295911	−0.062262	0.072981
	13	−0.332279	0.337432	−0.082858	0.101944	−0.311978	0.341101	−0.063754	0.074987
	14	−0.351090	0.356205	−0.082540	0.101854	−0.332096	0.334518	−0.063579	0.075100
	15	−0.370555	0.375518	−0.082665	0.102618	−0.352077	0.354327	−0.064127	0.075825

Table 2 Monte Carlo RMSEs and bias values of the entropy estimators $VE_{(m,n)}$ and $AE_{(m,n)}$ for the exponential distribution, $H(f) = 1$

n	m	SRS				RSS			
		$VE_{(m,n)}$		$AE_{(m,n)}$		$VE_{(m,n)}$		$AE_{(m,n)}$	
		Bias	RMSE	Bias	RMSE	Bias	RMSE	Bias	RMSE
10	1	−0.552032	0.677001	−0.495449	0.631471	−0.430553	0.505229	−0.376361	0.461201
	2	−0.442683	0.571820	−0.323532	0.483573	−0.337494	0.404667	−0.220406	0.315220
	3	−0.435444	0.561640	−0.265713	0.443276	−0.332760	0.401125	−0.159787	0.276197
	4	−0.451545	0.575390	−0.221689	0.424404	−0.348029	0.420617	−0.121584	0.266664
	5	−0.469437	0.597761	−0.179844	0.413541	−0.366628	0.445977	−0.080667	0.266812
20	1	−0.414064	0.490107	−0.384516	0.464796	−0.357765	0.398661	−0.333540	0.376843
	2	−0.285717	0.376086	−0.232518	0.338830	−0.234959	0.280262	−0.176512	0.232710
	3	−0.260773	0.351341	−0.175461	0.298406	−0.213397	0.261261	−0.125059	0.194705
	4	−0.256116	0.352810	0.141143	0.279706	−0.210620	0.259248	−0.098056	0.179990
	5	−0.262412	0.358638	0.118697	0.271887	−0.214122	0.265246	−0.072456	0.172661
	6	−0.265650	0.360325	0.090043	0.263318	−0.218028	0.272315	−0.048075	0.168086
	7	−0.266934	0.365008	−0.067175	0.260090	−0.224596	0.282196	−0.023128	0.173677
	8	−0.273952	0.377519	−0.041928	0.258647	−0.232629	0.293062	−0.000531	0.176806
	9	−0.280123	0.381968	−0.021108	0.262708	−0.236125	0.302083	0.027269	0.190739
	10	−0.285183	0.391290	0.004497	0.267634	−0.238413	0.310922	0.044912	0.203657
30	1	−0.367058	0.423423	−0.346283	0.406311	−0.332526	0.361491	−0.313657	0.343784
	2	−0.233677	0.306086	−0.198867	0.280012	−0.203455	0.236001	−0.163180	0.203230
	3	−0.202277	0.281503	−0.145618	0.241162	−0.170859	0.207468	−0.111717	0.161754
	4	−0.194424	0.275072	−0.115163	0.224526	−0.160246	0.199410	−0.084854	0.145930
	5	−0.191705	0.272356	−0.095073	0.217468	−0.159714	0.200465	−0.059819	0.134539
	6	−0.186870	0.272196	−0.070590	0.208597	−0.158702	0.202869	−0.043778	0.132887
	7	−0.191094	0.275374	−0.058550	0.205261	−0.161705	0.206226	−0.027194	0.130283
	8	−0.195662	0.280589	−0.036080	0.200329	−0.164468	0.212265	−0.010631	0.136358
	9	−0.196983	0.282040	−0.021144	0.202056	−0.165511	0.217222	−0.006685	0.138626
	10	−0.197171	0.283394	−0.005890	0.204787	−0.167152	0.220237	0.024904	0.145306
	11	−0.198853	0.286241	0.008492	0.207709	−0.173076	0.229318	0.039837	0.154215
	12	−0.204089	0.293653	0.022622	0.213445	−0.171555	0.232740	0.055108	0.163320
	13	−0.202908	0.298108	0.049154	0.220522	−0.176996	0.240454	0.070977	0.176787
	14	−0.205700	0.300842	0.061987	0.226574	−0.176922	0.244541	0.093001	0.193377
	15	−0.210699	0.305809	0.081431	0.238902	−0.177959	0.248760	0.109754	0.205539

$$c_i = \begin{cases} 1 + \dfrac{i-1}{m}, & 1 \leq i \leq m, \\ 2, & m+1 \leq i \leq n-m, \\ 1 + \dfrac{n-i}{m}, & n-m+1 \leq i \leq n. \end{cases}$$

They proved that $EE_{(m,n)}$ converges in probability to $H(f)$ as $n \to \infty$, $m \to \infty$ and $m/n \to 0$.

(Al-Omari AI (2012): Modified entropy estimators using simple random sampling, ranked set sampling and double ranked set sampling, Submitted) suggested a modified estimator of entropy of an unknown continuous pdf $f(x)$ as

Based on the simulation study, it is shown that this estimator has smaller bias and mean square error as compared to the Vasicek (1976) entropy estimator.

$$AE_{(m,n)} = \frac{1}{n} \sum_{i=1}^{n} \log\left(\frac{n}{c_i m}\left(X_{(i+m)} - X_{(i-m)}\right)\right), \quad (6)$$

Table 3 Monte Carlo RMSEs and bias values of the entropy estimators $VE_{(m,n)}$ and $AE_{(m,n)}$ for the standard normal distribution, $H(f) = 1.419$

n	m	SRS				RSS			
		$VE_{(m,n)}$		$AE_{(m,n)}$		$VE_{(m,n)}$		$AE_{(m,n)}$	
		Bias	RMSE	Bias	RMSE	Bias	RMSE	Bias	RMSE
10	1	−0.598925	0.676499	−0.538428	0.623068	−0.484489	0.549750	−0.429406	0.502967
	2	−0.521455	0.591007	−0.409842	0.496627	−0.422169	0.471157	−0.308706	0.375690
	3	−0.563002	0.623188	−0.386562	0.468471	−0.462240	0.504378	−0.291133	0.353844
	4	−0.610651	0.663364	0.388846	0.469519	−0.523019	0.557792	−0.292810	0.351636
	5	−0.671777	0.719069	−0.382242	0.461612	−0.584483	0.614209	−0.294820	0.349472
20	1	−0.435480	0.483459	−0.402721	0.452976	−0.382986	0.420310	−0.354315	0.393878
	2	−0.327145	0.375798	−0.267005	0.324501	−0.275716	0.313472	−0.218758	0.264068
	3	−0.317948	0.364927	−0.230598	0.292997	−0.268657	0.304811	−0.181588	0.230636
	4	−0.327070	0.372436	−0.214227	0.279269	−0.285331	0.318855	−0.168035	0.219922
	5	−0.352658	0.395796	−0.205782	0.272804	−0.305555	0.337744	−0.160392	0.213700
	6	0.375996	0.416964	−0.203268	0.269194	−0.335066	0.365185	−0.162263	0.216405
	7	−0.404050	0.442997	−0.200951	0.269828	−0.363782	0.391748	−0.162648	0.217866
	8	−0.439618	0.475094	−0.203704	0.270603	−0.395221	0.421583	−0.163443	0.217711
	9	−0.467134	0.500777	0.211872	0.276695	−0.428042	0.451680	−0.169841	0.224475
	10	−0.496926	0.527456	−0.209085	0.275281	−0.454818	0.477152	−0.171572	0.224804
30	1	−0.378860	0.413455	−0.359097	0.394766	−0.343626	0.370512	−0.328056	0.355718
	2	−0.259105	0.299687	−0.221750	0.266138	−0.226914	0.255947	−0.189446	0.223276
	3	−0.236758	0.277238	−0.177599	0.229027	−0.204698	0.234358	−0.147274	0.186797
	4	−0.234369	0.275867	−0.158560	0.213972	−0.204765	0.234413	−0.125487	0.169031
	5	−0.244288	0.283027	−0.148610	0.206988	−0.214434	0.243683	−0.117590	0.165087
	6	−0.255248	0.293332	−0.139542	0.200072	−0.227340	0.255901	−0.111407	0.161770
	7	−0.269724	0.305134	−0.132038	0.196792	−0.241325	0.268228	−0.105796	0.158654
	8	−0.285713	0.321039	−0.129915	0.193509	−0.254983	0.282376	−0.102504	0.157726
	9	−0.304064	0.337563	−0.131105	0.198239	−0.274697	0.301420	−0.103392	0.160749
	10	−0.320051	0.352764	−0.130086	0.196928	−0.295057	0.319933	−0.101392	0.160593
	11	−0.339131	0.369866	−0.127890	0.196985	−0.314201	0.339141	−0.102034	0.161378
	12	−0.361226	0.392070	−0.130212	0.197655	−0.333173	0.356224	−0.103026	0.163577
	13	−0.382347	0.410463	0.129885	0.199488	−0.353582	0.375170	−0.105978	0.165825
	14	−0.400618	0.428008	−0.131518	0.199794	−0.375752	0.397462	−0.109190	0.168154
	15	−0.423597	0.449968	−0.134062	0.200285	−0.394363	0.414605	−0.108705	0.167780

where

$$c_i = \begin{cases} 1 + \dfrac{1}{2}, & 1 \leq i \leq m, \\ 2, & m + 1 \leq i \leq n - m, \\ 1 + \dfrac{1}{2}, & n - m + 1 \leq i \leq n. \end{cases}$$

Alizadeh (2010) proposed a new estimator of entropy and studied its application in testing normality. Park and Park (2003) considered correcting moments for goodness-of-fit tests for two entropy estimates.

Inverse Gaussian distribution

A random variable X is said to have an inverse Gaussian distribution function $IG\ (x;\ \mu,\ \beta)$, if its pdf is of the following form

$$f(x) = \sqrt{\frac{\beta}{2\pi x^3}} \exp\left(-\frac{\beta}{2\mu^2 x}(x - \mu)^2\right), \quad \text{for } x > 0, \quad (7)$$

where $\mu > 0$ is the mean and $\beta > 0$ is the shape parameter. The variance of X is $\mu^3\beta$. Its characteristic function is

Table 4 Monte Carlo RMSEs and bias values of the entropy estimators $VE_{(m,n)}$ and $AE_{(m,n)}$ for the uniform distribution with $H(f) = 0$ and exponential distribution with $H(f) = 1$ using DRSS

| n | m | Uniform distribution and $H(f) = 0$ | | | | Exponential distribution and $H(f) = 1$ | | | |
| | | $VE_{(m,n)}$ | | $AE_{(m,n)}$ | | $VE_{(m,n)}$ | | $AE_{(m,n)}$ | |
		Bias	RMSE	Bias	RMSE	Bias	RMSE	Bias	RMSE
10	1	−0.327408	0.369593	−0.267924	0.318205	−0.365854	0.425279	−0.305667	0.379121
	2	−0.260621	0.278731	−0.145388	0.176159	−0.288898	0.340618	−0.173991	0.251460
	3	−0.296104	0.306116	−0.122180	0.144286	−0.300393	0.351750	−0.128545	0.223802
	4	−0.346305	0.352712	−0.115995	0.134276	−0.322839	0.377437	−0.089495	0.215854
	5	−0.404121	0.409902	−0.116805	0.135411	−0.335248	0.399189	−0.047170	0.219634
20	1	−0.308453	0.329353	−0.279902	0.302719	−0.329105	0.363241	−0.298237	0.335475
	2	−0.189231	0.202666	−0.132076	0.151177	−0.204908	0.240316	−0.150759	0.196279
	3	−0.182095	0.191163	−0.095961	0.112229	−0.191216	0.228320	−0.104346	0.163293
	4	−0.197693	0.204342	−0.082268	0.096978	−0.190904	0.229986	−0.075338	0.179771
	5	−0.220876	0.225845	−0.077708	0.091093	−0.197900	0.239789	−0.052175	0.145269
	6	−0.247733	0.251580	−0.075071	0.086966	−0.207032	0.251002	−0.026183	0.146832
	7	−0.275808	0.278919	−0.074331	0.085055	−0.209883	0.258152	−0.012044	0.152682
	8	−0.303823	0.306608	−0.073793	0.084202	−0.218701	0.271560	0.014201	0.161180
	9	−0.333903	0.336495	−0.075306	0.086127	−0.223692	0.278728	0.035069	0.173654
	10	−0.363272	0.365731	−0.075514	0.086480	−0.228126	0.290431	0.061574	0.189857
30	1	−0.298092	0.312767	−0.278830	0.293698	−0.308011	0.331033	−0.289677	0.314515
	2	−0.170745	0.180210	−0.133715	0.146379	−0.182416	0.207785	−0.143418	0.174632
	3	−0.146113	0.153646	−0.088998	0.100564	−0.152039	0.180708	−0.094799	0.136371
	4	−0.149143	0.154886	−0.072297	0.083848	−0.145325	0.176699	−0.071094	0.123270
	5	−0.159888	0.164564	−0.063874	0.074562	−0.146632	0.179028	−0.049250	0.114227
	6	−0.174419	0.178204	−0.060394	0.070784	−0.149443	0.184598	−0.030887	0.113500
	7	−0.191854	0.194940	−0.058041	0.067650	−0.150245	0.188158	−0.046556	0.115023
	8	−0.209886	0.212509	−0.056421	0.065369	−0.153441	0.194332	−0.001239	0.120306
	9	−0.229010	0.231261	−0.056053	0.064628	−0.157250	0.199936	0.012716	0.124585
	10	−0.248006	0.249993	−0.056843	0.064868	−0.162854	0.208891	0.029477	0.133242
	11	−0.267506	0.269188	−0.056931	0.064430	−0.163540	0.213175	0.045951	0.145582
	12	−0.287408	0.289018	−0.056982	0.064673	−0.167660	0.221482	0.063602	0.155340
	13	−0.307160	0.308699	−0.058363	0.066130	−0.171024	0.225764	0.079779	0.169499
	14	−0.327370	0.328890	−0.058038	0.065797	−0.170880	0.232977	0.096359	0.182124
	15	−0.346997	0.348439	−0.059523	0.067623	−0.169873	0.235173	0.115563	0.198755

given by

$$\phi_x(t) = exp\left(\frac{\beta}{\mu} - \sqrt{\beta}\sqrt{\frac{\beta}{\mu^2} - 2it}\right).$$

The $IG(x; \mu, \beta)$ has many applications in the field, for example see Seshadri (1999), and Folks and Chhikara (1998).

Method

The test procedure

Let $X_1, X_2, \ldots, X_n$ be a random sample of size n drawn from the pdf $f(x)$ and let $X_{(1)} \le X_{(2)} \le \cdots \le X_{(n)}$ be the order statistics of this sample. Our interest is to test that this random sample is coming from an inverse Gaussian population or not. Thus, the composite null hypothesis is $H_0: X \sim IG(x; \mu, \beta)$.

The following corollary is due to Mahdizaheh and Arghami (2010).

Table 5 Monte Carlo RMSEs and bias values of the entropy estimators $VE_{(m,n)}$ and $AE_{(m,n)}$ for the standard normal distribution and $H(f) = 1.419$ using DRSS

n	m	$VE_{(m,n)}$		$AE_{(m,n)}$	
		Bias	RMSE	Bias	RMSE
10	1	−0.415021	0.472162	−0.352434	0.416211
	2	−0.373395	0.412666	−0.262149	0.316029
	3	−0.427401	0.459119	−0.254450	0.303820
	4	−0.492911	0.518275	−0.264683	0.310442
	5	−0.554351	0.577281	−0.267798	0.312339
20	1	−0.350703	0.383160	−0.323780	0.359592
	2	−0.245907	0.277809	−0.190733	0.231106
	3	−0.246496	0.276941	−0.158832	0.201924
	4	−0.262789	0.290545	−0.148107	0.194728
	5	−0.291340	0.317967	−0.145734	0.191755
	6	−0.316105	0.341597	−0.147800	0.195946
	7	−0.349246	0.373132	−0.150312	0.199934
	8	−0.384526	0.406764	−0.152801	0.203493
	9	−0.416151	0.436696	−0.156902	0.205954
	10	−0.445901	0.465518	0.159050	0.207883
30	1	−0.321940	0.345223	−0.307781	0.332609
	2	−0.206709	0.231560	−0.169564	0.198438
	3	−0.187163	0.212774	−0.129694	0.163913
	4	−0.190073	0.215577	−0.114103	0.152713
	5	−0.199843	0.224569	−0.103570	0.145964
	6	−0.214636	0.239021	−0.100510	0.146417
	7	−0.231613	0.255278	−0.095517	0.143483
	8	−0.247340	0.271084	−0.094560	0.145579
	9	−0.268298	0.291044	−0.091548	0.145394
	10	−0.286538	0.308661	−0.094236	0.149024
	11	−0.305310	0.326485	−0.093843	0.150300
	12	−0.324892	0.346062	−0.096171	0.152896
	13	−0.343097	0.363236	−0.096892	0.153854
	14	−0.369990	0.388586	−0.100541	0.155029
	15	−0.387740	0.406081	−0.101202	0.156143

Critical points at significance level 0.05 of the test statistic are given in Table 6. The optimal choice of the window size for a given sample size in the estimation of entropy using spacing's is still open problem for testing goodness-of-fit. The bold fonts in Table 6 are the largest critical values based on SRS, RSS and DRSS. For the suggested test, the optimal window size values are summarized in Table 7.

Corollary 1: Assume that X is a random variable has an inverse Gaussian distribution IG $(x; \mu, \beta)$ and let $Y = 1/\sqrt{X}$ Then the entropy of Y is given by $H(f(y)) = \log(0.5\,\phi\sqrt{2\pi e})$, where $\phi^2 = 1/\beta = E(Y^2) - 1/E(Y^{-2})$.

The following corollary is due to Mudholkar and Tian (2002).

Corollary 2: The random variable X with inverse Gaussian distribution IG $(x; \mu, \beta)$ is characterized by the property that $1/\sqrt{X}$ attains the maximum entropy among all nonnegative, absolutely continuous random variables Y with a given value at $E(Y^2) - 1/E(Y^{-2})$.

Let $VE_{(m,n)}(f_y)$ be the sample estimate of $VE(f_y)$ for the distribution of $Y = 1/\sqrt{X}$ defined as

$$VE_{(m,n)}(f_y) = \frac{1}{n}\sum_{i=1}^{n}\text{Log}\left(\frac{n}{2m}\left(y_{(i+m)} - y_{(i-m)}\right)\right), \quad (8)$$

where $y_{(i)} = \left(x_{(n-i+1)}\right)^{-1/2} (i = 1, 2, \ldots, n)$.

Mahdizaheh and Arghami (2010) followed Vasicek (1976) and proposed rejecting the null hypothesis H_0: $X \sim IG$ $(x; \mu, \beta)$ if

$$K_{(m,n)}(f_y) = \frac{2exp\left(VE_{(m,n)}(f_y)\right)}{\psi} \leq K^*_{(m,n,\alpha)}(f_y), \quad (9)$$

where ψ^2 is a uniform minimum variance unbiased (UMVU) estimate of $\emptyset^2$ defined as

$$\psi^2 = \frac{1}{n-1}\sum(1/x_i - 1/\bar{x})$$
$$= \frac{1}{n-1}\left(\sum_{i=1}^{n}y_i^2 - n^2\left(\sum_{i=1}^{n}yi^{-2}\right)^{-1}\right). \quad (10)$$

Suggested test

Let $X_{i(i)}$ denote the ith order statistic from the ith sample $(i = 1, 2, \ldots, n)$. Then, the measured RSS units are denoted by $X_{1(1)}, X_{2(2)}, \ldots, X_{n(n)}$. The cumulative distribution function of $X_{i(i)}$ is given by

$$F_{(i)}(x) = \sum_{j=i}^{n}\binom{n}{j} F^j(x)(1 - F(x))^{n-j}, -\infty < x < \infty,$$

with probability density function defined as

$$f_{(i)}(x) = n\binom{n-1}{i-1}F^{i-1}(x)(1 - F(x))^{n-i}f(x), \quad -\infty < x < \infty.$$

The mean and the variance of the ith order statistic, $X_{i(i)}$ can be written respectively as

$$\mu(i) = \int_{-\infty}^{\infty}xf_{(i)}(x)dx, \text{ and } \sigma_{(i)}^2$$
$$= \int_{-\infty}^{\infty}\left(x - \mu_{(i)}\right)^2 f_{(i)}(x)dx.$$

Table 6 Critical values of the test statistics at significance level $a = 0.05$ using SRS, RSS and DRSS

					$n = 30$			
n	m	SRS	RSS	DRSS	m	SRS	RSS	DRSS
10	1	1.77481	1.92014	2.11693	1	2.45932	2.50879	2.57507
	2	2.32375	2.49737	2.73051	2	3.00586	3.06976	3.15363
	3	2.55582	2.70474	2.87862	3	3.19857	3.25881	3.33729
	4	2.67573	2.81527	2.91803	4	3.27582	3.35586	3.42156
	5	**2.73289**	**2.83557**	**2.91884**	5	3.32359	3.39547	3.45623
20	1	2.24771	2.35314	2.42654	6	3.35015	3.42129	3.47623
	2	2.79602	2.88869	3.02510	7	3.36693	3.43050	**3.47907**
	3	2.97493	3.08786	3.19524	8	3.37529	3.43391	3.47352
	4	3.04798	3.15706	3.25697	9	3.37021	**3.43604**	3.47057
	5	3.09802	3.19645	**3.28312**	10	3.38831	3.43064	3.47215
	6	3.13033	3.21615	3.28262	11	**3.39279**	3.42939	3.45317
	7	3.15950	**3.22789**	3.27655	12	3.38330	3.41772	3.44495
	8	3.15719	3.21777	3.26882	13	3.37597	3.42184	3.44197
	9	**3.16680**	3.21856	3.26432	14	3.36220	3.41612	3.44014
	10	3.15824	3.21474	3.25051	15	3.38366	3.41508	3.43684

The ranked set sampling method was suggested by McIntyre (1952) for estimating the mean of pasture and forage yields. The RSS can be described as follows:

Step 1: Select n simple random samples each of size n from the target population.

Step 2: Without cost, visually rank the units within each sample with respect to the variable of interest.

Step 3: For actual measurement, from the ith $(i = 1, 2, \ldots, n)$ sample of n units, select the ith smallest ranked unit. The method is repeated h times if needed to increase the sample size to hn units.

Al-Saleh and Al-Kadiri (2000) suggested double ranked set sampling (DRSS) method for estimating the population mean. The DRSS can be described as in the following steps:

Step 1 Randomly select n^2 samples each of size n from the target population.

Step 2 Apply the RSS method on the n^2 samples obtained in Step 1. This step yields n samples each of size n.

Step 3 Reapply the RSS method again on the n samples obtained on Step 2 to obtain a sample of size n from the DRSS data. The cycle can be repeated h times if needed to obtain a sample of size hn units.

Table 7 Optimal window sizes

n	SRS	RSS	DRSS
10	5	5	5
20	9	7	5
30	11	9	7

The SRS estimator of the population mean is given by $\hat{\mu}_{SRS} = \sum_{i=1}^{n} X_i/n$, with variance $Var(\hat{\mu}_{SRS}) = \sigma^2/n$. The RSS estimator of the population mean is defined as $\hat{\mu}_{RSS} = \sum_{i=1}^{n} X_{i(i)}/n$, with variance given by $Var(\hat{\mu}_{RSS}) = \frac{\sigma^2}{n} - \frac{1}{n^2}\sum_{i=1}^{n}\left(\mu_{(i)} - \mu\right)^2$. The relative precision (RP) of RSS relative to SRS for estimating the population mean is

$$RP = \mathrm{Var}\mu^{SRS}\mathrm{Var}\mu^{RSS} = 1 - i = 1n\mu i - \mu2n\sigma^2.$$

Takahasi and Wakimoto (1968) showed that the parent pdf $f(x)$ and the population mean can be expressed as $f(x) = \frac{1}{n}\sum_{i=1}^{n} f_{(i)}(x)$, and $\mu = \frac{1}{n}\sum_{i=1}^{n}\mu_{(i)}$, respectively. Also, they showed that $1 \leq RP \leq \frac{m+1}{2}$, where the lower bound is attained if and only if the underlying distribution is degenerate, while the upper bound is attained if and only if the underlying distribution of the data is rectangular.

Al-Saleh and Al-Omari (2002) extended the DRSS for multistage RSS method to increase the efficiency of the estimators for fixed value of the sample size, Al-Omari and Raqab (2012) suggested truncation RSS method for estimating the population mean and median, Al-Omari (2011) suggested double robust extreme RSS for estimating the population mean, Haq and Shabbir (2010) proposed a family of ratio estimators of the population mean using extreme RSS based on two auxiliary variables.

Table 8 Power comparison for the entropy tests at the significance level $\alpha = 0.05$

n	m	Exponential (1)			Uniform (0,1)			Weibull (2,1)		
		SRS	RSS	DRSS	SRS	RSS	DRSS	SRS	RSS	DRSS
10	1	0.1869	0.2330	0.2559	0.4089	0.5078	0.5921	0.1059	0.1238	0.1346
	2	**0.2167**	**0.2776**	**0.3610**	**0.4874**	**0.6422**	0.8381	**0.1269**	0.1640	**0.2240**
	3	0.1960	0.2562	0.3242	0.4796	0.6398	**0.8455**	0.1261	**0.1659**	0.2230
	4	0.1366	0.1875	0.1981	0.3735	0.5284	0.6825	0.0961	0.1391	0.1593
	5	0.0629	0.0750	0.0780	0.1897	0.2481	0.3011	0.0460	0.0574	0.0622
20	1	0.3805	0.4530	0.4682	0.7665	0.8704	0.9186	0.1874	0.2311	0.2351
	2	0.4584	0.5375	0.6152	0.8661	0.9528	0.9930	0.2566	0.3062	0.3597
	3	**0.4713**	**0.5680**	**0.6360**	**0.8873**	**0.9716**	**0.9970**	**0.2625**	**0.3341**	**0.3890**
	4	0.4179	0.5201	0.6027	0.8711	0.9680	0.9968	0.2299	0.2964	0.3552
	5	0.3829	0.4685	0.5284	0.8346	0.9484	0.9944	0.2095	0.2648	0.3106
	6	0.3094	0.3855	0.4221	0.8024	0.9211	0.9802	0.1682	0.2106	0.2364
	7	0.2377	0.2899	0.3074	0.7229	0.8564	0.9312	0.1368	0.1611	0.1635
	8	0.1660	0.1827	0.1942	0.5806	0.7019	0.7954	0.0877	0.0955	0.0963
	9	0.1022	0.1131	0.1132	0.4095	0.4875	0.5456	0.0600	0.0581	0.0633
	10	0.0538	0.0615	0.0638	0.2145	0.2585	0.2627	0.0297	0.0328	0.0346
30	1	0.5400	0.5913	0.6094	0.9188	0.9660	0.9851	0.2729	0.3091	0.3125
	2	0.6402	0.7097	0.7585	0.9724	0.9960	0.9997	0.3776	0.4276	0.4669
	3	**0.6734**	**0.7431**	0.7941	**0.9832**	0.9982	0.9999	**0.4116**	0.4605	0.5075
	4	0.6510	0.7374	**0.7959**	0.9804	**0.9989**	**1.0000**	0.3941	**0.4650**	**0.5156**
	5	0.6252	0.7048	0.7711	0.9800	0.9979	0.9999	0.3636	0.4324	0.4829
	6	0.5763	0.6583	0.7229	0.9690	0.9978	0.9998	0.3109	0.3757	0.4322
	7	0.5170	0.6015	0.6531	0.9558	0.9940	0.9995	0.2795	0.3274	0.3575
	8	0.4526	0.5237	0.5565	0.9392	0.9875	0.9982	0.2166	0.2672	0.2778
	9	0.3843	0.4356	0.4609	0.8973	0.9730	0.9949	0.1768	0.2066	0.2134
	10	0.3102	0.3424	0.3547	0.8673	0.9445	0.9823	0.1421	0.1438	0.1592
	11	0.2440	0.2528	0.2585	0.7882	0.8763	0.9285	0.1066	0.1070	0.1020
	12	0.1772	0.1788	0.1785	0.6678	0.7474	0.8160	0.0713	0.0697	0.0660
	13	0.1117	0.1218	0.1141	0.5201	0.6034	0.6372	0.0447	0.0501	0.0502
	14	0.0697	0.0774	0.0800	0.3516	0.4083	0.4327	0.0269	0.0363	0.0288
	15	0.0477	0.0448	0.0522	0.2284	0.2458	0.2411	0.0231	0.0261	0.0197

Goodness-of-fit test for the $IG\,(x;\,\mu,\,\beta)$ distribution is considered using SRS, RSS and DRSS methods. Our composite null hypothesis is H_0: $X \sim IG\,(x;\,\mu,\,\beta)$. Following Mudholkar and Tian (2002), we reject H_0 if

$$K_{(m,n)}\left(f_y\right) = \frac{2\exp\left[AE_{(m,n)}\left(f_y\right)\right]}{\psi} \leq K^*_{(m,n,\alpha)}\left(f_y\right), \qquad (11)$$

where

$$AE_{(m,n)} = \frac{1}{n}\sum_{i=1}^{n} \mathrm{Log}\left(\frac{n}{c_i m}\left(X_{(i+m)} - X_{(i-m)}\right)\right) \qquad \text{and}$$

$$c_i = \begin{cases} 1 + \dfrac{1}{2}, & 1 \leq i \leq m, \\ 2, & m+1 \leq i \leq n-m, \\ 1 + \dfrac{1}{2}, & n-m+1 \leq i \leq n. \end{cases}$$

Note that, $AE_{(m,n)}\left(f_y\right)$ is the sample estimate of $AE\left(f_y\right)$. Since the entropy estimators are functions of order statistics, then the entropy estimation using RSS and DRSS involves ordering the RSS units.

Results and discussion

In this section, a Monte Carlo experiment is presented to investigate the performance of the entropy estimators i.e. $AE_{(m,n)}$ as well as $VE_{(m,n)}$ and as well as to study the

Table 9 Power comparison for the entropy tests at the significance level $a = 0.05$

n	m	Lognormal (0,2)			Beta (2,2)			Beta (5,2)		
		SRS	RSS	DRSS	SRS	RSS	DRSS	SRS	RSS	DRSS
10	1	0.1347	0.1595	0.1806	0.1758	0.1990	0.2343	0.1436	0.1667	0.1823
	2	**0.1576**	**0.1849**	**0.2383**	0.2208	0.2925	0.4210	0.2027	0.2649	0.3855
	3	0.1177	0.1532	0.1853	**0.2341**	**0.3255**	**0.4670**	**0.2443**	0.3276	**0.5106**
	4	0.0667	0.0894	0.0936	0.1871	0.2774	0.3626	0.2303	**0.3554**	0.4872
	5	0.0262	0.0267	0.0241	0.0910	0.1194	0.1480	0.1644	0.2462	0.3241
20	1	0.2802	0.3461	0.3535	0.3543	0.4343	0.4556	0.2923	0.3556	0.3693
	2	0.3447	0.4144	**0.4731**	0.4954	0.5982	0.7032	0.4418	0.5150	0.6393
	3	**0.3504**	**0.4282**	0.4726	**0.5214**	**0.6633**	**0.7879**	**0.4817**	0.6162	0.7499
	4	0.3037	0.3743	0.4325	0.5056	0.6472	0.7819	0.4799	0.6238	**0.7869**
	5	0.2402	0.3071	0.3379	0.4875	0.6170	0.7554	0.4742	**0.6288**	0.7809
	6	0.1870	0.2164	0.2338	0.4256	0.5471	0.6569	0.4546	0.5935	0.7156
	7	0.1251	0.1346	0.1326	0.3672	0.4858	0.5137	0.4299	0.5452	0.6399
	8	0.0669	0.0671	0.0720	0.2603	0.3153	0.3578	0.3735	0.4543	0.5274
	9	0.0324	0.0317	0.0323	0.1594	0.1886	0.2044	0.3094	0.3651	0.4164
	10	0.0116	0.0126	0.0136	0.0868	0.0973	0.0967	0.2227	0.2661	0.2867
30	1	0.4096	0.4578	0.4737	0.5287	0.5856	0.6167	0.4344	0.4767	0.5121
	2	0.5141	0.5748	0.6309	0.7055	0.7838	0.8603	0.6237	0.7156	0.7936
	3	**0.5292**	**0.6032**	**0.6622**	**0.7543**	0.8437	0.9182	0.6911	0.7996	0.8857
	4	0.5187	0.6013	0.6542	0.7542	**0.8670**	**0.9382**	0.6993	0.8376	**0.9258**
	5	0.4831	0.5571	0.5990	0.7308	0.8530	0.9339	**0.7030**	**0.8398**	0.9240
	6	0.4209	0.4965	0.5441	0.7038	0.8338	0.9185	0.6877	0.8228	0.9141
	7	0.3574	0.4220	0.4439	0.6584	0.7854	0.8702	0.6559	0.7989	0.8874
	8	0.2916	0.3275	0.3447	0.5932	0.7100	0.7995	0.6239	0.7564	0.8375
	9	0.2172	0.2460	0.2466	0.5197	0.6383	0.7055	0.5672	0.7001	0.7779
	10	0.1442	0.1705	0.1664	0.4502	0.5295	0.5999	0.5433	0.6273	0.7271
	11	0.1055	0.1037	0.0977	0.3810	0.4140	0.4532	0.4848	0.5615	0.6114
	12	0.0549	0.0555	0.0599	0.2764	0.2975	0.3117	0.4196	0.4751	0.5126
	13	0.0311	0.0288	0.0285	0.1922	0.2188	0.2187	0.3449	0.4049	0.4171
	14	0.0129	0.0148	0.0148	0.1130	0.1356	0.1376	0.2720	0.3261	0.3560
	15	0.0067	0.0070	0.0070	0.0824	0.0830	0.0822	0.2466	0.2687	0.2729

powers of the suggested tests under different alternatives hypotheses. The root mean square errors (RMSEs) and the bias values are obtained for the estimators based on 10,000 samples of sizes $n = 10$, 20, 30 with window sizes $1 \leq m \leq 5$, $1 \leq m \leq 10$ and $1 \leq m \leq 15$, respectively.

Comparison between $VE_{(m,n)}$ and $AE_{(m,n)}$

The samples are selected from the uniform, exponential and the standard normal distributions using SRS, RSS and DRSS methods. From Tables 1, 2, 3, 4, 5, 6, and 7 we can see that $AE_{(m,n)}$ is more efficient than $VE_{(m,n)}$ for all cases considered in this study. Also, the DRSS is superior to SRS and RSS. For more details about this comparison see (Al-Omari AI (2012): Modified entropy estimators using simple random sampling, ranked set sampling and double ranked set sampling, Submitted).

We can see that these optimal values are different from Mahdizaheh and Arghami (2010) values where their suggested test is based on Vasicek (1976) entropy estimator. Here, we can conclude that the optimal window size depends on the entropy estimator used for the goodness-of-fit test.

Power of the tests

The power of the suggested goodness-of-fit tests using SRS, RSS and DRSS is considered here relative to the same alternatives considered by Mahdizaheh and Arghami (2010) for the distributions, exponential(1), uniform(0,1), Weibull(2,1), lognormal(0,2), beta(2,2), and beta(5,2). 10000 samples of sizes $n = 30, 20, 30$ are generated for each method at the significance level 0.05.

Based on Tables 8 and 9, we can conclude that gain in the performance of the new suggested tests using different methods considered in this paper is obtained. However, we found that the DRSS is superior to both RSS and SRS methods based on the sample size. Also, the RSS performs better than SRS for all cases considered here. The bold fonts in Tables 8 and 9 are the optimal power values for each design with the same sample size. These optimal power values are $< n/2$. However, the optimal values of the window size are 2, 3, 4, 5. For fixed n, the power values decreases as m increases, while it increases in n.

Conclusion

In this paper, new goodness-of-fit tests for the inverse Gaussian distribution are suggested using SRS, RSS and DRSS based on the maximum entropy characterization. It is found that the new tests are more powerful under RSS and DRSS, and the test under DRSS is superior to the tests under RSS and SRS methods. We recommend using the suggested goodness-of-fit tests for the inverse Gaussian distribution. As the DRSS is better than RSS, the current work can be extended to multistage RSS design and for some other probability distributions.

Competing interests
Both authors declared that they have no competing.

Authors' contribution
The work presented here was carried out in collaboration between authors. AA carried out the theoretical and discussion of this paper. AH carried out the Monte Carlo simulations. All authors read and approved the final manuscript.

Acknowledgment
The authors are grateful to the editors and the anonymous reviewers for their valuable comments and suggestions.

Author details
[1]Department of Mathematics, Faculty of Science, Al al-Bayt University, Mafraq 25113, Jordan. [2]Department of Statistics, Quaid-i-Azam University, Islamabad 45320, Pakistan.

References
Alizadeh HN (2010) A new estimator of entropy and its application in testing normality. J Stat Comput Simul 80:1151–1162
Al-Omari AI (2011) Estimation of mean based on modified robust extreme ranked set sampling. J Stat Comput Simul 81(8):1055–1066
Al-Omari AI, Raqab MZ (2012) Estimation of the population mean and median using truncation-based ranked set samples. Accepted in J Stat Comput Simul. doi:10.1080/00949655.2012.662684
Al-Saleh MF, Al-Kadiri MA (2000) Double ranked set sampling. Stat probability lett 48(2):205–212
Al-Saleh MF, Al-Omari AI (2002) Multistage ranked set sampling. J Stat Planning and Inference 102(2):273–286
Ebrahimi N, Pflughoeft K, Soofi E (1994) Two measures of sample entropy. Stat Probability Lett 20:225–234

Folks JL, Chhikara RS (1998) The inverse Gaussian distribution and its statistical application-a review. J R Soc, Series B 40:263–289
Haq A, Shabbir J (2010) A family of ratio estimators for population mean in extreme ranked set sampling using two auxiliary variables. SORT 34(1):45–64
Mahdizaheh M, Arghami NR (2010) Efficiency of ranked set sampling in entropy estimation and goodness-of-fit testing for the inverse Gaussian law. J Stat Comput Simul 80(7):761–774
McIntyre GA (1952) A method for unbiased selective sampling using ranked sets. Australian J Agricultural Res 3:385–390
Mudholkar GS, Tian L (2002) An entropy characterization of the inverse Gaussian distribution and related goodness-of-fit test. J Stat Planning and Inference 102:211–221
Park S, Park D (2003) Correcting moments for goodness of fit tests based on two entropy estimates. J Stat Comput Simul 73(9):685–694
Seshadri V (1999) The inverse Gaussian distribution: Statistical theory and applications. Springer, New York
Shannon CE (1948) A mathematical theory of communications. Bell System Technical J 27(379–423):623–656
Takahasi K, Wakimoto K (1968) On the unbiased estimates of the population mean based on the sample stratified by means of ordering. Annals of the Institute of Statistical Mathematics 20:1–31
Van Es B (1992) Estimating functionals related to a density by class of statistics based on spacing's. Scand J Stat 19:61–72
Vasicek O (1976) A test for normality based on sample entropy. J Royal Stat Soc B 38:54–59

15

A decision support system for benchmarking the energy and waste performance of schools in Toronto

Julian Scott Yeomans[*]

Abstract

Background: The Toronto District School Board (TDSB) oversees the largest school district in Canada and has been spent more than one third of its annual maintenance budget on energy and waste. This has directed attention toward system-wide reductions to both energy consumption patterns and waste generation rates. In this paper, a decision support system (DSS) that can process unit-incompatible measures is used for rating, ranking, and benchmarking the schools within the TDSB.

Results: The DSS permits the ranking of any set of schools by contextually evaluating their relative attractiveness to other identified school groupings. Consequently, the DSS was used to explicitly rank each school's performance within the district and to determine realistic energy improvement targets. Achieving these benchmarks would reduce system-wide energy costs by twenty-five percent.

Conclusions: The TDSB study demonstrates that this DSS provides an extremely useful approach for evaluating, benchmarking and ranking the relative energy and waste performance within the school system, and the potential to extend its much broader applicability into other applications clearly warrants additional exploration.

Keywords: Benchmarking Performance, Data Envelopment Analysis, Decision Support Systems, Energy & Waste

Background

Representing more than 600 schools, the Toronto District School Board (TDSB) oversees the largest education constituency within Canada. In the 2003 fiscal year, the TDSB allocated $48 million of its annual budget to the energy and waste requirements of the school system, but within three years, found that the energy and waste expenditures had escalated to more than $69 million (Christie 2003, 2007; Christie & Coppinger 2006). This rapid forty percent increase in spending during a period of public economic retrenchment necessitated stringent attention toward system-wide reductions to both energy consumption patterns and waste generation rates (Christie 2003, 2007; Christie & Coppinger 2006).

While the need to decrease energy consumption and minimize waste was universally acknowledged, to achieve long-term success in these efforts, the TDSB resolutely believed that any successful systematic reduction efforts would need to adequately address three critical questions: (i) what initiatives would achieve the most effective results; (ii) how could these "most effective reduction initiatives" actually be identified; and, (iii) how effective would these initiatives prove to be once implemented (Christie 2003) Additional "essential components" to effectively implement any proposed reduction schemes necessitated that improvement initiatives had to include: (i) the establishment of realistic benchmarks for each school to strive toward; (ii) the setting of achievable annual performance targets for each school; (iii) a commitment to rational and objective management of the system data; and, (iv) an assurance of transparency and neutrality in policy-setting and decision-making (Christie 2003, 2007; Christie & Coppinger 2006). Unfortunately, it proves to be an extremely difficult process to evaluate system-wide performance and to establish benchmarks when there are multiple incommensurate criteria measurements present (Camp 1995), which was the case within the

Correspondence: syeomans@schulich.yorku.ca
Operations Management & Information Systems Area, Schulich School of Business, York University, 4700 Keele Street, Toronto, Ontario M3J 1P3, Canada

budgetary-constrained and politically-charged environment of the TDSB (Christie 2003, 2007; Christie & Coppinger 2006).

Decision support systems (DSS) are intelligent information systems based on decision models that can be used to extract large quantities of data from databases, to provide interfaces and methods for effectively processing it, and for deriving meaningful decisions of managerial/economic significance from it. DSS have been used to analyze a wide variety of performance information and to provide a readily-accessible medium for distributing any knowledge generated to a wide variety of stakeholders (Lin *et al.* 2008). In this paper, a DSS that can simultaneously combine unit-incompatible energy and waste performance measures is introduced for evaluating the system-wide energy and waste performance of the schools within the TDSB. This DSS incorporates several data envelopment analysis (DEA) modules that had been developed in Yeomans (2004) and threads these modules together using commonly available software (Albright 2010; Seref et al. 2007; Zhu 2003). The underlying DEA methodology has been shown to hold advantages over many other multi-criteria methods by providing an objective decision-making tool that does not require variables to have the same scale or conversion weights applied to them (Cook & Zhu 2008; Zhu 2003), while permitting a simultaneous combination of both quantitative and qualitative measures (Cook *et al.* 1996).

The DSS developed can be used to rate and rank each school according to its energy and waste performance relative to the other schools in the system by recursively partitioning the schools into sub-groups of relatively superior and inferior performers. In addition, the DSS can establish realistically achievable improvement targets for each school by benchmarking their performance against the operations of schools in higher efficiency categories. This relative performance comparison between the schools is important since it ensures that the underlying analysis involves the use of energy and waste values actually occurring at peer institutions within the TDSB and not through "externally-generated", potentially unrepresentative values.

The DSS developed directly addresses the TDSB's three requisite "critical questions" while simultaneously addressing most of the key issues in the "essential components" identified above. Furthermore, the most practical contribution from this approach is that, since DEA can be readily implemented using common spreadsheet software linked together by relatively straightforward VBA programming (Albright 2010; Seref et al. 2007; Zhu 2003), this entire methodology can be implemented on virtually any computer. Consequently, practitioners in any organization could easily modify and extend this methodology to match their own very specific multi-

criteria applications and circumstances. An illustrative example of an application of the DSS is provided through an analysis of a subset of the high schools in the TDSB.

Results and Discussion

The primary purpose in the performance evaluations of most organizational systems is to appraise the current operations of individual entities and to benchmark these against peer entities to identify best practices. While individual performance measure, or "gap", analysis has often provided the fundamental basis for performance evaluation and benchmarking (Zhu 2003), it remains a difficult task to satisfactorily combine multiple disparate, unit-incompatible, single-criteria measurements into an overall conclusion (Camp 1995). Since a complex entity's actual performance generally represents multifaceted phenomena, the use of single measures in gap analysis explicitly ignores all interactions, substitutions, and tradeoffs between the various performance measures. Therefore, it is a rare occurrence when a one-measure-at-a-time gap-analysis can suffice for the purposes of an effective performance evaluation of organizational systems (Camp 1995; Zhu 2003).

Clearly it is difficult to evaluate an organization's performance or to establish benchmarks when there are multiple measurements present (Camp 1995). If the specific algebraic functional relationship between performance measures is known, then established multi-criteria techniques can be used to estimate best-practice levels of performance. These algebraic functional forms cannot be specified without *a priori* information on the corresponding tradeoffs and, unfortunately, such information is generally unavailable in most practical situations. However, when the best practices of similar types of operations can be identified empirically at a specific point in time, it becomes possible to empirically estimate the resulting best-practice level of performance, or efficient frontier, using these observations (Zhu 2003).

DEA has proved to be an effective empirical tool for identifying multi-criteria efficient frontiers, for subsequently evaluating relative performance efficiencies, and for implicitly estimating the tradeoffs inherent within the empirically designed frontier. DEA's empirical orientation and absence of *a priori* assumptions establish it as the ideal analytical instrument for application to a wide variety of practical situations, since its underlying theoretical basis is consistent with the practice of rating entities by concurrently examining the relative efficiencies of their multiple performance measures. Furthermore, DEA allows performance comparisons to be made between numerous entities that have been evaluated by multiple, unit-incomparable measures without employing any *a priori* weightings or conversion factors typically required in other methods. Since the process is adaptable and invariant to data type, DEA permits the inclusion and comparison of non-numeric environmental-type variables that might prove incomparable using many other techniques (Zhu 2003). By incorporating DEA into its DSS, the focus of performance evaluation for the TDSB application could shift from a characterization of each school's energy and waste usage in terms of single measures to evaluating performance from a mathematically rational, multidimensional system perspective (Linton *et al.* 2007; Yeomans 2004). Consequently, inherent energy and waste relationships and their role in performance ranking can be explicitly brought into the analysis in a rational, transparent, and neutral fashion.

An overview of the energy and waste DSS for the TDSB

The DSS, itself, contains a series of specific DEA modules (more fully explained in subsequent sections) for conducting the performance evaluation of the TDSB. The first module recursively partitions the selected schools into sub-groups of relatively superior and inferior performers according to their energy and waste performance. The second module then determines realistically achievable energy and waste improvement targets for each school by benchmarking them against the system-wide operations for schools in selected higher performance categories. If desired, the DSS can also execute a third module to establish an explicit rank ordering of each school relative to any desired set of schools within the TDSB.

The set of DEA modules was created using readily available spreadsheet optimization software linked together by a combination of straightforward programming. In order to make the entire analysis process readily accessible to the various system users, the DSS was implemented using Microsoft Access and Microsoft Excel together with VBA (contained in all Microsoft Office products) as the programming language. By using these specific computer packages, the entire set of modules for evaluating the performance of the TDSB was created using software residing on most current personal computers. In the following sections, each of the individual DEA modules used in the DSS is explained in detail and illustrated using data from 65 high schools in the TDSB.

DEA relative performance rating module

The first module in the DSS partitions any selected group of schools into relatively superior and inferior performers based upon their energy and waste performance. In DEA terminology a decision-making unit (DMU) designates the specific entity being studied (Zhu 2003). For the example, the set of DMUs consists of the 65 high schools in the TDSB. Each DMU possesses a set of inputs and outputs that represent its multiple measures of performance and, for the high schools, these inputs

and outputs consist of various observed energy and waste measures. While a considerable number of different combinations of energy and waste performance indicators could have been selected from the available data, for illustrative purposes the example in this paper contains only two inputs and four outputs. The inputs considered were (i) school enrolments measured in terms of the number of students and (ii) school sizes measured in square metres (sqm), while the set of outputs consisted of (i) total energy consumption measured in gigajoules (GJ), (ii) total energy costs in dollars ($), (iii) total waste in kilograms (Kg), and (iv) waste diversion percentage (%).

One analytical requirement for DEA is that all inputs have to be measured in units where "less is better", while all outputs have to be expressed in units in which "more is better". Hence, prior to implementing any of the procedures, the DSS transforms all selected inputs and outputs into a format consistent with this analytical requirements. Table 1 provides a complete list of the transformed inputs and outputs for the 65 schools.

The analytical approach of DEA evaluates the data by "enveloping" the entities being studied based upon the values of the performance measures. The underlying concept of DEA requires an evaluation of each DMU through a projection onto an empirically constructed, multi-dimensional efficient frontier. The enveloping process determines the efficiency of DMUs by: (i) creating an m + s dimensional surface, or "efficient frontier", of the efficient DMUs (where m represents the number of inputs and s represents the number of outputs); (ii) assigning an efficiency score of $\theta = 1$ to any DMU on the efficient frontier; (iii) determining the distance from the frontier for all inefficient DMUs; and, (iv) calculating the value of θ for inefficient DMUs as its proportional, multi-dimensional distance from the efficient frontier. The efficiency score of any inefficient DMU is always a value of $\theta < 1$. The enveloping constructs an efficient frontier of the best-practice entities and also shows how any inefficient DMU can be improved by providing the amounts and directions for improvement to its specific measures (Zhu 2003).

In assessing the high schools, the overall goal is to identify the system's best performers by contrasting each school's observed performance metrics relative to those of every other school considered. However, DEA only determines whether a school is efficient ($\theta = 1$) or inefficient ($\theta < 1$). The magnitude of θ cannot be used to establish relative degrees of inefficiency between non-efficient schools. Because the relative performance of any school can be contrasted only to an identified best-practice frontier, actual measures of relative inefficiency would change only when the best-practice frontier is altered (that is, when one or more of the efficient

Table 1 Each high school's energy and waste input/output measurements transformed into appropriate format for DEA usage

FACILITY NAME	Floor Area sqm	Student Enrollment	Transformed Total Energy GJ	Transformed total Energy Cost	Transformed Total Waste Kg	Diversion (%)
Agincourt CI	19,554	1,533	19,826	$254,168	3,488	10.25
Albert Campbell CI	22,964	2,115	3,275	$1	3,546	9.59
Bendale BTI	14,693	855	15,794	$227,171	5,169	15.56
Birchmount Park CI	16,826	1,276	23,304	$287,084	5,150	41.52
Bloor CI	13,656	797	28,974	$314,772	3,924	18.07
Cedarbrae CI	23,668	1,557	14,999	$202,965	4,415	40.02
Central Commerce Collegiate	20,729	1,034	20,843	$298,930	6,114	100
Central Etobicoke HS	11,086	492	21,900	$291,602	5,025	35.99
CW Jefferys CI	16,401	1,057	22,486	$279,963	5,189	12.24
David & Mary Thomson CI	21,576	1,628	14,098	$206,145	5,856	8.95
Downsview SS	21,483	877	9,661	$176,278	5,514	13.05
Dr Norman Bethune CI	14,254	1,130	23,011	$286,933	5,682	56.31
Earl Haig SS	24,849	2,394	21,630	$146,889	1,159	69.46
Eastdale CI	5,501	197	34,717	$407,993	5,834	81.85
Eastern HS of Commerce / Subway Academy I	18,330	1,155	26,755	$342,930	5,520	53.13
Emery CI	22,306	1,570	16,563	$190,444	6,139	100
Etobicoke CI	19,367	1,502	22,215	$232,432	4,776	30.52
Etobicoke School of the Arts	12,537	889	22,467	$319,969	5,270	37.14
Frank Oke SS	4,322	154	34,371	$413,854	5,755	25.6
George Harvey CI	25,025	1,183	13,911	$179,360	4,635	30.21
Georges Vanier SS	23,721	1,045	1	$15,462	5,859	0
Greenwood SS / School of Life Experience	7,847	404	32,836	$379,072	5,982	69.35
Harbord CI	18,437	1,040	22,577	$213,196	4,992	33.39
Heydon Park SS	7,475	220	36,678	$441,925	6,122	0
Humberside CI	17,655	1,150	24,382	$304,460	5,226	63.09
Inglenook Community School	1,607	128	40,091	$488,102	6,317	89.74
Jarvis CI	21,783	1,313	19,217	$265,879	5,066	66.76
Kipling CI	12,276	729	27,813	$362,378	5,504	21.96
Lakeshore CI	16,208	920	22,371	$302,844	5,276	38.44
Lawrence Park CI	15,634	1,026	27,163	$336,863	5,149	49.39
Leaside HS	13,560	1,163	27,529	$339,132	5,369	37.07
Malvern CI	14,331	1,046	28,028	$345,788	4,982	41.88
Maplewood HS	10,728	523	25,715	$312,227	5,914	100
Martingrove CI	14,737	1,041	32,751	$331,622	5,691	48.38
Nelson A Boylen CI	9,708	611	28,556	$325,182	6,033	0
Newtonbrook SS	18,230	1,789	14,985	$170,648	5,443	0
North Albion CI	15,961	1,110	26,433	$332,279	3,934	20.16
North Toronto CI	16,046	1,114	28,736	$355,959	2,802	20.76
Northern SS	29,471	1,998	16,571	$223,581	1	21.05
Northview Heights SS	23,864	1,444	17,419	$220,008	3,644	18.95
Oakwood CI	18,588	983	28,163	$349,240	6,155	38.56
Parkdale CI	14,435	631	28,357	$342,029	5,639	64.45
RH King Academy	17,796	1,469	18,416	$228,731	4,751	36.07

Table 1 Each high school's energy and waste input/output measurements transformed into appropriate format for DEA usage (Continued)

Richview CI	11,030	992	27,608	$344,498	5,335	39.36
Riverdale CI	23,418	1,217	21,011	$210,624	3,854	29.29
Rosedale Heights SS	16,271	680	22,744	$289,208	5,503	50.75
Runnymede CI	13,491	806	29,109	$366,070	5,504	43.94
Scarlett Heights Entrepreneurial Academy	11,528	749	28,293	$348,189	369	3.31
School of Experiential Education	2,525	86	37,662	$459,735	3,814	79.81
Silverthorn CI	16,537	1,263	20,870	$242,216	4,868	38
Sir John A Macdonald CI	17,324	1,576	12,097	$105,763	4,444	10.15
Sir Robert L Borden BTI	13,246	722	21,546	$268,610	4,820	23.52
Sir William Osler HS	11,010	359	22,425	$304,830	5,725	39.89
Thistletown CI	15,540	1,103	21,448	$296,117	5,002	24.78
Ursula Franklin Academy	19,001	405	26,430	$320,347	4,395	63.15
Vaughan Road Academy	17,021	839	26,630	$324,903	5,610	67.77
Victoria Park SS	20,525	1,295	16,123	$211,734	5,823	76.02
West Hill CI	20,161	1,338	7,304	$105,948	2,922	15.03
West Toronto CI	19,852	813	18,979	$152,372	4,665	61.43
Western Technical-Commercial School / The Student School	44,367	1,220	6,691	$73,962	4,191	57.35
Weston CI	18,317	1,271	21,145	$206,202	4,546	37.46
Westview Centennial SS	25,323	1,385	11,118	$101,969	3,430	21.95
William Lyon Mackenzie CI	11,619	1,231	25,392	$316,995	6,039	100
Woburn CI	20,126	1,458	22,884	$264,443	3,887	69.23
York Mills CI	16,207	1,373	23,638	$274,777	3,012	73.49

schools is removed). However, a recursive enveloping module can be created that stratifies the schools into numerous levels, or groupings, of relative best-practice frontiers, rather than the single DEA frontier.

The stratifying module of the DSS proceeds by removing all of those schools placed onto the original best-practice frontier and then using the original DEA enveloping methodology to form a new second-level best-practice frontier from the set of remaining schools. That is, once a first efficient frontier is calculated, all of the associated efficient schools are removed from further consideration and a new efficient frontier based only upon the remaining, initially inefficient schools is calculated. The schools on this second-level efficiency frontier are subsequently removed, permitting a third-level frontier to be constructed, then a fourth-level frontier, and so on, until no schools remain. The final result from this stratification is the creation of a series of efficient frontiers. The recursive procedure stratifies the original set of schools into L groupings of school efficiencies, with the specific value for L algorithmically determined, a posteriori, by an "empty-set" stopping rule.

When the stratification module was applied to the high school data, the 65 schools were partitioned into $L = 12$ distinct groups with 2, 2, 3, 6, 9, 10, 10, 5, 7, 5, 3, and 3 schools assigned to each respective efficiency stratum (see Table 2). This stratification effectively partitions the high schools into distinct groupings of comparably-performing schools based upon the multi-criteria measurements of their energy and waste performance.

Benchmark module for generating realistic energy and waste performance targets

Benchmarking is widely used for the identification and adoption of best practices and as a means for improving performance and increasing productivity. Benchmarking can be thought of as the process of defining valid measures of performance comparison among peer schools, using these to determine the relative standing of the peer schools, and ultimately in establishing standards of excellence for performance improvement. The TDSB had stated in their "essential component" requirements that, in order to improve their system-wide energy and waste usage, it was crucial for them to be able to determine attainable performance benchmarks and to establish achievable annual performance targets for each school (Christie 2003). Clearly the satisfaction of these components would necessitate the creation of multi-

Table 2 Changes required to each high school's current energy and waste measures to advance into the next higher efficiency group

DMU No.	DMU Name	Changes required to original outputs to attain efficiency at the next higher level			
		Change to Total Energy GJ	Change to Total Energy Cost $	Change to Total Waste Kg	Change to Diversion(%)
Level 1					
26	Inglenook Community School	0	0	0	0
49	School of Experiential Education	0	0	0	0
Level 2					
14	Eastdale CI	−5,374	−80,109	−483	8
19	Frank Oke SS	−5,720	−74,248	−562	64
Level 3					
22	Greenwood SS / School of Life Experience	−7,256	−109,030	−335	20
24	Heydon Park SS	−3,413	−46,177	−195	90
33	Maplewood HS	−5,334	−77,604	−265	55
Level 4					
35	Nelson A Boylen CI	−6,411	−98,118	−96	8
42	Parkdale CI	−4,340	−36,153	−348	4
44	Richview CI	−5,423	−46,343	−718	5
53	Sir William Osler HS	−11,658	−96,559	−317	2
55	Ursula Franklin Academy	−8,319	−88,192	−1,443	17
63	William Lyon Mackenzie CI	−7,230	−73,073	−584	7
Level 5					
7	Central Commerce Collegiate	−4,022	−47,268	−1,321	5
8	Central Etobicoke HS	−1,757	−23,397	−620	3
12	Dr Norman Bethune CI	−389	−13,090	−30	0
28	Kipling CI	−5,553	−39,705	−603	2
34	Martingrove CI	−1,720	−73,754	−299	3
47	Runnymede CI	−5,031	−34,249	−515	4
48	Scarlett Heights Entrepreneurial Academy	−55	−671	−5,337	14
56	Vaughan Road Academy	−6,589	−60,058	−342	4
65	York Mills CI	−11,079	−133,216	−2,822	8
Level 6					
5	Bloor CI	−1,359	−14,768	−1,924	9
16	Emery CI	−2,444	−33,938	−401	5
18	Etobicoke School of the Arts	−6,006	−23,782	−392	3
25	Humberside CI	−2,360	−21,550	−386	4
30	Lawrence Park CI	−1,642	−19,472	−410	3
31	Leaside HS	−1,369	−15,909	−252	2
32	Malvern CI	−1,400	−17,269	−539	2
46	Rosedale Heights SS	−1,224	−12,518	−238	2
52	Sir Robert L Borden BTI	−1,970	−33,561	−441	17
59	West Toronto CI	−3,580	−127,440	−551	7
Level 7					
3	Bendale BTI	−6,958	−61,342	−350	36
11	Downsview SS	−4,154	−4,813	−151	18
15	Eastern HS of Commerce / Subway Academy I	−1,860	−6,006	−97	1

Table 2 Changes required to each high school's current energy and waste measures to advance into the next higher efficiency group (Continued)

29	Lakeshore CI	−2,506	−11,754	−205	10
37	North Albion CI	−1,868	−17,433	−206	14
38	North Toronto CI	−624	−7,736	−2,715	23
41	Oakwood CI	−11,928	−138,862	−162	51
54	Thistletown CI	−1,156	−2,288	−39	9
57	Victoria Park SS	−17,854	−184,878	−70	1
64	Woburn CI	−929	−19,192	−158	3
Level 8					
4	Birchmount Park CI	−1,966	−33,642	−434	13
9	CW Jefferys CI	−189	−17,568	−44	27
21	Georges Vanier SS	−28,555	−309,719	−174	0
23	Harbord CI	−485	−88,334	−107	4
27	Jarvis CI	−3,704	−17,101	−326	4
Level 9					
13	Earl Haig SS	−1,326	−122,171	−2,108	4
20	George Harvey CI	−8,662	−36,385	−364	2
36	Newtonbrook SS	−3,185	−72,412	−333	31
45	Riverdale CI	−1,451	−14,550	−1,151	2
50	Silverthorn CI	−1,998	−52,234	−341	2
60	Western Technical-Commercial School / The Student School	−12,969	−166,149	−624	9
61	Weston CI	−2,175	−81,412	−595	4
Level 10					
1	Agincourt CI	−2,368	−30,352	−1,648	6
6	Cedarbrae CI	−4,810	−39,285	−533	5
10	David & Mary Thomson CI	−12,958	−105,347	−152	0
17	Etobicoke CI	−117	−1,224	−226	0
43	RH King Academy	−2,774	−8,051	−141	1
Level 11					
40	Northview Heights SS	−3,353	−17,186	−285	2
51	Sir John A Macdonald CI	−10,389	−174,200	−746	2
62	Westview Centennial SS	−8,051	−110,998	−1,294	8
Level 12					
2	Albert Campbell CI	−18,940	−232,431	−1,230	21
39	Northern SS	−2,571	−18,254	−4,099	2
58	West Hill CI	−15,242	−130,494	−2,138	11

criteria benchmark targets to improve each school's performance and it would be imperative that the various competing stakeholders within the system felt that these targets had been set fairly, objectively, and transparently (Christie 2003, 2007). As observed in the previous section, the stratification module partitioned the 65 high schools into 12 sets of comparably performing groups in which all schools within a higher grouping were better performers than any school in each of the lower groupings. In this sense, the stratification module could be viewed as a type of benchmarking, since the module creates various different performance groupings of peer-efficient schools to which any underperforming schools could be benchmarked.

However, by using the stratification module's output, it becomes possible to establish performance goals for less efficient schools by benchmarking their current energy and waste usage against the more efficient operations of the schools in any of the higher performing strata. For an inefficient DMU, it is possible to calculate the changes needed to its inputs/outputs for it to become efficient relative to the DMUs on the efficient frontier. These efficiency targets would be the specific values that the inputs and outputs of the inefficient DMU would need to attain in order to move onto the efficiency frontier. For the benchmarking and target setting required in the TDSB case, the set of DMUs in the following procedure would consist of one specific, inefficient DMU under evaluation (i.e. DMU_0) and all of the DMUs in the next higher contextual grouping. Hence, assume that there exists a set of n DMUs, with each DMU_j, $j = 1,\ldots, n$, consisting of m input measures x_{ij}, $i = 1,\ldots, m$, and s output measures y_{rj}, $r = 1,\ldots, s$. Suppose that DMU_0 is being evaluated with x_{i0} and y_{r0} representing its ith input and rth output measures. Let s_i^-, $i = 1,\ldots, m$, and s_r^+, $r = 1,\ldots, s$, represent the ith input and r^{th} output slack variables, respectively, and let ε be some non-Archimedean scalar. Then by solving the optimization model:

$$\min \theta - \varepsilon \left(\sum_{i=1}^{m} s_i^- + \sum_{r=1}^{s} s_r^+ \right)$$

subject to:

$$\sum_{j=1}^{n} \lambda_j x_{ij} + s_i^- = \theta x_{i0} \; i = 1, \ldots, m$$

$$\sum_{j=1}^{n} \lambda_j y_{rj} - s_r^+ = y_{r0} \; r = 1, \ldots s$$

$$\sum_{j=1}^{n} \lambda_j = 1$$

$$\lambda_j \geq 0 \; j = 1, \ldots, n$$

efficiency targets $\hat{x}_{i0} = \theta^* x_{i0} - s_1^-$, $i = 1,\ldots, m$, and $\hat{y}_{ro} = y_{r0} + s_y^{+*}$, $r = 1,\ldots, s$, can be calculated for each of the inputs and outputs of DMU_0. The presence of the non-Archimedean ε in the objective function effectively allows the minimization over θ to pre-empt the optimization involving the slacks, s_i^- and s_r^+. This creates a two-stage optimization process with the maximal reduction of the inputs being achieved in the first stage via the optimal θ^*, followed by the movement onto the efficient frontier achieved in the second stage via the subsequent optimization of the slack variables. This approach to multi-criteria benchmarking proves particularly suitable to practical situations in which no objective or pre-existing engineered standards are available to define efficient or effective performance.

For illustrative purposes, an incremental goal for each school to progress only into the next highest level of efficiency was set. While a longer-term perspective might seek to advance all schools into the very highest level of performers, these intermediate targets establish more attainable improvements that each inefficient school would need to undertake in order to move into the next higher level of efficiency. Hence, the benchmarking module was used to calculate the specific efficiency targets required for each school to proceed into the next higher category of energy and waste performance and Table 2 shows the specific improvements required in each measure in order to reach the calculated targets. The specific changes shown in Table 2 represent: (i) annual reductions in energy use; (ii) annual reductions in energy expenses; (iii) annual reductions in the quantity of solid wastes generated; and, (iv) annual increases in the percentage diversion of solid wastes. While the target-setting module actually produces multi-criteria measures for improvement, if the changes in Table 2 were to be achieved, then one significant subset of these improvements would be the 25 % percent reduction in annual energy costs. This reduction, alone, represents a direct annual cost savings of $3.7 million from the current energy budget of $14.9 million allocated to the TDSB's high school. Furthermore, if the established targets were achieved, the table shows potential reductions of over 300,000 GJ of energy and of 44,000 Kg of waste. It should be noted that an analogous percentage cost, energy and waste improvements would be obtained from the targets set for the entire set of the more than 600 schools of the TDSB.

Rank ordering module

While the stratification module partitions the schools into distinct levels of energy and waste performers, it does not rank order the standings of any of the schools within each grouping. If only a small number of schools were under consideration in an analysis or if a large number of groupings each containing only a very small number of schools had been produced, then the stratification, itself, might be sufficient for actually ranking the specific schools. For the general case, however, the stratification might not prove restrictive enough to permit sufficient degrees of preference discrimination. If this situation proves to be the case, then a contextual attractiveness concept can be incorporated into the DSS to permit an explicit ranking of the schools (Simonson & Tversky 1992; Tversky & Simonson 1993). Obtaining relative attractiveness scores for the schools requires that relative performance be defined with respect to some particular evaluation context and the L partitions from the stratification module can be used to supply these contexts.

Define $H_q^*(d)$ to be the d-degree, $d = 1,\ldots, L - l_0$, contextual attractiveness of $DMU_q = (x_q, y_q)$ from some specific level $\mathbf{E}^{l_0}$, $l_0 \in \{1,\ldots, L - 1\}$. $H_q^*(d)$ can be calculated by solving the model:

$$H_q^*(d) = \min H_q(d) \quad d = 1,\ldots,L - l_0$$

subject to:

$$\sum_{j \in F(E^{l_0+d})} \lambda_j x_j \leq H_q(d) x_q$$

$$\sum_{j \in F(E^{l_0+d})} \lambda_j y_j \geq y_q$$

$$\sum_{j \in F(E^{l_0+d})} \lambda_j = 1$$

$$\lambda_j \geq 0 \quad j \in F(\mathbf{E}^{l_0+d})$$

DMU_q is viewed as a more attractive option than another DMU if it possesses a larger value for its contextual attractiveness measure $H_q^*(d)$. Hence, it is possible to rank each school within each stratum using a direct sorting of these contextual attractiveness scores. Since all schools within a contextual grouping are considered better performers than any school in a lower contextual group, when the rankings within each grouping are subsequently concatenated, a complete, rank ordering of the entire set of selected schools will be produced, *de facto*.

Using the next-lower stratification grouping as the appropriate context, the attractiveness scores $H_q^*(d)$, for each high school were calculated. Since any school is considered more efficient than the other schools within its grouping if it possesses a larger contextual attractiveness score, it now becomes possible to rank order the schools within each stratum by sorting these scores. Concatenating these individually sorted groupings produces a comprehensive rank ordering from 1 to 65 based upon the energy and waste usage in all of the schools. The results of this contextual scoring system, the subsequent sorting within each partition, and the overall efficiency ranking of each high school within the TDSB are shown in Table 3.

If a large number of schools existed within any particular level after the stratification stage, then a situation might occur in which two or more of the schools each received exactly the same attractiveness measure. This tied ranking problem can be alleviated by incorporating a lexicographic ranking modification into the procedure described above and could be accomplished in the following manner. In order to reduce the likelihood of tied rankings, the lexicographic procedure would be used to calculate attractiveness measures for each school by using *every* level lower than it as the contextual basis. Hence, each school in the highest level would receive L-1 separate contextual attractiveness scores, each school in the second highest level would receive L-2 separate

attractiveness scores, and so on. A lexicographical rank ordering of schools within each specific grouping would then be performed with greater emphasis given to attractiveness scores calculated from closer contextual groupings (i.e. through an alphabetical or dictionary style of sorting). This multi-scoring, lexicographic rank ordering process would produce additional discriminating powers by reducing the likelihood of ties occurring between schools within any specific grouping. However, since no tied-scores occurred in the contextual attractiveness calculations already performed, this lexicographic ranking produces the same rank ordering as Table 3.

Conclusions

In this paper, a DEA-based DSS for analyzing, rating, ranking, and benchmarking the multi-criteria energy and waste system of the schools in the TDSB has been studied. Several benefits of this DSS were demonstrated through an illustrative investigation of 65 Toronto high schools. The DSS stratified the schools into similarly-efficient groupings based upon their energy and waste usage. The DSS then generated realistic energy and waste improvement targets for any relatively inefficient school against another grouping of benchmarked schools. Amongst other findings, it was shown that achieving these target reductions would produce system-wide energy cost savings of twenty-five percent. The DSS also permits the ranking of any set of schools by contextually evaluating their relative attractiveness to other identified school groupings. The findings with respect to the high schools have been extended in an analysis of all 600 schools within the TDSB. Based upon the TDSB study, DEA has shown itself to be an extremely useful approach for evaluating, benchmarking and ranking the relative energy and waste performance within the school system, and the potential for its much broader applicability to other applications clearly warrants additional exploration.

Methods

The mathematical models of each of the DEA modules were created as separate computer models within Microsoft Excel worksheets (Albright 2010; Seref et al. 2007; Zhu 2003). The optimization of the spreadsheets of these DEA models was performed using the standard, built-in Excel Solver function. The data for all 600 schools in the TDSB can be found in Christie (2003) and the data for the 65 schools used in the analysis appears in Table 1. All data was stored in a Microsoft Access database. The programming language used to link together the various components was VBA which is contained in all Microsoft Office products. Hence, the mathematical and computer-based DSS for evaluating

Table 3 Overall rankings of high schools based upon contextual attractiveness scores calculated relative to the next lower partitioning group

DMU No.	DMU Name	Attractiveness Score Hq(d)	Overall Ranking
Level 1			
49	School of Experiential Education	0.406	1
26	Inglenook Community School	0.251	2
Level 2			
19	Frank Oke SS	0.539	3
14	Eastdale CI	0.422	4
Level 3			
33	Maplewood HS	0.540	5
22	Greenwood SS / School of Life Experience	0.510	6
24	Heydon Park SS	0.378	7
Level 4			
44	Richview CI	0.862	8
42	Parkdale CI	0.798	9
35	Nelson A Boylen CI	0.725	10
53	Sir William Osler HS	0.640	11
63	William Lyon Mackenzie CI	0.541	12
55	Ursula Franklin Academy	0.531	13
Level 5			
12	Dr Norman Bethune CI	0.889	14
34	Martingrove CI	0.870	15
65	York Mills CI	0.848	16
56	Vaughan Road Academy	0.844	17
47	Runnymede CI	0.821	18
48	Scarlett Heights Entrepreneurial Academy	0.802	19
28	Kipling CI	0.773	20
7	Central Commerce Collegiate	0.728	21
8	Central Etobicoke HS	0.707	22
Level 6			
52	Sir Robert L Borden BTI	0.914	23
25	Humberside CI	0.895	24
30	Lawrence Park CI	0.861	25
16	Emery CI	0.826	26
32	Malvern CI	0.795	27
31	Leaside HS	0.763	28
5	Bloor CI	0.757	29
18	Etobicoke School of the Arts	0.745	30
59	West Toronto CI	0.712	31
46	Rosedale Heights SS	0.642	32
Level 7			
54	Thistletown CI	0.896	33
64	Woburn CI	0.859	34
57	Victoria Park SS	0.825	35

Table 3 Overall rankings of high schools based upon contextual attractiveness scores calculated relative to the next lower partitioning group *(Continued)*

3	Bendale BTI	0.823	**36**
37	North Albion CI	0.820	**37**
11	Downsview SS	0.816	**38**
38	North Toronto CI	0.769	**39**
15	Eastern HS of Commerce / Subway Academy I	0.760	**40**
29	Lakeshore CI	0.716	**41**
41	Oakwood CI	0.671	**42**
Level 8			
4	Birchmount Park CI	0.852	**43**
23	Harbord CI	0.769	**44**
9	CW Jefferys CI	0.724	**45**
21	Georges Vanier SS	0.699	**46**
27	Jarvis CI	0.647	**47**
Level 9			
36	Newtonbrook SS	0.905	**48**
45	Riverdale CI	0.851	**49**
20	George Harvey CI	0.840	**50**
61	Weston CI	0.792	**51**
50	Silverthorn CI	0.787	**52**
13	Earl Haig SS	0.725	**53**
60	Western Technical-Commercial School / The Student School	0.547	**54**
Level 10			
10	David & Mary Thomson CI	0.756	**55**
1	Agincourt CI	0.709	**56**
17	Etobicoke CI	0.524	**57**
6	Cedarbrae CI	0.510	**58**
43	RH King Academy	0.418	**59**
Level 11			
62	Westview Centennial SS	0.706	**60**
40	Northview Heights SS	0.541	**61**
51	Sir John A Macdonald CI	0.534	**62**
Level 12			
58	West Hill CI	0.923	**63**
39	Northern SS	0.889	**64**
2	Albert Campbell CI	0.657	**65**

the performance of the TDSB was created using software residing on essentially all personal computers.

Abbreviations
DSS: Decision support system; DEA: Data envelopment analysis; TDSB: Toronto district school board; DMU: Decision making unit; VBA: Visual basic for applications.

Competing interests
The author declares that there are no competing interests.

Acknowledgments
This work was supported in part by grant OGP0155871 from the Natural Sciences and Engineering Research Council. The author would like to thank Richard Christie of the Toronto District School Board for providing access to the energy and waste data of the various schools.

Author's contributions
JSY is the sole author of this paper, and read and approved the final manuscript.

References

Albright SC (2010) VBA for Modelers: Developing Decision Support Systems. Cengage Learning, Mason, OH

Camp RC (1995) Business Process Benchmarking, Finding and Implementing Best Practices. ASQC Quality Press, Milwaukee, WI

Christie R (2003) The Case for an Annual Environment Report at the Toronto District. Department of Environmental Education, School Services, Toronto District School Board

Christie R (2007) Environmental Report. Department of Ecological Literacy and Sustainable Development, Toronto District School Board

Christie R, Coppinger F (2006) Toronto District School Board Energy Conservation Report. Department of Ecological Literacy and Sustainable Development, Toronto District School Board

Cook W, Zhu J (2008) Data Envelopment Analysis: Modeling Operational Processes and Measuring Productivity. CreateSpace, Charleston, SC

Cook W, Kress M, Seiford L (1996) Data Envelopment Analysis in the Presence of both Quantitative and Qualitative Factors. J Oper Res Soc 47:945–953

Lin QG, Huang GH, Bass B, Chen B, Zhang BY, Zhang XD (2008) CCEM: A City-cluster Energy Systems Planning Model. Energy Sources, Part A: Recovery, Utilization, and Environmental Effects 31:1–14

Linton J, Morabito J, Yeomans JS (2007) An Extension to a DEA Support System Used for Assessing R&D Projects. R&D Management 37:29–36

Seref MMH, Ahuja RA, Winston WL (2007) Developing Decision Spreadsheet-Based Decision Support Systems. Dynamic Ideas, Boston

Simonson I, Tversky A (1992) Choice in Context: Tradeoff Contrast and Extremeness Aversion. J Mark Res 29:281–295

Tversky A, Simonson I (1993) Context-Dependent Preferences. Manag Sci 39:1179–1189

Yeomans JS (2004) Rating and Evaluating the Combined Financial and Environmental Performance of Companies in the Metals and Mining Sector. Journal of Environmental Informatics 2004(3):95–105

Zhu J (2003) Quantitative Models for Performance Evaluation and Benchmarking. Kluwer Academic Publishers, Norwell, MA

16

Factors affecting the value of environmental predictions to the energy sector

Matt Davison[1*], Ozgur Gurtuna[2], Claude Masse[3] and Brian Mills[4]

Abstract

Background: Energy extraction, production, and transmission systems are highly sensitive to states of the natural environment such as temperature, wind speed, and even ice cover. Forecasts of such state variables are termed environmental predictions. How much value can such environmental predictions provide to the operator of a given energy system? This paper presents three illustrative Canadian case studies, selected to provide a good cross section across sectors, forecast types, and decision time scales, to provide insights into this important question.

Results: Using these case studies this paper examines what distinguishes economically valuable forecasts from economically less valuable forecasts. It is found that the risk aversion of the decision makers, the degree to which the decision has multiple inputs, and the certainty of the forecasts, together with the sensitivity of the system to the environmental variable in question, all play important roles.

Conclusions: To the extent that risk aversion results from government regulations and organizational guidelines, the conclusions suggest changes in public policy and industry practices that could help unlock value from currently underused forecasts.

Keywords: Weather information, Energy, Electricity, Value, Decision, Economics

Background

Energy is a fundamental input to the global economy and modern society. Despite technological advances, many aspects of the energy web-demand, production, transmission and distribution-remain sensitive and vulnerable to the fluctuating states and elements of the natural environment, as thoroughly reviewed in (Larsen, 2006). For example, temperature and humidity strongly influence electricity and natural gas consumption (Teisberg et al., 2005) while the rate of inflow of water to hydroelectric generating facilities affects the production of electricity (Hamlet et al. 2002). Tropical storms (Considine et al., 2002) and the trajectory of pack ice and icebergs in northern waters can both have a huge impact on natural gas and oil production.

Science and its institutions have developed a degree of skill in predicting temperature, wind speed, hurricane tracks, and other environmental variables and their relationships with aspects of the energy system. While most people think of this as the domain of public meteorological and hydrometeorological agencies, the specialized private weather forecasting sector is occupying a greater share of this niche. Environmental predictions (EPs) provided by these enterprises become valuable to the extent that they reduce uncertainty in ways meaningful to decisions or actions taken by stakeholders in the energy system. The value of meteorological forecasts to the entire British economy is quantified by (Teske & Robinson, 1994), a broader goal than that presented here. Efforts to quantify the value of EP to the energy sector are surprisingly absent in Canada, an energy-rich nation exposed to variable weather patterns.

The focus of this paper is on demonstrating the economic value of environmental prediction information to selected agents or enterprises involved in the production, transmission and distribution of energy. The aim is not to make an exhaustive estimate of the value of every possible EP to every participant in the Canadian energy sector but rather, by means of case studies, to identify factors allowing greater or lesser value to be extracted

* Correspondence: mdavison@uwo.ca
[1]Departments of Applied Mathematics and Statistical & Actuarial Sciences and the Richard Ivey School of Business, The University of Western Ontario, London, ON, Canada
Full list of author information is available at the end of the article

from EP. Thus, while the goals of this paper are narrower than (Teske & Robinson, 1994), they are broader than the results reported by (National Renewable Energy Laboratory, 1995).

While the main objective of this work is to quantify the economic value of EP, broader notions of value, extending beyond the realm of cost reductions and profit maximization, should also include societal benefits such as reductions in adverse health effects from air pollution reductions. While not explicitly treated here, these broader metrics are valid and necessary elements of determining the societal value of better environmental predictions.

An overview of the environmental prediction valuation methodology is described. Several case studies highlighting specific weather and climate-sensitive aspects of the Canadian energy sector are then used to illustrate some of the non-environmental factors governing how the economic value of a prediction can significantly vary. The case studies are selected so much as to be representative of the Canadian energy environment but to illustrate the impact of various external criteria on forecast value. We agree with (Considine et al., 2002) that non-risk averse or "expected value" decision makers are able to extract more value from (necessarily) imperfect weather forecasts than are safety oriented, worst case scenario minimizing decision makers. This study makes the added point that, even in the absence of significant risk to human life and welfare, the nature of regulatory constraints on energy system operators can influence whether they behave as expected value or worst case decision makers, and hence on their ability to extract value from EP. The paper concludes with implications for policy and future research. A complete account of this research is reported in (Turquoise Technology Solutions 2007).

Methodology for valuing an environmental prediction

Environmental predictions may allow entities in the energy sector to make better decisions. For example, knowing it will be hot tomorrow allows energy producers to plan for the resulting high demand. However, the forecast only has value if it reduces uncertainty: the forecast of a hot July day will be much less valuable than the forecast of an unexpected May heat wave. The "net benefit" value of the forecast can be quantified by determining the added financial value of these better informed decisions (Mason, 1966). In addition, entities must be able to integrate weather forecasts into their decision structures in order to benefit from a particular forecast (Stewart et al., 2004). Of course, for a forecast to be useful it must save more money than it costs to produce, with the caveat that there may be multiple uses for the same forecast.

A variety of approaches have been developed to value EP-like information as reviewed in (Turquoise Technology Solutions 2007). Early efforts to quantify the value of weather information are synthesized by (Macauley, 2005) which also sheds light on the cost/benefit decision of purchasing additional pieces of information. The latter decision problem is also described by several governmental bodies in (Danish Ministry of Transport, 2006; WMO Secretariat, 2007; NAV Canada, 2002). Early papers applying a classical decision science approach include (Baquet et al., 1976), who quantified the value of frost forecasts to orchard operators deciding whether or not to run heaters. A more general theoretical framework for this valuation problem was provided by (Hilton, 1981) while (Kite-Powell 2005) applied the theory to an ocean environmental data application.

The expected value is a cornerstone in many of these valuation frameworks. For example, suppose a power company must decide how to bid tomorrow's production from their various generation assets into tomorrow's market. Because this process is repeated every day and the possibility of bid leading to a catastrophic loss is very low, it makes sense to measure the effectiveness of a bidding strategy by its average performance, or expected value. In such a situation a slight edge in predicting temperature, hence demand, and hence electricity price, might have a small daily value but a large annual value.

However, the expected value is not always an appropriate metric. Hurricane forecasts to drilling rig operators in the Gulf of Mexico are valued by (Considine et al., 2002) who suggest that, unless a forecast brings near certainty, it is worth little. Oil rigs will be evacuated even at a slight chance of catastrophic weather conditions because the costs of a missed evacuation dwarf that of an unnecessary evacuation. An incremental improvement in the accuracy of the forecast is unlikely to change this decision.

Results

The current paper aims to draw qualitative conclusions about the characteristics of situations leading to valuable environmental forecasts. Its goal is not to quantify the value of all forms of EP to all sectors of the energy system. It proceeds by analyzing three representative cases. For more information about how these cases were selected and constructed, please see the Methods section.

Electricity demand and temperature

The framework for this case study summarizes the U.S. focused studies of (Teisberg et al., 2005) and (Hobbs et al., 1999) and use similar stylized facts about the general behaviour of energy systems. Although these works are not focused on Canada, the similar power systems and climate conditions of the two countries allow useful comparisons.

Because economically meaningful quantities of electricity cannot be stored, generation must match consumption in a dynamic manner. Consumption varies tremendously due to weather (i.e., air conditioning during heat waves and heating during extreme cold) and more regular patterns of use (e.g., lower demand during the night and on weekends or holidays). Electricity is most economically generated in large plants which produce it at a constant rate. Generating electricity at a moment's notice is much more expensive. For this reason, accurate demand forecasts, which in turn require accurate temperature forecasts are very valuable. This case study further develops the short term supply and demand aspects of this issue and outlines how the economic value of forecasts can be approximated.

Power demand is low at night and high during the day and tends to be lower on weekends and public holidays. Because of heating and cooling requirements, exterior temperature is a key power use driver. Figure 1 shows a plot of Ontario power demand against the temperature at Toronto's Pearson airport at two different hours of the day. The figure shows that when days are cold, power demand rises when temperature falls, while on hot days power demand rises with temperature. Since Ontario does not have a spatially uniform temperature, this relationship could be improved by use of temperature forecasts at various locations, but the figure is nonetheless instructive. It is also important to note that most of the other power use drivers, such as number of daylight hours and timing of weekends and public holidays are predictable and so do not require sophisticated forecasts

Matching this fluctuating demand requires an understanding of the economic and engineering features of the commonly used generation sources. A typical electrical system includes several generating technologies with different marginal costs of production. These plants are dispatched using solutions of the "unit commitment

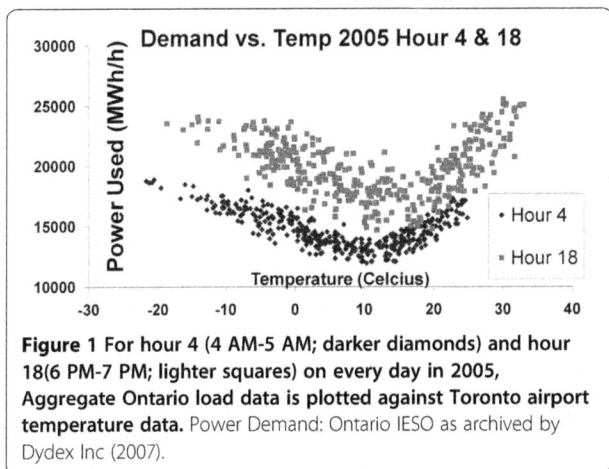

Figure 1 For hour 4 (4 AM-5 AM; darker diamonds) and hour 18(6 PM-7 PM; lighter squares) on every day in 2005, **Aggregate Ontario load data is plotted against Toronto airport temperature data.** Power Demand: Ontario IESO as archived by Dydex Inc (2007).

problem" (Baldick 1995) that, for a given level of production, use the cheapest sources of power first. Power demand varies over the day, with predictable and unpredictable components, so unit commitment must be performed repeatedly to find the best variable generation mix over the course of the day. The dynamic nature of this process creates a complication since the output of some generation types cannot be adjusted quickly, if at all. The need to take these flexibility characteristics into account can result in plans which no longer use the absolute low-cost producer. For example, nuclear power must typically always be dispatched.

Highly responsive plants are therefore required to enable the variation of these plans. Two main methods can be used for this purpose: spinning reserve and Automatic Generation Control (AGC). A unit is part of the spinning reserve if, though ready to immediately do so, it is not generating any power. Plants running in AGC surrender operational control to the system operator, which operates them to balance system load. The technical and environmental issues around spinning reserve and AGC are reviewed in (NOAA/NESDIS 2002). AGC units are able to rapidly increase or decrease their output, as commanded, to respond to load changes. Thermal power plants operating in spinning reserves still consume significant amounts of fuel to "keep their boilers hot". Therefore improvements in temperature (and hence load) forecast accuracy can create both economic and environmental value by reducing the unnecessary use of spinning reserves. In a U.S. study, (Teisberg et al., 2005) found that accurate 24-hour temperature forecasts are worth US$166 million per year in optimizing the generation mix.

It should be noted that the high value reported here results from the impact of small daily profits repeated day after day. This repeated decision framework means that small statistical improvements in decisions will, over time, contribute to significant value.

In contrast to the simple single variable/single repeated decision setting of this case study, there are situations when multi-factor forecast information may be valuable for tactical, operational and strategic decisions. These are discussed in the following section within the context of hydroelectric power.

Hydrology case study

Hydroelectric generation depends strongly on water inflows and water levels (Zhao & Davison 2009a). Both flow rates and water levels vary over time, and this variability has a significant impact on the generation capacity of utilities. Indeed, (Hydro Québec, 2006) identifies hydrological inflow risk as its largest single financial risk factor, surpassing interest rate risk and foreign exchange risk.

This case study quantifies the potential value of sophisticated hydrological EP to hydroelectricity companies across operational and tactical decision scales. The use of the term "potential" signifies the curious absence of EP in current industry practices.

The value of hydrological EP to operational decisions

On hourly time scales, accurate hydrological inflow predictions can be used to buffer electricity demand fluctuations much as temperature predictions were used in 3.1. Such forecasts can be obtained by accurately measuring upstream rainfall forecasts and using these as inputs to river flow models. This value can be quantified using the framework of (Hobbs et al., 1999).

Quick-response hydro allows the amount of water sent through the turbines to be changed on short notice. In jurisdictions with little hydro, the equivalent balancing services must be provided by more expensive spinning reserve, as discussed in section 3.1. The more accurate the short-term prediction of water inflows, the more prominent will be hydro's balancing role.

The value of accurate short term forecasts of water inflows can be captured by treating the power generated by a turbine as a negative source of demand, reducing the power required from the rest of the system. We ignore any diversification effect inherent in combining inflow and load uncertainty and consider only the savings attributable to EP by requiring less backup for uncertain hydro inflows. Since the more uncertainty the more valuable the forecast which reduces this uncertainty, including diversification would reduce this forecast value. This approach to valuing short term hydrological forecasts lies within the theoretical framework of (Hobbs et al. 1999) which allows the value of a decrease in mean absolute percentage error (MAPE) of a load forecast to be assessed. Specifically, a 1% reduction in MAPE is reported to decrease variable generation costs by 0.1%–0.3% when MAPE is in the range of 3–5%, while (Ranaweera et al. 1997) suggest that MAPE is between 2% and 5%.

In Alberta, (very seasonal) hydroelectric power generation accounts for about 1000 MW of an 11,500 MW market. During peak flow (freshet) season, when power output is highest, the output variability is negligible. At the end of the fall, when natural water flows are lowest, variability is also negligible as flows are the result of planned reservoir withdrawals. The following analysis is based on dams operated by TransAlta corporation on the Bow River and South Saskatchewan river systems and uses information provided by a TransAlta hydrologist (Lin 2006). The two rivers have appreciable variability in water inflows for 3 months of the year. This variability is measured by a capacity factor, the ratio of average annual river energy production to the amount which would be produced if the generating stations were always working at

full output). This capacity factor is 31% and 21% for the Bow and North Saskatchewan river systems respectively, so we assume a capacity factor of 25% in the following calculation. During this season, the hydrological inflows uncertainty at a given dam is about 10% over the 2-day time window. As the TransAlta corporation alone has 13 dams on the Bow and North Saskatchewan River rivers systems we can suppose that the actual variability over the entire system is scaled down, by diversification and adopting an independence assumption, to $10\%/\sqrt{(13)}$, or about 3%. If hydrological forecasts could reduce uncertainty in operational scale river inflow forecasting by an estimate of about 1/3 they would therefore result in a further 1% improvement of hydro system MAPE.

This improvement saves about 0.2% of the variable generation cost during that same time period (Hobbs et al. 1999). Saving 0.2% of the $50/MWh marginal cost natural gas fuelling Alberta's variable generation is $0.10/MWh. Over three months, a 1,000 MW facility working at 25% capacity factor generates 546,000 MWh of electricity, implying overall cost savings of $55,000 per year for Alberta. While this conservative estimate is based on strict assumptions, even multiplying by ten would not generate material savings.

Hydrological EP can also play an interesting role in the timing decisions of hydro operators when it comes to generating power now or storing water for later. The operator of a hydroelectric facility with water storage ability can profit from fluctuations in the price for electrical power. If a power system contains a mix of hydroelectric and other variable output fossil fuel generators, during periods of low power prices water can be held back and the amount of hydro power generated can be decreased, being replaced by fossil fuel generated power. The amount of water saved can then be channelled through the turbine during high power price periods. Implementing this strategy requires overcoming challenges in predicting both power prices and water inflows.

In regulated markets with time-of-day pricing, the price is deterministic. Even in deregulated markets, (Davison et al. 2002) show that the price results from the balance between the load and the supply of power. Over short time horizons, the power supply is largely predictable, barring unlikely unplanned outages. The short term power demand is almost entirely determined by foreseeable time-of-day and day-of-week effects and the temperature, which can be predicted accurately over 1–3 days (see section 3.1).

In order to isolate the value of improved hydrological EP in this decision problem, (Zhao & Davison 2009b) consider a simple model of a hydroelectric facility operating in a market with time-varying but deterministic power prices and hourly water inflows following a simple random model. They compute the optimal facility operating strategy and the resulting facility expected value using a dynamic programming approach. By modelling the problem with a 48-hour time horizon, (Zhao & Davison 2009b) assume that, while the form of the stochastic model governing the future inflows is known, the actual inflows are unknown. Stochastic dynamic programming is used to calculate the optimal control and the associated value of the pump storage facility. Next, they assume that the future inflows, although generated from sampling within the same stochastic model, are known and use deterministic dynamic programming to obtain the optimal controls and associated values for each set of inflows. Finally, using Monte Carlo techniques they compute the average of these two values. The first optimal control assumes random inflows, the second assumes deterministic inflows simulated from the same distribution as the random inflows. They conclude that an accurate 48-hour flow forecast increases the value of the hydroelectric facility by about 2%.

In order to determine the value of this prediction capability for the Alberta setting, we use the same base case scenario, but this time with storage as an additional model feature. If we take the 2% savings at face value, then the annual value of perfect 48-hour load forecasts to Alberta will be approximately $3 million per year. Since for 9 months of the year water storage is not an option and the best strategy available is "use it or lose it", even perfect forecasts cannot yield value. Hence, on an annualized basis, $750,000 is the appropriate prediction value.

The value of hydrological EP to tactical decisions

In large watersheds longer hydrological forecasts are possible. Reservoir levels can be used to manage seasonal fluctuations in demand as well as to profit from seasonal price fluctuations. This is done by selling hydro production to neighbouring markets when prices are high and, when prices are low, conserving water in the reservoirs for future use.

The ability of such "water trading" entities to generate profits from time varying electricity prices in neighbouring markets is limited by the ability to store water. Even the largest reservoirs cannot prudently be filled above an upper threshold nor drawn down below a lower threshold. If future water inflows are uncertain, these "safety limits" must be computed by taking into account certain risk factors. If, on the other hand, future water inflows can be well characterized, water supplies can be managed much more aggressively. This approach has significant potential economic value as described by (Hamlet et al. 2002).

Watersheds are typically used by many stakeholders. Rivers provide hydroelectric generation, recreational opportunities, and unique habitats for fish and other flora

and fauna, and the tension between these demands constrains the use of the rivers. Moreover, regulators tend to be very risk averse and prefer erring on the more conservative side when it comes to managing water levels. Such conservative protocols, while ensuring adequate water levels at all times and for all purposes, also have the negative effect of leaving a lot of unused water in most years-water which could have been used to generate valuable hydroelectric power. If a better forecast of extreme levels is possible, and if this forecast is accepted by regulating bodies, it can actually provide significant economic value.

The information provided by TransAlta hydrologist (Lin 2006) together with data on variable Alberta electricity market prices, was used to obtain a rough estimate of the value of annual EP projections for the Alberta market. In addition to the system capacity factor described in section 3.2.1, an important parameter for a hydroelectric system is the storage capacity which denotes the ratio of the amount of water that can be stored in the reservoir to the average annual flow rate. This storage capacity is 24% for the Bow River system and 44% for the North Saskatchewan river. TransAlta's reservoirs are drawn down to their minimum capacity each spring to prepare them to accept as much of the spring runoff as possible.

In Alberta the power price fluctuates according to a deregulated market. According to the Alberta Electric System operator (2007), in 2005–2006 power could be sold for $116.31 per MWh during low flow October-December times ($142.14 during "on peak:" daytime weekday hours), while in high flow April-June average prices were much lower at $50.06, $61.14 for on peak power. This price variability makes the ability to save water for use in October-December very valuable: we suppose that such storage could allow power to be sold for $100/MWh rather than $50/MWh.

The Columbia river study of (Hamlet et al. 2002) suggests that the difference between a dry-and a wet-year is substantial-this is also the case in the intermittent Alberta market. Assume that an EP application for annual Alberta flow rate allowed for an additional 10% of the high capacity spring runoff water to be stored for release in the late summer or winter high price season. In this case, the value of EPs for the Alberta market is very significant: about $3.6 million (accounting for the difference between the roughly $100/MWh available during the high demand, low flow months and the roughly $50/MWh available during the spring months). Despite this large potential value, our interviews did not reveal significant use of seasonal forecasts in Alberta, a point of discussion we leave for section 4.

In summary, the value of temperature forecasts to the electricity industry relies to a great extent on the fact that expected value decision making is possible. An agent like a utility can use weather information to support its decisions and take actions that will reduce costs or hedge against financial risk. This is not always the case, as shown in the case study presented in the next section.

Sea ice environmental predictions for offshore oil and gas production

Not unlike the case of hurricanes affecting oil and gas production in the Gulf of Mexico (Considine et al. 2002), both drifting sea ice (pack ice) and icebergs can be a threat to offshore oil platforms in northern waters, such as those tapping the Jeanne D'Arc basin located below the Grand Banks, East of Newfoundland (Hibernia, Terra Nova, and White Rose). These three sites respectively produce about 314,000, 150,000 and 100,000 barrels per day of light crude oil (Offshore technology, 2006). All three locations are at risk not only from icebergs, which given the shallow 80 m depths can scour the ocean floor and damage underwater infrastructure, but also from floating sea ice at thicknesses of 1–2 m. It is interesting to note that different oil platform technology has been selected for these sites.

The Hibernia Oil Platform is a fixed "gravity based structure" (GBS) which is designed to be nearly invulnerable to the encroachment of sea ice or icebergs. This invulnerability comes at the heavy financial cost of $1 billion for the additional reinforcement (Ralph, 2007). In contrast, the extraction at the Terra Nova and White Rose sites is accomplished by means of a floating production, storage, and offloading vessel (FPSO) that can be disengaged in response to a severe ice threat. In the latter case, the timing of the decision to disengage is critical. If the vessel is moved in response to every threat, production will be delayed at a high cost. On the other hand, if the vessel never leaves, it will be at risk to damage from ice. Coupled with this set of decision processes is a complicated threat management environment which allows sea ice and icebergs to be diverted from the structure using icebreakers and even water cannons. Environmental predictions of the trajectory of floating sea ice and icebergs are an integral part not only of the threat management system but also of the evacuation system.

To obtain insight into the challenges of incorporating forecasts into sea ice/iceberg management for offshore operations, simple expected value decision experiments for the FPSO case are discussed below. However, while the risk to life and limb from floating sea ice is low (Ralph, 2007; Kirby 2007) icebergs present significant catastrophic loss risk. The need to heavily weigh such catastrophic events in decisions will decrease the following estimates.

The decision may be framed by assuming that the platform is surrounded by three concentric circles, denoting various threat levels. Each successive breach by an iceberg or sea ice reduces the time remaining to the decision maker and forces an intermediate shutdown so that, if necessary, there is enough time to evacuate.

Assume that the FPSO services an oil well, modelled on White Rose, with a daily production of 100,000 barrels of light crude (sold on the international market), and a production lifetime of 10 years. An ice threat (sea ice or iceberg) arriving at the outer perimeter of a given concentric circle will take a given time to hit the oil platform.

The following analysis is constructed to model how a decision maker would incorporate ice forecast information in FPSO movement decisions. It does not represent detailed industry practice. In the absence of forecasts, the probability of ice breaching a given area is assumed proportional to that area. Additional information about the direction from which ice threats typically come will change this scaling relationship (if ice always comes from the north, the probability of ice breaching a circular region is proportional to its diameter).

For indicative purposes only, assume that the outer circle has a radius three times and the middle circle a radius twice that of the inner circle. The resulting probability of ice breaches must follow in the progression 9:4:1. For the purpose of this analysis, we assume that, during an assumed 30-day ice season, the outer circle is breached 9% of days. Countermeasures involving serious economic loss need only be initiated as the second circle is breached, so this corresponds to an assumption of approximately 1.33 times per year of second circle events and approximately once every 3 years of extremely serious, inner circle breach events.

A threat protocol to safeguard the oil platform workers and the natural environment is instituted so that when the outer circle is breached by ice, the steps of a shutdown and evacuation sequence are initiated. We consider a three-level threat model. When ice is outside the outer concentric circle, a threat level of zero exists. When ice breaches the outermost circle, the FPSO makes ready for sea, incurring a modest cost in fuel burned. If the ice leaves the outermost circle again, this step can be reversed. If, on the other hand, ice continues to breach the second circle, the FPSO must shut down oil production, at a much heavier cost. Even ignoring any petroleum engineering consequences, the result is to defer production of oil from the present to the end of the reservoir's life. This incurs a known cost of 100,000 x today's oil price, with a countervailing benefit of 100,000 x the oil price 10 years from now, discounted to the present day at an appropriate interest rate (spread over the London Interbank Offered Rate, LIBOR). If we assume that today's oil price is US$70 per barrel, discounted at 6%, and that the value of oil grows simply with a 2% inflation rate, the cost of deferring a day's production is about $2.25 million. Naturally, this result is very sensitive to the underlying oil price assumptions.

Further, assume that the ice leaves the middle circle after a single day (either to breach the inner circle or to retreat to the outer circle), and that if the ice retreats to the outer circle, production can be restarted immediately. If the ice breaches the innermost circle, the FPSO needs to disconnect from the wellhead and steam away. The cost of this is estimated to be an entire week of lost production-to cover the time required to steam to safety, return, and perform the delicate operation of re-connecting to the wellhead.

A mobile escort vessel armed with water cannon and having tug and icebreaking capabilities can provide safety countermeasures. Such vessels cost about $25,000/day (Ralph, 2007) and are employed to redirect (with cannon or towing) icebergs out of the threat area and to break up floating sea ice.

We assume that employing such a vessel without using forecasts clears the threat at each stage with a fixed probability. Employing a vessel will be nearly worthwhile if its use for a single day reduces the probability of a middle circle breach lasting one day by even 1%. However, the vessel cannot come on a day-by-day basis. It must be engaged monthly at a cost of $750,000. Even though the Hibernia, White Rose, and Terra Nova oil platforms lie within a few hours steaming of one another so perhaps could share a single vessel, the cost of engaging it may not be worthwhile if the probability of an ice-related shutdown is small.

The expected value of environmental prediction information is assessed by carrying through two sets of calculations, one with and one without ice forecasts. The base case computes the expected annual cost of ice to an offshore platform similar to the White Rose without any use of countermeasures such as escort vessels nor any use of environmental prediction. This provides a baseline against which the other strategies can be judged. Four percent of the time the middle circle is breached. A middle circle breach requires a one day production shutdown at a cost of one day's deferred production, estimated to be $2.25 million. As we assume 30 ice risk days in a year so the total cost of this middle circle breach is $30 \times 0.04 \times \$2.25 = \$2.7$ million. Conditioned on the middle circle being breached, one-quarter of the time the inner circle is breached next. An inner circle breach requires a shutdown for 7 days (after which it is assumed that the ice has left the outermost threat circle) at a cost of 7 days of lost production. Again, there are 30 ice days in a year so the cost of an inner circle breach is $4.725 million. So the total

expected annual cost of ice, with no countermeasures and no use of environmental prediction, is about $7.4 million.

A more realistic variation of the base case assumes that an escort vessel is employed for the 30 days. The cost of the escort vessel is shared between 3 platforms at a daily cost of $25,000 for a total cost of $0.25 million per platform. We assume that 10% of the time, the escort vessel is able to stop a floating ice threat from moving from one threat level to the next higher level. When the escort vessel is present, the middle circle is breached only $4\% \times (0.9)^2 = 3.24\% =$ of the time, and the inner circle is breached just $(0.9)^3 \times 1\% = 0.729\%$ of the time. So the expected loss due to ice decreases to $5.6 million. As the escort vessel costs $250,000, the total expected loss with escort vessel is $5.85 million. This represents a savings of $1.55 million relative to the original base case, suggesting that with the crude assumptions employed here, an escort vessel is well worth the money even in the absence of EP.

What is the effect of EP on the latter variation of the base case? In essence, good EP allows the threat regions to be shrunk, reducing the probability of having to engage in costly countermeasures. The better the EP, the more comfortable we can be with nearby ice and the more confident we can be about the escort vessel's ability to avert threats. Under reasonable assumptions quantifying these effects (Turquoise Technology Solutions 2007), the net effect is to save $3.98 million, a substantial improvement from the $1.55 million saved by the escort vessel alone or of the $740,000 saved by passive EP alone. We can therefore attribute a significant savings of $3.98 million - $1.55 million = $2.4 million to the incremental benefit of EP in this context.

Discussions with stakeholders (Dewhurst, 2007) emphasized that weather forecasts were used by stakeholders on an operational level, but mostly to minimize danger due to high winds and high seas rather than to forecast ice movement. Because of the time pressure under which such decisions are taken and their multifactor nature, it is difficult to isolate the value of a decision framework such as that described here.

The case studies presented here suggest that predictions of environmental variables ranging from temperature to water inflows and levels to the motion of floating ice in the ocean all have value to different participants in the Canadian energy industry. However, a recurring theme arising in our discussions with many stakeholders was that, despite this potential value, the full possibilities of environmental predictions were rarely being used. In the next section of this paper we address the reasons for this apparent failure and present some policy recommendations which could release more value from these forecasts.

Discussion

Why isn't EP used where results suggest it would be very beneficial? For example, before 2007, the Alberta utility we interviewed was not using hydrological EP to manage their hydroelectric assets, although they were at that time considering the purchase of some hydrological decision support tools. One reason for this lies in the extreme complexity of making hydrological forecasts in the highly variable Rocky Mountain fed waterways. The priorities of other river stakeholders must be respected, and legal structures forbidding water levels below or above given thresholds must be obeyed. The addition of these constraints means that even if information allowing positive expected value decisions exists, the potential for the worst case arising from these decisions to breach legal limits eliminates the value to act on this information.

This dynamic was seen to be operating in the (Hamlet et al. 2002) analysis of hydrological prediction strategy on the Columbia river watershed. The U.S. Army Corps of Engineers restricts outflows on the Columbia River watershed using very conservative "observed worst year" or only slightly less conservative "observed third worst year" data. Worst case or "minimax" decision makers are often in the poorest position to make use of probabilistic EP, since the reduction of a remote, though possible, event to an even more remote, yet still possible event does not change worst case scenarios at all. Therefore, for conservative organisations such as the Corps of Engineers which are more concerned with mitigating ill effects than in optimizing frequent effects, minimax decision making is a natural tendency. This factor came up repeatedly in decisions involving physical as well as financial danger, negatively impacting the practical value of EP (for example in the case of sea ice forecasts).

Another reason for not using environmental forecasts which arose in discussions with both hydrological and ocean rig stakeholders was the multifaceted nature of decisions. When a decision depends on many variables, reducing uncertainty in just one of the variables may not be that valuable, even if reducing uncertainty in all of them might be. Such an example impacts the subject of the current study since the value of an improved forecast for just one of two uncertain inputs to a decision is often hard to realize. Even if the forecast is important, quantifying its value to the decision can be very difficult, especially after the fact.

These insights suggest some policy implications. Although extreme caution must be used when taking decisions involving human life and injury, it is possible that some laws could be rewritten in terms of exceedence frequency (e.g., the water level can fall below this limit a maximum of 5 minutes per day) or by replacing outright prohibition by fines. This would allow less than perfect forecasts to be used even by expected value decision makers, with the case-by-case potential to unlock a great deal of value. Of course, care must be taken to change regulations only when the impact on non-economic aspects of value is both acceptable and well understood.

For multi-faceted decisions, other financial tools such as forwards markets and hedging programs can, where appropriate, be used to limit exposure to market fluctuations, allowing the operators of businesses subject to environmental risk to extract more value from long range climate forecasts while controlling the risks related to economic and financial factors.

Conclusions

This study presented an analysis, in a Canadian context, of the value of various environmental predictions or forecasts to various participants in energy markets. The case studies presented show that value is dynamic and varies by decision time scale, location, and situation. The case studies described here were based on expert opinions, plausible assumptions or parameters obtained from the literature, however more research is required to substantiate the effect of improved hydrometeorological information on decisions.

The case studies demonstrated that expected value decision makers who can key on a single factor when making their decisions are much more likely to extract value from probabilistic forecasts than worst case avoiding decision makers. To support this conclusion, we saw that while temperature forecasts are routinely used by power generators in solving the unit commitment problem, use of hydrological forecasts was less common in the operation of Alberta hydro dams. While weather forecasts were used in oil rig operations, they were not yet being used to forecast the motion of floating pack ice and icebergs. To some extent this is a chicken and egg problem – until forecasts get better, they will not be valuable to worst case focused decision makers, but they won't get sufficiently customized until they are used on an operational basis. As a result, framing regulations to enable expected value decisions might facilitate various agents to improve bespoke forecasts and hence extract more economic value from advanced scientific forecasts.

Finally, expanding on the traditional 'economic' definition of value to include social, cultural, and environmental factors might entirely change the final numbers; however, until such definitions become internalized in business or regulatory decision-making, considering them will yield little or no real impact.

Methods

The case studies reported here are selected from those analyzed for the broader study reported in (Turquoise Technology Solutions 2007), which were selected to represent decisions taken over operational (less than one week), tactical (one day to one year), and strategic (more than one year) time scales and were chosen to highlight various EP modalities and various important aspects of the Canadian energy system including energy production, transmission, distribution and consumption.

This paper presented three cases: the value of in the first case temperature and in the second case hydrological forecasts to the electricity generation sector followed by a case developing the value of sea ice and iceberg predictions to the offshore oil and gas production sector. The cases were constructed by incorporating results from U.S. and European studies and developing simple, yet plausible expected value decision experiments and analyses. To the extent possible, real agent experience and expert opinions were used to elucidate relevant aspects of the decision problem such as environmental and economic thresholds.

Competing Interests

The authors declare that they have no competing interests.

Authors' Contributions

This paper arose from a consulting project performed by OG and MD for Environment Canada. BM and CM, who work for Environment Canada, were on the steering panel for this project and were instrumental not only in selecting the case studies presented here from a broader menu but also in introducing OG and MD to valuable contacts and in providing feedback on the project. MD and OG developed the case studies, reviewing the published literature, performing the interviews and creating the models presented here. MD drafted this paper from a more comprehensive report co-authored by MD and OG; all authors gave detailed feedback on this paper is it progressed through the various publication stages. All authors read and approved the final manuscript

Acknowledgements

The authors would like to thank anonymous referees for their insightful comments which have improved this paper. MD thanks the Canadian Natural Science and Engineering Research Council and the Canada Research Chairs Program for financial support.

Author details

[1]Departments of Applied Mathematics and Statistical & Actuarial Sciences and the Richard Ivey School of Business, The University of Western Ontario, London, ON, Canada. [2]Turquoise Technology Solutions Inc, Montreal, Canada. [3]Environment Canada, Montreal, Canada. [4]Adaptation and Impacts Research, Environment Canada, Waterloo, Canada.

References

Alberta Electric Systems Operator: http://www.aeso.org, accessed May 2007

Baldick R (1995) The generalized unit commitment problem. IEEE Trans Power Syst 10:465–475

Baquet AE, Halter AN, Conklin FS (1976) The Value of Frost Forecasting: A Bayesian Appraisal. Am J Agric Econ 58:511–520

Considine T, Jablonowski C, Posner B, Bishop CH (2002) The Value of Hurricane Forecasts to Oil and Gas Producers in the Gulf of Mexico. J Appl Meteorol 43:1270–1281

Danish Ministry of Transport (2006) Meteorology: A Revenue Generating Science: A Mapping of Meteorological Services With An Economic Assessment Of Selected Cases. The Ministry of Transport and Energy, Denmark

Davison M, Anderson CL, Marcus B, Anderson K (2002) Development of a Hybrid Model for Electrical Power Spot Prices. IEEE Trans Power Syst 17:257–264

Dewhurst M (2007) Oceans Inc meteorologist, personal communication

Hamlet AF, Huppert D, Lettenmaier DP (2002) Economic Value of Long-Lead Streamflow Forecasts for Columbia River Hydropower. J Water Resour Plann Manage 128:91–101, March/April

Hilton RW (1981) The Determinants of Information Value: Synthesizing Some General Results. Manage Sci 27:57–64

Hobbs BF, Jitprapaikulsarn S, Konda S, Chankong V, Loparo K, Maratukulam DJ (1999) Analysis of the Value for Unit Commitment of Improved Load Forecasts. IEEE Trans Power Syst 14:1342–1348

Hydro Québec (2006) Annual Report

Kirby N (2007) Personal communication with Mr. Kirby, seaman

Kite-Powell H (2005) Estimating Economic Benefits from NOAA PORTS Installations. A Value of Information Approach, NOAA Technical Report NOS CO-OPS 044, pp 3–5

Larsen PH (2006) An Evaluation of the Sensitivity of the U.S. Economy Sectors to Weather. M.Sc. Thesis, Cornell University, http://hdl.handle.net/1813/2957

Lin LL (2006) (TransAlta Inc) personal communication October 2006

Macauley MK (2005) Discussion Paper 05–26, Resources for the Future. In: The Value of Information: Measuring the Contribution of Space-Derived Earth Science Data to National Resource Management., , http://www.rff.org

Mason BJ (1966) The Role of Meteorology in the National Economy. Weather 21:383–393

National Renewable Energy Laboratory (1995) Estimating the Economic Value of Wind Forecasting to Utilities

NAV Canada (2002) Assessment of Aerodrome Forecast (TAF) Accuracy Improvement

NOAA/NESDIS (2002) Geostationary Operational Environmental Satellite System GOES-R Sounder and Imager Cost/Benefit Analysis

Offshore Technology (2006), http://www.offshore-technology.com/projects, accessed July 2006

Power load data: Ontario Independent Electricity System Operator, http://www.ieso.ca, accessed May 2007

Ralph F (2007) Personal communication with C-CORE. Engineer, July, 2007

Ranaweera DK, Karady GK, Farmer RG (1997) Economic Impact Analysis of Load Forecasting. IEEE Trans Power Syst 12:1388–1392

Stewart TR, Pielke R, Nath R (2004) Understanding User Decision Making and the Value of Improved Precipitation Forecasts: Lessons from a case study. Bull Am Meteorol Soc 85:223–235

Teisberg TJ, Weiner RF, Khotonozad A (2005) The Economic Value of Temperature Forecasts in Electricity Generation. Bull Am Meteorol Soc 86:1765–1780

Temperature data: Environment Canada: 2007. http://www.climate.weatheroffice.ec.gc.ca/ClimateData/canada_e.html, accessed May 2007

Teske S, Robinson P (1994) The Benefit of the UK Met Office to the National Economy. Conf Econ Benefits Meteorol Hydrol Serv 630:21–24, WMO/TD

Turquoise Technology Solutions (2007) Environmental Predictions and the Energy Sector: A Canadian Perspective. Report commissioned by Environment Canada, In

WMO Secretariat (2007) Sustainable Living-Reducing Risks and Increasing Opportunities: The Social and Economic Benefits of Meteorological and Hydrological Services: Issues and Actions

Zhao G, Davison M (2009a) When does variable power pricing alter the behavior of hydroelectric facility operators? Renew Energy 34:1064–1077

Zhao G, Davison M (2009b) Valuing Hydrological Forecasts for a Pump Storage Facility. J Hydrol 373:453–462

17

A hybrid fuzzy stochastic analytical hierarchy process (FSAHP) approach for evaluating ballast water treatment technologies

Liang Jing, Bing Chen[*], Baiyu Zhang and Hongxuan Peng

Abstract

Background: Environmental decisions can be complex because of the inherent trade-offs among environmental, social, ecological, and economic factors. This paper presents a novel hybrid fuzzy stochastic analytical hierarchy process (FSAHP) approach to aid decision making by incorporating fuzzy and stochastic uncertainty into the traditional analytic hierarchy process (AHP). A case study related to ballast water management is used to demonstrate the applicability of the proposed approach. Nine experts from government ministries and academic institutions are invited to evaluate five treatment technologies (i.e., heat treatment, ultraviolet, ozone, ultrasound, and biocide) based on a number of criteria such as efficacy, capital cost, and human risk.

Results: The experts' preferences over the set of alternatives are represented as linguistic terms instead of numerical values. The beta-PERT distribution is adopted to approximate the probability density functions of the values of their inputs. Statistical analysis indicates that ultraviolet has the highest score (0.22–0.24) in most replications and its overlap with the second-best alternative is statistically negligible. Ozone, ultrasound, and heat treatment are mostly found as the second-, third-, and fourth-best alternatives with considerable overlaps that may be reduced if more experts are involved.

Conclusions: As compared with the traditional AHP, the proposed FSAHP approach can not only take into account linguistic information but also capture the uncertainty associated with insufficient information and biased opinions in group decision-making problems.

Keywords: FSAHP approach; Fuzzy; Stochastic; Ballast water management; Group decision-making

Background

Environmental decisions can be complex because of the inherent trade-offs among environmental, social, ecological, and economic factors [1,2]. Many multi-criteria decision making (MCDM) approaches have been developed to facilitate decision making under uncertainty [3-5]. Kornyshova and Salinesi [6] classified them into categories such as outranking methods, analytic hierarchy process, multiattribute utility theory, weighting methods, fuzzy methods, and multiobjective programming. Among them, the analytic hierarchy process (AHP), first proposed by Saaty [7], is one of the most widely used MCDM approaches. It structures the rational analysis of decision making by dividing a problem into hierarchies including goal, criteria, subcriteria (if any), and decision alternatives. One of the most important features or, in other words, the strength of the AHP revolves around the possibility of evaluating quantitative as well as qualitative criteria and alternatives on the same preference scale. Pairwise comparison judgments are given by decision makers using numerical, verbal or graphical scales and are subsequently synthesized to obtain the overall priorities. This comparison enables the AHP to capture subjective and quantitative judgment made by decision makers. Many attempts have been reported in the literature to apply the AHP in problems with high complexity and uncertainty, especially in the environmental sector [8-13].

However, the AHP has been criticized for its inability to quantify the uncertainty associated with decision

* Correspondence: bchen@mun.ca
Faculty of Engineering and Applied Science, Memorial University of Newfoundland, St. John's, NL A1B 3X5, Canada

making [14]. Banuelas and Antony [15] highlighted that the basic theory of the AHP does not allow any statistical conclusion to be drawn. Rosenbloom [16] stated that a small difference in the utilities of alternatives may not be appropriate to conclude that one alternative is superior to the other. Carlucci and Schiuma [17] argued that the AHP is not able to address the interactions and feedback dependencies between the elements of a decision problem. In addition, in many real-world applications, the available information is imprecise, incomplete and occasionally unreliable due to the unquantifiable nature of data or lack of knowledge. Human experts tend to use linguistic terms (e.g., good, poor, excellent) to express their judgments which can not be handled effectively using crisp scales.

To overcome the aforementioned limitations, much research effort has therefore been directed towards taking uncertainties (e.g., fuzzy sets and probability distributions) into account in the AHP. On one hand, to capture linguistic information, Yu [18] employed an absolute term linearization technique and a fuzzy rating expression into a GP-AHP model for solving fuzzy AHP problems. Tolga et al. [13] combined the use of fuzzy set theory with the AHP to address the uncertainty of assigning crisp concepts in decision-making topics. Tesfamariam and Sadiq [19] incorporated uncertainty into the AHP using fuzzy arithmetic operations for environmental risk management. Chowdhury and Husain [8] integrated fuzzy set theory, the AHP, and the concept of entropy to select the best management plan for a drinking water facility. Kaya and Kahraman [10] proposed a hybrid fuzzy AHP-ELECTRE approach for modeling the uncertainty of linguistic expression. On the other hand, to deal with insufficient information and opinion difference in group decision-making processes, pairwise comparison elements were suggested to be viewed as random variables and computed via Monte Carlo simulation by Rosenbloom [16], Eskandari and Rabelo [20] and Jing et al. [21]. To date, triangular distribution is the most commonly used distribution for modeling expert judgment in the AHP [15,22]. However, it may place too much emphasis on the most likely value at the expense of the values to either side [23]. It is possible to overcome this disadvantage of the triangular distribution by using the beta-PERT distribution. The beta-PERT distribution has also been widely used for modeling expert judgments and providing a close fit to normal distributions with less demand for data [24,25]. Although various types of uncertainty have been discussed in the literature, there has been no study investigating the feasibility of incorporating both fuzzy and stochastic uncertainty into the AHP.

In response to this, in this paper, a hybrid fuzzy stochastic analytical hierarchy process (FSAHP) approach is developed by integrating the beta-PERT distribution, fuzzy set theory, pairwise comparison and Monte Carlo simulation. A real-world case study for ballast water management is presented to test the feasibility and efficiency of the proposed approach in a group decision-making environment. Ballast water is carried by ships to acquire the optimum operating depth of the propeller and to maintain maneuverability and stability [26]. It is recognized as the principle source of invasive species and pollutants in coastal freshwater and marine ecosystems, causing severe negative effects on the environment and human health [27-30]. To address the associated concerns, the International Maritime Organization (IMO) has adopted many legal instruments whereby ships will be required to establish a ballast water management system between 2009 and 2016 [31]. Many treatment technologies such as filtration, heat treatment, hydrocyclone, ultraviolet, ozonation, oxidization, electric pulse, and deoxygenation have been tested and applied to remove unwanted species and pollutants from ballast water [29]. However, Gregg and Hallegraeff [32] argued that no treatment option had been shown fully biologically effective, environmentally friendly, safe and practical for onboard applications. In addition, the performance of most treatment processes is likely to be affected by the cold environment and unpredictable weather conditions [26,29]. The evaluation of their applicability and associated risk is of paramount importance and lacks in-depth research. How to choose the best technology from a sustainability metrics perspective still exists as a challenge to the government and other public bodies with environmental responsibilities.

Methods
Fuzzy sets and fuzzy numbers
Zadeh [33] first introduced the concept of fuzzy set theory which was oriented to the rationality of uncertainty due to imprecision or vagueness. Fuzzy set theory is an extension of the classical set theory in which elements have grades of membership ranging from 0 to 1. A triangular fuzzy number (TFN) is defined by its membership function $\mu(x)$ as

$$\mu(x) = \begin{cases} (x-a)/(b-a), & a \leq x \leq b \\ (x-c)/(b-c), & b \leq x \leq c \\ 0, & otherwise \end{cases} \quad (1)$$

where a, b and c denote the minimum, most likely, and maximum values, respectively. Fuzzy numbers are well suited to represent the imprecise nature of judgments, such as linguistic terms used by human experts. Some basic arithmetic operations of fuzzy numbers can be found at Kaufmann et al. [34].

Stochastic programming
Monte Carlo simulation, which applies probability theory to address variable and uncertain phenomena, relies

on statistical representation of available information. It has been widely applied to obtain more detailed information for systems that are too complex to be solved analytically. Monte Carlo simulation in its simplest form involves random sampling from a probability distribution. Various probability distributions (e.g., uniform, normal, beta, and lognormal) have been used in connection with Monte Carlo simulation to model the uncertainty of environmental systems. Banuelas and Antony [15] presented a modified analytic hierarchy process with triangular probability distribution to include uncertainty in the judgments. Li and Chen [35] developed a fuzzy-stochastic-interval linear programming (FSILP) approach for supporting municipal solid waste management by tackling uncertainties expressed in normal probability distributions, fuzzy membership functions and discrete intervals. Jing et al. [25] proposed a Monte Carlo simulation aided analytic hierarchy process (MC–AHP) approach by employing the beta-PERT distribution to prioritize nonpoint source pollution mitigation strategies. Jing et al. [21] further integrated the uniform distribution with interval judgment to a hybrid stochastic-interval analytic hierarchy process (SIAHP) framework for group decision making on wastewater reuse. In this paper, the beta-PERT distribution is employed to model expert judgment. It uses the most likely, minimum, and maximum values of expert estimates to generate a distribution that more closely resembles realistic probability distribution.

Fuzzy stochastic analytic hierarchy process (FSAHP)

The proposed FSAHP approach is capable of capturing not only a human's appraisal of ambiguity but also the uncertainty introduced by the lack of information or scattered opinions. Experts' linguistic assessments are aggregated to approximate a series of beta-PERT distributions for randomized fuzzy pairwise comparisons.

Monte Carlo simulation is then used to generate random fuzzy pairwise comparison matrices (FPCMs), calculate the fuzzy weights, and produce the final scores for each decision alternative. The detailed steps are summarized as follows:

Step 1: Structure the decision problem into a hierarchy of interrelated subproblems that can be analyzed independently. The hierarchy usually includes a main goal, criteria, and alternatives, from the top to the bottom. Each criterion may be further decomposed to a number of lower-level subcriteria as a new level. The goal, criteria, subcriteria (if any), and alternatives can be determined through literature reviews and collective discussions.

Step 2: Linguistic judgments on each alternative and criterion with respect to the elements on the level immediately above can be obtained from experts through questionnaires, surveys, interviews, expert panels, and direct observations. Instead of using a crisp ratio scale, seven TFNs (Figure 1) are used to represent linguistic terms with the expectation that experts will feel more comfortable using such terms in their assessment. It should be noted that such a verbal clarification becomes impractical when too many rating scales (e.g., 10-point format) are involved because the level of agreement becomes too fine to be easily expressed in words [36].

Step 3: For the assessment of each alterative and criterion, the number of TFNs should be equal to the number of experts. The minimum (a), most likely (b) and maximum (c) values of the TFNs are aggregated into three individual groups. In order to generate random TFNs, Equations 2–5 are used to approximate an independent beta-PERT distribution for each group.

$$mean = \frac{min + 4modal + max}{N} \qquad (2)$$

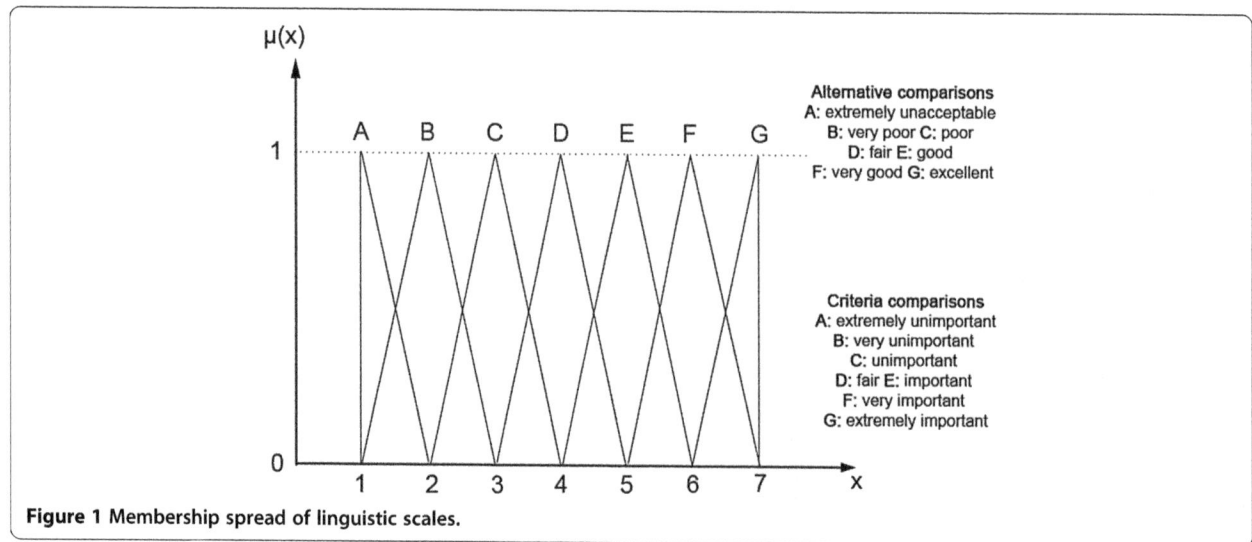

Figure 1 Membership spread of linguistic scales.

$$stdev = \frac{max-min}{N} \quad (3)$$

$$\alpha = \left(\frac{mean-min}{max-min}\right)\left(\frac{(mean-min)(max-mean)}{stdev^2}-1\right) \quad (4)$$

$$\beta = \left(\frac{max-mean}{mean-min}\right)*\alpha \quad (5)$$

where *mean, min, modal, max, stdev* denote the mean, smallest, most probable, largest values, and standard deviations of a, b, and c, respectively; N is the number of experts; α and β are the shape factors. Equations 6–8 are then used to generate random numbers (i.e., $random_a$, $random_b$, $random_c$) that follow the beta-PERT distributions for a, b, and c, respectively. It is noteworthy that the triangular shape needs to be verified to validate these random numbers.

$$random_a = min_a + betarnd(\alpha_a, \beta_a) \\ * (max_a - min_a) \quad (6)$$

$$random_b = min_b + betarnd(\alpha_b, \beta_b) \\ * (max_b - min_b) \quad (7)$$

$$random_c = min_c + betarnd(\alpha_c, \beta_c) \\ * (max_c - min_c) \quad (8)$$

where *betarnd* denotes standard Matlab function (i.e., beta distribution) which returns a random number between 0 and 1.

Step 4: Set up fuzzy pairwise comparison matrices (FPCMs) for each hierarchy level based on fuzzy arithmetic. For example, when m alternatives ($C_1...C_m$) on a given level are evaluated against each other with regard to the pth criterion ($p = 1, 2, 3...n$) on the preceding level, an $m \times m$ FPCM is obtained as below

$$\begin{array}{c} \\ C_1 \\ C_2 \\ C_3 \\ \vdots \\ C_m \end{array} \begin{array}{ccccc} C_1 & C_2 & C_3 & \cdots & C_m \\ \left[\begin{array}{ccccc} (1,1,1) & \tilde{x}_{12} & \tilde{x}_{13} & \cdots & \tilde{x}_{1m} \\ 1/\tilde{x}_{12} & (1,1,1) & \tilde{x}_{23} & \cdots & \tilde{x}_{2m} \\ 1/\tilde{x}_{13} & 1/\tilde{x}_{23} & (1,1,1) & \cdots & \tilde{x}_{3m} \\ \vdots & \vdots & \vdots & \vdots & \vdots \\ 1/\tilde{x}_{1m} & 1/\tilde{x}_{2m} & 1/\tilde{x}_{3m} & \cdots & (1,1,1) \end{array}\right] \end{array} \quad (9)$$

To calculate each non-diagonal fuzzy element (e.g., $\tilde{x}_{13}$), the dominance of one alternative or criterion over another is determined by the division of two TFNs. For example, if the random TFNs for C_1 and C_3 are (a_1, b_1, c_1) and (a_3, b_3, c_3), respectively, then $\tilde{x}_{13} = (a_1/c_3, b_1/b_3, c_1/a_3)$ and $1/\tilde{x}_{13} = (a_3/c_1, b_3/b_1, c_3/a_1)$.

Step 5: Calculate the fuzzy weights of each FPCM. For example, in Equation 9, the geometric means of each row and the corresponding fuzzy weights are obtained using Equations 10–13. The weight assessing method by geometric mean is applied because of its simplicity and ease when dealing with fuzzy matrices [10].

$$a_i = \left[\prod_{j=1}^{m} a_{ij}\right]^{1/m} ; \quad b_i = \left[\prod_{j=1}^{m} b_{ij}\right]^{1/m} ; \quad c_i \\ = \left[\prod_{j=1}^{m} c_{ij}\right]^{1/m} \quad for \; i = 1, 2, ..., m \quad (10)$$

$$a_{sum} = \sum_{i=1}^{m} a_i ; b_{sum} = \sum_{i=1}^{m} b_i ; c_{sum} = \sum_{i=1}^{m} c_i \quad (11)$$

$$\tilde{w}_{ip} = \left(\frac{a_i}{c_{sum}}, \frac{b_i}{b_{sum}}, \frac{c_i}{a_{sum}}\right) \quad for \; i = 1, 2, ..., m \quad (12)$$

where a_{ij}, b_{ij}, and c_{ij} are the minimum, most likely, and maximum values of each non-diagonal fuzzy element $\tilde{x}_{ij}$, respectively; m is the size of the FPCM or the number of decision alternatives; a_i, b_i, and c_i are the geometric means of the minimum, most likely, and maximum values of the fuzzy elements on the ith row, respectively; a_{sum}, b_{sum}, and c_{sum} are the sum of a_i, b_i, and c_i, respectively; and $\tilde{w}_{ip}$ are the fuzzy weights of the ith alternative against the pth criterion. Repeating this step to obtain all other $\tilde{w}_{ip}$ and $\tilde{w}_p$, which are the fuzzy weights of the pth criterion in terms of the goal.

Step 6: As with the traditional AHP, the proposed FSAHP approach also measures the inconsistency of each FPCM. Due to the presence of fuzzy numbers, the traditional consistency algorithms are not effective in addressing such uncertainties. Hence, in this paper, a new inconsistency index (CI_F) based on the distance of the matrix to a specific consistent matrix is adopted from Ramík and Korviny [37].

$$s_i^L = \min_i\left\{\frac{b_i}{a_i}\right\} \cdot \frac{a_i}{b_{sum}} \quad (13)$$

$$s_i^M = \frac{b_i}{b_{sum}} \quad (14)$$

$$s_i^U = \max_i\left\{\frac{b_i}{c_i}\right\} \cdot \frac{c_i}{b_{sum}} \quad (15)$$

$$CI_F = \gamma \cdot \max_{ij}\left\{ \max\left\{ \left|\frac{s_i^L}{s_j^U}-a_{ij}\right|, \left|\frac{s_i^M}{s_j^M}-b_{ij}\right|, \left|\frac{s_i^U}{s_j^L}-c_{ij}\right| \right\} \right\} \quad (16)$$

$$\gamma = \frac{1}{\max\left\{\sigma - \sigma^{(2-2m)/m}, \sigma^2\left(\left(\frac{2}{m}\right)^{2/(m-2)} - \left(\frac{2}{m}\right)^{m/(m-2)}\right)\right\}}$$

$$if \ \sigma < \left(\frac{m}{2}\right)^{m/(m-2)}$$

$$\gamma = \frac{1}{\max\{\sigma - \sigma^{(2-2m)/m}, \sigma^{(2m-2)/m} - \sigma\}} \quad if \ \sigma \geq \left(\frac{m}{2}\right)^{m/(m-2)}$$

$$(17)$$

where s_i^L, s_i^M, and s_i^U are the minimum, most likely, and maximum values of the optimal solution that has the minimal measure of fuzziness, respectively; σ is the linguistic scale (i.e., [1/7, 7] in this study); γ is the normality constant; CI_F is the inconsistency index of a FPCM such that a value of 0.1 or less is considered to be acceptable, otherwise the FPCM should be revised.

Step 7: The overall fuzzy priorities $\tilde{w}_i$ of the ith alternative can be calculated by aggregating the weights throughout the hierarchy:

$$\tilde{w}_i = \sum_{p=1}^{n} \tilde{w}_{ip} \times \tilde{w}_p \qquad (18)$$

where $\tilde{w}_{ip}$ are the fuzzy merits of the ith alternative with regard to the pth criterion, respectively; $\tilde{w}_p$ are the fuzzy weights of the pth criterion against the goal; and n is the number of evaluation criteria.

Step 8: Defuzzify $\tilde{w}_i$ by using the center of gravity (COG) method and rank the decision alternatives based on their normalized crisp overall scores w_i.

$$w_i^* = \frac{\int_a^c x\mu_{\tilde{w}_i}(x)dx}{\int_a^c \mu_{\tilde{w}_i}(x)dx} \qquad (19)$$

$$w_i = \frac{w_i^*}{\sum_{i=1}^{m} w_i^*} \qquad (20)$$

where w_i^* are the crisp overall scores of the ith alternative; a and c denote the support of $\tilde{w}_i$; $\mu_{\tilde{w}_i}(x)$ are the corresponding membership functions of $\tilde{w}_i$; and w_i are the normalized crisp overall scores of each decision alternative and are sequenced from high to low in the order of 1 to 5. To validate this ranking scheme, or in other words, the defuzzification results, Chen's fuzzy ranking method is also employed to further compare the overall fuzzy priorities $\tilde{w}_i$ and rank them from the highest to the lowest [38].

Step 9: Repeat Steps 4 to 8 for a number of iterations (e.g., 1000, 5000), the overall scores of alternatives can be obtained and plotted as probability density functions rather than as point values.

Case study

This case study was conducted to demonstrate the applicability and effectiveness of the proposed FSAHP approach in addressing uncertainty in the context of group decision-making. A cargo shop was assumed to be required for an onboard ballast water treatment system in order to operate in the North Atlantic. The decision alternatives and evaluation criteria were determined based on literature review and discussion with experts from governmental ministries and academic institutions. The experts were further invited to fill out the questionnaire on the basis of linguistic terms. Their opinions were analyzed and interpreted to facilitate the implementation of the FSAHP approach.

Hierarchy structure

As depicted in Figure 2, the goal was to select the best onboard treatment technology in order to eliminate invasive microorganisms and to remove water soluble organics from ballast water, particularly in the harsh environments. Five treatment technologies including heat treatment, ultraviolet (UV), ozone, ultrasound, and biocide were chosen [26-30]. Heat treatment is capable of killing invasive species embedded in sediment that has accumulated at the bottom of the ballast tanks. It should be pointed out that discharging warm water potentially threatens biological communities and a complete treatment process may take hours or days, which is not always practical. Despite the potential threats posed by mercury contamination and genetic mutation, UV manages to eliminate microorganisms by breaking chemical bonds in DNA and RNA molecules and cell proteins [29]. Recently, ozone has been widely employed in removing microorganisms from ballast water. The often-cited disadvantages of using ozone as a disinfectant have been reported as the possible formation of toxic byproducts, low solubility, and high instability [39]. Ultrasound can induce the collapse of microscopic gas bubbles in the exposed liquid and lead to the rupture of cell membranes, yet it is less effective in killing some microorganisms such as bacteria [40]. Many chemical biocides have been documented as possible treatment options to the problem of ballast-mediated invasive species. However, some concerns, such as risks from storage and handling, high operational and material cost, and possible discharge of toxic residues need to be taken into account [32]. Based on the recommendations from literature [29,41,42] and expert opinions, in this study, eight evaluation criteria including efficacy on microorganisms, efficacy on organic pollutants, adaptability to harsh environment, capital cost, operation and maintenance (O&M) cost, human health risk, ecological risk, and waste production were chosen.

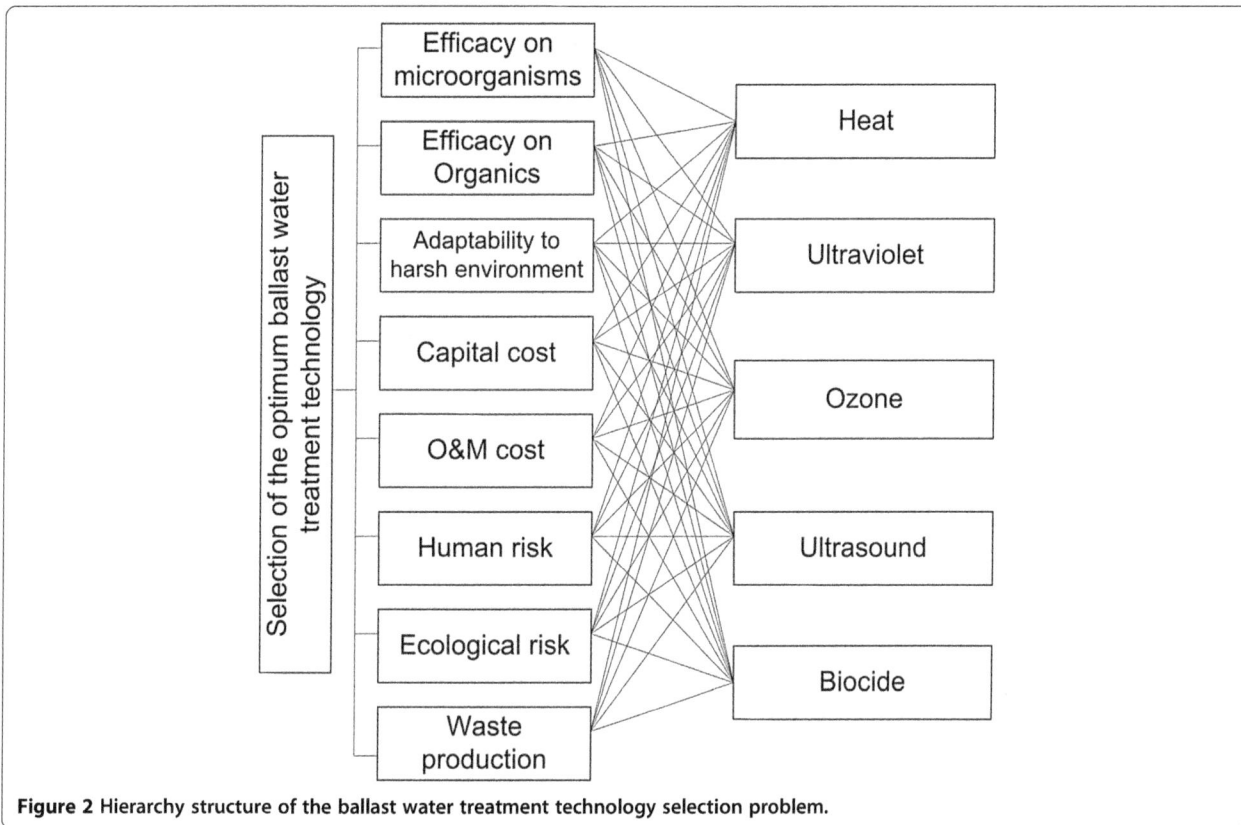

Figure 2 Hierarchy structure of the ballast water treatment technology selection problem.

Data acquisition

In the absence of quantitative data about each alternative, experts' qualitative judgments were used to measure the priorities of alternatives. The linguistic assessments for the qualitative attributes were provided by nine local experts from governmental ministries (environmental division) and academic institutions (professors and graduate students at Memorial University of Newfoundland). They were asked to rate the performance of each alternative and the importance of each criterion using the linguistic scales provided in Figure 1. Tables 1 and 2 summarize the linguistic assessments made by each participating expert. These assessments were aggregated in groups such that the beta-PERT distributions of each group can be estimated to generate random TFNs. For example, the performance of heat, ultraviolet, ozone, ultrasound and biocide with respect to their efficacy on microorganisms was randomly generated as C (2, 3, 4), G (6, 7, 7), G (6, 7, 7), E (4, 5, 6) and F (5, 6, 7), respectively. To obtain the corresponding FPCM (Equation 21), elements in the first row were given by the fuzzy comparisons between the performance of heat (2, 3, 4) and all the others, respectively. The consistency of this FPCM was less than 0.1, which was acceptable, and the fuzzy weights of each alternative were able to be calculated. It should be noted that the number of Monte Carlo iterations used for this case study was determined as 1000 by taking time constraints and the efficiency of convergence into account [22].

$$
\begin{array}{c}
& \begin{array}{ccccc} \textit{Heat} & \textit{Ultraviolet} & \textit{Ozone} & \textit{Ultrasound} & \textit{Biocide} \end{array} \\
\begin{array}{c} \textit{Heat} \\[20pt] \textit{Ultraviolet} \\[18pt] \textit{Ozone} \\[18pt] \textit{Ultrasound} \\[18pt] \textit{Biocide} \end{array}
&
\begin{array}{ccccc}
(1,1,1) & \dfrac{(2,3,4)}{(6,7,7)} & \dfrac{(2,3,4)}{(6,7,7)} & \dfrac{(2,3,4)}{(4,5,6)} & \dfrac{(2,3,4)}{(5,6,7)} \\[14pt]
\left[\dfrac{(2,3,4)}{(6,7,7)}\right]^{-1} & (1,1,1) & \dfrac{(6,7,7)}{(6,7,7)} & \dfrac{(6,7,7)}{(4,5,6)} & \dfrac{(6,7,7)}{(5,6,7)} \\[14pt]
\left[\dfrac{(2,3,4)}{(6,7,7)}\right]^{-1} & \left[\dfrac{(6,7,7)}{(6,7,7)}\right]^{-1} & (1,1,1) & \dfrac{(6,7,7)}{(4,5,6)} & \dfrac{(6,7,7)}{(5,6,7)} \\[14pt]
\left[\dfrac{(2,3,4)}{(4,5,6)}\right]^{-1} & \left[\dfrac{(6,7,7)}{(4,5,6)}\right]^{-1} & \left[\dfrac{(6,7,7)}{(4,5,6)}\right]^{-1} & (1,1,1) & \dfrac{(4,5,6)}{(5,6,7)} \\[14pt]
\left[\dfrac{(2,3,4)}{(5,6,7)}\right]^{-1} & \left[\dfrac{(6,7,7)}{(5,6,7)}\right]^{-1} & \left[\dfrac{(6,7,7)}{(5,6,7)}\right]^{-1} & \left[\dfrac{(4,5,6)}{(5,6,7)}\right]^{-1} & (1,1,1)
\end{array}
\end{array}
\tag{21}
$$

Table 1 Expert assessment for ballast water treatment technologies

Criteria	Alternatives	Expert assessment								
		1	2	3	4	5	6	7	8	9
Efficacy on microorganisms	Heat	C	D	B	B	C	C	D	C	C
	Ultraviolet	G	F	G	G	G	G	F	F	F
	Ozone	G	G	E	G	G	F	F	G	F
	Ultrasound	E	E	F	E	F	D	E	E	E
	Biocide	F	G	G	F	F	G	E	F	G
Efficacy on organics	Heat	B	A	A	C	C	B	C	C	B
	Ultraviolet	F	E	G	F	G	F	F	G	F
	Ozone	F	G	E	D	F	E	F	G	E
	Ultrasound	E	E	D	E	D	E	D	E	C
	Biocide	B	B	A	B	B	B	C	B	A
Adaptability to harsh environment	Heat	C	B	D	C	B	C	C	B	C
	Ultraviolet	F	E	G	F	F	E	F	E	F
	Ozone	F	E	E	F	G	D	E	F	G
	Ultrasound	E	E	F	D	E	D	E	D	D
	Biocide	D	D	F	E	D	E	D	C	D
Capital cost	Heat	F	D	E	E	F	E	F	E	E
	Ultraviolet	D	E	D	E	D	B	D	D	C
	Ozone	C	D	C	D	C	C	C	B	C
	Ultrasound	B	C	B	C	D	C	D	B	D
	Biocide	G	F	F	E	G	F	E	F	E
O&M cost	Heat	F	G	F	E	E	G	E	F	G
	Ultraviolet	D	E	E	D	D	E	D	E	E
	Ozone	C	C	C	D	C	E	C	D	C
	Ultrasound	C	B	C	C	D	D	B	C	D
	Biocide	D	E	D	F	F	E	D	E	F
Human risk	Heat	G	F	G	E	F	F	E	E	F
	Ultraviolet	C	C	D	C	D	C	D	C	B
	Ozone	C	D	B	D	E	D	E	C	D
	Ultrasound	B	D	D	C	D	D	E	D	C
	Biocide	B	B	C	C	D	C	C	D	B
Ecological risk	Heat	C	D	B	C	D	E	C	E	D
	Ultraviolet	D	F	D	E	F	G	F	D	D
	Ozone	F	E	D	E	E	F	F	D	E
	Ultrasound	E	D	E	D	E	D	E	D	D
	Biocide	C	B	E	D	C	C	D	C	C
Waste production	Heat	E	D	D	D	C	D	E	D	C
	Ultraviolet	F	G	F	E	F	G	G	F	E
	Ozone	D	E	D	D	E	E	E	F	F
	Ultrasound	D	C	C	C	D	D	E	D	E
	Biocide	C	B	C	B	D	C	E	D	B

Table 2 Expert assessment for evaluation criteria

Goal	Criteria	Expert assessment								
		1	2	3	4	5	6	7	8	9
Best treatment technology	Efficacy on microorganisms	G	F	F	E	F	E	G	E	F
	Efficacy on organics	E	C	D	E	F	F	D	F	D
	Adaptability to harsh environment	F	F	G	F	E	D	E	C	D
	Capital cost	F	E	B	E	C	D	F	D	C
	O&M cost	C	B	E	D	C	F	F	E	F
	Human risk	F	G	E	F	D	E	D	E	E
	Ecological risk	D	E	F	D	E	C	D	F	D
	Waste production	B	D	D	A	D	B	E	G	C

Results and discussion

The results and statistics were obtained by following the proposed FSAHP approach. Figure 3, for example, depicts the probability density of the scores of each alternative with respect to the criterion of human health risk after 1,000 iterations. The histogram bar plot clearly demonstrates that heat treatment (0.26–0.33) appeared to be the most attractive solution in terms of the lowest health risk, followed by ozone (0.16–0.25) without any overlap. Ultrasound, biocide, and UV were seen as the least preferable option with considerable overlaps between each other, implying that experts were not confident about ranking one over the others. The correlation coefficients between the scores of ultrasound and biocide, biocide and UV, and ultrasound and UV were -0.201, -0.476, and 0.308, respectively. A negative correlation coefficient between two variables usually implies that the increase of one variable is associated with the decrease of the other. One the other hand, a positive correlation coefficient means that two variables increase (or decrease) simultaneously in the same

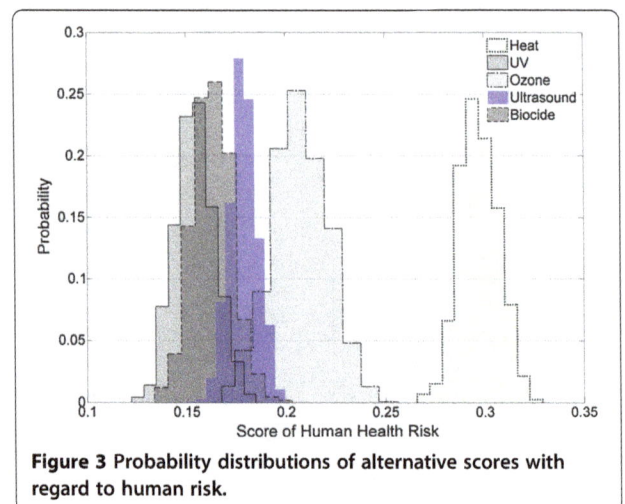

Figure 3 Probability distributions of alternative scores with regard to human risk.

direction. These principles become more prominent as the absolute value of a correlation coefficient close to 1. In this case study, negative correlation coefficients can be interpreted as larger overlaps as compared to positive correlation coefficients based on the fact that the scores were closely distributed (Figure 3). Tables 3 and 4 further validate these conclusions by showing the ranking of alternative priorities based on the COG and Chen's defuzzification methods, respectively. A statistical test of the null hypothesis that heat treatment was not the probabilistic optimal alternative (versus the alternate assumption that it was) was conducted to examine if the difference between it and the second best option (i.e., ozone) was statistically significant. Heat treatment was ranked first by both methods with the confidence level exceeding 95%, indicating the null assumption that it is not probabilistic optimal (versus the alternate assumption that it is) is rejected. Ultrasound took the third place in more than 75% of the iterations while UV had the least preference in over 70% of the cases. From the technical perspective, the results were expected because heat sources such as waste heat from the engine jacket coolers and additional auxiliary boiler are usually not accessible by most crew members. On the other hand, short-term exposure to high level ozone can temporarily influence lung function and respiratory tract; meanwhile, some by-products (e.g., bromate) produced from ozonation may also pose risks to human health. UV was ranked as the least preferable alternative because excessive human exposure to UV is positively associated with severe health problems including photoaged skin, ocular diseases, and skin cancers.

The probability density distributions of criteria weights using the kernel-smoothing method are plotted in Figure 4. It reveals that efficacy on microorganisms, adaptability to harsh environments, O&M cost, and human health risk were the most influential criteria that need to be prioritized in the decision making process. The overall scores of each alternative towards the goal are shown in Figure 5 as histograms. Another statistical test of the null hypothesis that UV was not the probabilistic optimal alternative (versus the alternate assumption that it

Table 4 Ranking with regard to human risk based on Chen's method

Treatment technology	Rank				
	1	2	3	4	5
Heat	1000	0	0	0	0
UV	0	2	20	277	701
Ozone	0	929	57	11	3
Ultrasound	0	64	752	180	4
Biocide	0	5	171	532	292
Total	1000	1000	1000	1000	1000

was) was conducted. Tables 5 and 6 reveal that UV was ranked with the highest overall score at 100% confidence level, indicating that the null assumption that it was not probabilistic optimal (versus the alternate assumption that it is) was rejected. Ozone, heat treatment, and ultrasound had the second, third, and fourth places at the confidence levels of 61.0–71.4%, 56.0–68.4%, and 78.4–84.6%, respectively. Figure 6 further supports this ranking scheme by using box plot to graphically illustrate the minimum, lower quartiles, medians, upper quartiles, and maximum of the overall scores. It indicates that the score distribution of ozone has a remarkable overlap with that of ultrasound as their medians, lower percentiles, and upper percentiles are close to each other. Nonetheless, ozone has a wider spread of scores as compared to ultrasound, suggesting that the experts were more unanimous on the performance of ultrasound during their assessment. Another interesting point to note is that both COG and Chen's methods produced similar defuzzification results, which demonstrated their applicability in the proposed FSAHP approach. In addition, the results also depicted that the proposal approach can well address linguistic inputs in group decision making

Table 3 Ranking with regard to human risk based on the COG method

Treatment technology	Rank				
	1	2	3	4	5
Heat	1000	0	0	0	0
UV	0	3	6	260	731
Ozone	0	943	51	6	0
Ultrasound	0	50	798	150	2
Biocide	0	4	145	584	267
Total	1000	1000	1000	1000	1000

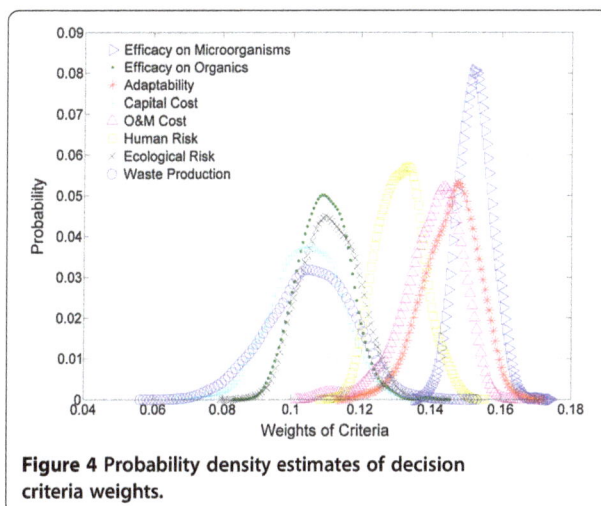

Figure 4 Probability density estimates of decision criteria weights.

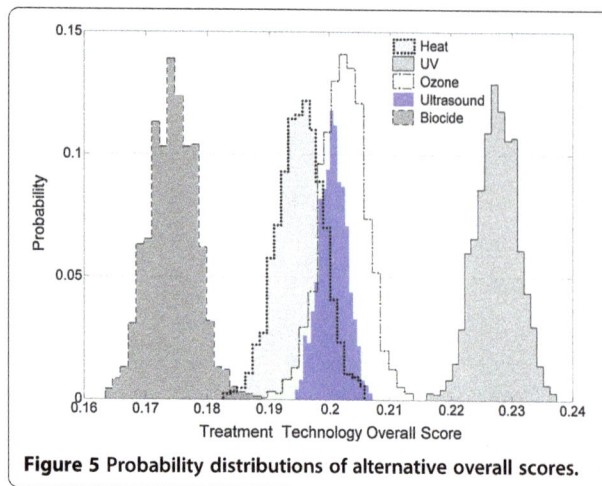

Figure 5 Probability distributions of alternative overall scores.

Table 6 Summary of the simulation results for the final ranking based on Chen's method

Treatment technology	Rank				
	1	2	3	4	5
Heat	0	25	98	876	1
UV	1000	0	0	0	0
Ozone	0	746	218	36	0
Ultrasound	0	229	684	87	0
Biocide	0	0	0	1	999
Total	1000	1000	1000	1000	1000

processes. The decision makers would be more comfortable and confident to give vague judgments rather than evaluating pairwise comparisons using single numeric values. Verbal assessments were collected and compared against with each other wherein the priorities of each alternative were determined. The use of the beta-PERT distribution was also able to lessen the uncertainty caused by insufficient information or biased opinions.

Conclusions

As one of the most widely exploited multi-criteria decision making (MCDM) approaches, the analytic hierarchy process (AHP) has been well documented in the literature. However, it has been criticized for its inability to quantify the uncertainty associated with decision making. In this paper, a hybrid fuzzy stochastic analytical hierarchy process (FSAHP) approach was developed in order to assist decision making with more confidence by integrating fuzzy set theory, probabilistic distribution, pairwise comparison and Monte Carlo simulation. A case study related to ballast water management was carried out to verify the feasibility and efficiency of the proposed approach. Five treatment technologies were evaluated against a number of environmental, economic,

and technical criteria by nine experts. The results revealed that UV was ranked with the highest overall score at 100% confidence level, indicating that the null assumption that it was not probabilistic optimal (versus the alternate assumption that it is) was rejected. Ozone, heat treatment, and ultrasound had the second, third, and fourth places at the confidence levels of 61.0–71.4%, 56.0–68.4%, and 78.4 – 84.6%, respectively. Considerable overlaps existed among these three alternatives which may be attributed to the irreducible uncertainty caused by subjective judgments or lack of knowledge. The results also revealed that both COG and Chen's defuzzification methods were able to provide the decision makers with reliable decision references. The proposed FSAHP approach can offer a number of benefits such as the capability of capturing human's appraisal of ambiguity and addressing the effects of uncertain judgment when dealing with insufficient information or biased opinions. However, this approach is highly sensitive to expert dependence whereby any misjudgment may affect its reliability and efficiency. As a complex methodology, it requires more computational efforts in assessing composite priorities than the traditional AHP.

Table 5 Summary of the simulation results for the final ranking based on the COG method

Treatment technology	Rank				
	1	2	3	4	5
Heat	0	71	144	784	1
UV	1000	0	0	0	0
Ozone	0	610	296	94	0
Ultrasound	0	319	560	121	0
Biocide	0	0	0	1	999
Total	1000	1000	1000	1000	1000

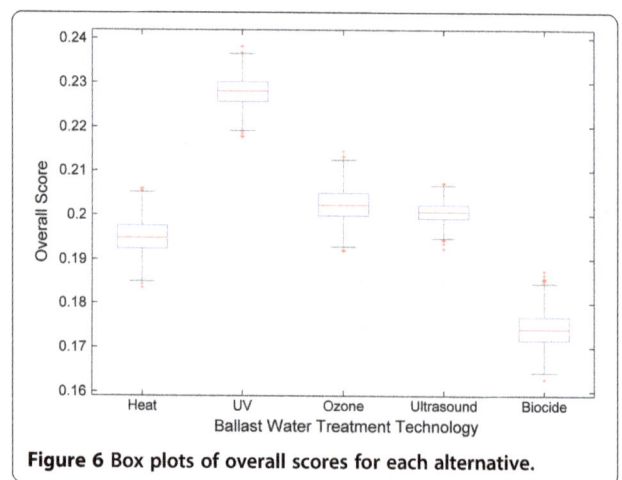

Figure 6 Box plots of overall scores for each alternative.

Competing interests

The authors declared that they have no competing interest.

Authors' contributions

LJ and BC co-developed the FSAHP method for group decision-making and conducted the design of the study. BZ and HP participated in the design of questionnaire, acquisition of data and data analysis. LJ performed data analysis and drafted the manuscript, which BC, BZ and HP helped edit and polish. All authors have read and approved the final manuscript.

Acknowledgements

Special thanks go to American Bureau of Shipping Harsh Environment Technology Centre (ABS-HETC), Research & Development Corporation Newfoundland and Labrador (RDC NL), Natural Sciences and Engineering Research Council of Canada (NSERC), and Memorial University of Newfoundland for funding this work.

References

1. Kiker GA, Bridges TS, Varghese A, Seager TP, Linkov I: **Application of multicriteria decision analysis in environmental decision making.** *Integr Environ Assess Manage* 2005, 1(2):95–108.
2. Matott LS, Babendreier JE, Purucker ST: **Evaluating uncertainty in integrated environmental models: a review of concepts and tools.** *Water Resour Res* 2009, **45**. doi:10.1029/2008WR007301.
3. Steele K, Carmel Y, Cross J, Wilcox C: **Uses and misuses of multicriteria decision analysis (MCDA) in environmental decision making.** *Risk Anal* 2008, 29(1):26–33.
4. Yeh CH, Chang YH: **Modeling subjective evaluation for fuzzy group multicriteria decision making.** *Eur J Oper Res* 2009, 194(2):464–473.
5. Yu L, Wang S, Lai KK: **An intelligent-agent-based fuzzy group decision making model for financial multicriteria decision support: the case of credit scoring.** *Eur J Oper Res* 2009, 195(3):942–959.
6. Kornyshova E, Salinesi C: **MCDM techniques selection approaches: state of the art.** In *Proceedings of the 2007 IEEE Symposium on Computational Intelligence in Multi-Criteria Decision-Making (MCDM), Honolulu.* 2007.
7. Saaty TL: *The Analytic Hierarchy Process: Planning, Priority Setting and Resource Allocation.* New York, NY, USA: McGraw-Hill; 1980.
8. Chowdhury S, Husain T: **Evaluation of drinking water treatment technology: an entropy-based fuzzy application.** *J Environ Eng-ASCE* 2006, 132(10):1264–1271.
9. Jablonsky J: **Measuring the efficiency of production units by AHP models.** *Math Comput Modell* 2007, 46:1091–1098.
10. Kaya T, Kahraman C: **An integrated fuzzy AHP–ELECTRE methodology for environmental impact assessment.** *Expert Syst Appl* 2011, 38:8553–8562.
11. Sadiq R, Tesfamariam S: **Environmental decision-making under uncertainty using intuitionistic fuzzy analytic hierarchy process (IF-AHP).** *Stoch Env Res Risk A* 2009, 23(1):75–91.
12. Tiryaki F, Ahlatcioglu B: **Fuzzy portfolio selection using fuzzy analytic hierarchy process.** *Inform Sci* 2009, 179:53–69.
13. Tolga E, Demircan ML, Kahraman C: **Operating system selection using fuzzy replacement analysis and analytic hierarchy process.** *Int J Prod Econ* 2005, **97**:89–117.
14. Deng H: **Multicriteria analysis with fuzzy pair-wise comparison.** *Int J Approximate Reasoning* 1999, 21:215–231.
15. Banuelas R, Antony J: **Modified analytic hierarchy process to incorporate uncertainty and managerial aspects.** *Int J Prod Res* 2004, 42(18):3851–3872.
16. Rosenbloom ES: **A probabilistic interpretation of the final rankings in AHP.** *Eur J Oper Res* 1996, **96**:371–378.
17. Carlucci D, Schiuma G: **Applying the analytic network process to disclose knowledge assets value creation dynamics.** *Expert Syst Appl* 2009, 36(4):7687–7694.
18. Yu CS: **A GP-AHP method for solving group decision-making fuzzy AHP problems.** *Comput Oper Res* 2002, 29:1969–2001.
19. Tesfamariam S, Sadiq R: **Risk-based environmental decision-making using fuzzy analytic hierarchy process (F-AHP).** *Stoch Env Res Risk A* 2006, 21:35–50.
20. Eskandari H, Rabelo L: **Handling uncertainty in the analytic hierarchy process: a stochastic approach.** *Int J Inf Tech Decis* 2007, 6(1):177–189.
21. Jing L, Chen B, Zhang BY, Li P: **A hybrid stochastic-interval analytic hierarchy process (SIAHP) approach for prioritizing the strategies of reusing treated wastewater.** *Math Probl Eng* 2013. doi:10.1155/2013/874805.
22. Hsu T, Pan FFC: **Application of Monte Carlo AHP in ranking dental quality attributes.** *Expert Syst Appl* 2009, 36:2310–2316.
23. Phanikumar CV, Maitra B: **Valuing urban bus attributes: an experience in Kolkata.** *J Publ Transport* 2006, 9(2):69–87.
24. Coates G, Rahimifard S: **Modelling of post-fragmentation waste stream processing within UK shredder facilities.** *Waste Manage* 2009, 29(1):44–53.
25. Jing L, Chen B, Zhang BY, Li P, Zheng JS: **A Monte Carlo simulation aided analytic hierarchy process (MC–AHP) approach for best management practices assessment in nonpoint source pollution control.** *J Environ Eng-ASCE* 2012, 139(5):618–626.
26. Endresen Ø, Behrens HL, Brynestad S, Andersen AB, Skjong R: **Challenges in global ballast water management.** *Mar Pollut Bull* 2004, 48:615–623.
27. Cangelosi AA, Mays NL, Balcer MD, Reavie ED, Reid DM, Sturtevant R, Gao X: **The response of zooplankton and phytoplankton from the North American Great Lakes to filtration.** *Harmful Algae* 2007, 6:547–566.
28. Galil BS, Nehring S, Panov V: **Waterways as invasion highways – impact of climate change and globalization.** *Ecol Stud* 2007, 193(2):59–74.
29. Jing L, Chen B, Zhang BY, Peng HX: **A review of ballast water management practices and challenges in harsh and arctic environments.** *Environ Rev* 2012, 20:83–108.
30. Parmesan C: **Ecological and evolutionary responses to recent climate change.** *Annu Rev Ecol Evol Syst* 2006, 37:637–669.
31. Gollasch S, David M, Voigt M, Dragsund E, Hewitt C, Fukuyo Y: **Critical review of the IMO international convention on the management of ships' ballast water and sediments.** *Harmful Algae* 2007, 6:585–600.
32. Gregg MD, Hallegraeff GM: **Efficacy of three commercially available ballast water biocides against vegetative microalgae, dinoflagellate cysts and bacteria.** *Harmful Algae* 2007, 6:567–584.
33. Zadeh LA: **Fuzzy Sets.** *Inform Contr* 1965, 8:338–353.
34. Kaufmann A, Gupta MM, Kaufmann A: *Introduction to fuzzy arithmetic: theory and applications.* New York: Van Nostrand Reinhold Company; 1985.
35. Li P, Chen B: **FSILP: Fuzzy-stochastic-interval linear programming for supporting municipal solid waste management.** *J Environ Manage* 2011, 92:1198–1209.
36. Dawes J: **Do data characteristics change according to the number of scale point used?** *Int J Market Res* 2007, 50(1):61–77.
37. Ramík J, Korviny P: **Inconsistency of pair-wise comparison matrix with fuzzy elements based on geometric mean.** *Fuzzy Sets Syst* 2010, 161:1604–1613.
38. Chen SJ, Hwang CL: *Fuzzy multiple attribute decision making.* Heidelberg: Springer; 1992.
39. Herwig RP, Cordell JR, Perrins JC, Dinnel PA, Gensemer RW, Stubblefield WA, Ruiz GM, Kopp JA, House ML, Cooper WJ: **Ozone treatment of ballast water on the oil tanker S/T Tonsina: chemistry, biology and toxicity.** *Mar Ecol Prog Ser* 2006, 324:37–55.
40. Holm ER, Stamper DM, Brizzolara RA, Barnes L, Deamer N, Burkholder JM: **Sonication of bacteria, phytoplankton and zooplankton: Application to treatment of ballast water.** *Mar Pollut Bull* 2008, 56:1201–1208.
41. de Lafontaine Y, Despatie SP, Veilleux É, Wiley C: **Onboard ship evaluation of the effectiveness and the potential environmental effects of PERACLEAN® Ocean for ballast water treatment in very cold conditions.** *Environ Toxicol* 2009, 24(1):49–65.
42. Tsolaki E, Diamadopoulos E: **Technologies for ballast water treatment: a review.** *J Chem Technol Biotechnol* 2010, 85:19–32.

Permissions

List of Contributors

Kebede Wolka Wolancho
Wondo Genet College of Forestry and Natural Resources, School of Natural Resources and Environmental Studies, Hawassa University, P.O. Box 128, Shashemene, Ethiopia

Qing Chen
School of Applied Mathematics, Xiamen University of Technology, Ligong Road, 361024 Xiamen, China

Zhong Tan
School of Mathematical Sciences, Xiamen University, Siming South Road, 361005 Xiamen, China

Tianhong Li
College of Environmental Sciences and Engineering, The Key Laboratory for Water and Sediment Sciences, Peking University, Beijing 100871, China
Shenzhen Graduate School of Peking University, Shenzhen 518055, China

Wenkai Li
Geography and Planning School, Sun Yat-sen University, Guangzhou 510275, China

Habitamu Taddese
Hawassa University, Wondo Genet College of Forestry and Natural Resources, P. O. Box 128, Shashemene, Ethiopia

Wei Hu
Department of Soil Science, University of Saskatchewan, Saskatoon, SK S7N 5A8, Canada

Jing Xie
Department of Soil Science, University of Saskatchewan, Saskatoon, SK S7N 5A8, Canada

Henry Wai Chau
Department of Soil and Physical Science, Lincoln University, PO Box 84Lincoln, Christchurch 7647, New Zealand

Bing Cheng Si
Department of Soil Science, University of Saskatchewan, Saskatoon, SK S7N 5A8, Canada

Xiaosheng Qin
School of Civil & Environmental Engineering, Nanyang Technological University, 50 Nanyang Avenue 639798 Singapore
Earth Observatory of Singapore (EOS), Nanyang Technological University, 50 Nanyang Avenue 639798 Singapore

Ye Xu
S-C Research Academy of Energy & Environmental Studies, North China Electric Power University, Beijing 102206, China

Jianjun Yu
DHI-NTU Water & Environment Research Centre and Education Hub, Nanyang Technological University, 50 Nanyang Avenue 639798, Singapore

Masayuki Otaki
Institute of Social Science, University of Tokyo, 7-3-1 Hongo, Bunkyo, Tokyo, Japan

Bin Lu
Department of Applied Mathematics, University of Western Ontario, London, ON N6A 5B7, Canada

Matt Davison
Department of Applied Mathematics, Department of Statistical and Actuarial Science, Richard Ivey School of Business, University of Western Ontario, London, ON N6A 5B7, Canada

Xiaowen Ding
Key Laboratory of Urban Stormwater System and Water Environment, (Beijing University of Civil Engineering and Architecture), Ministry of Education, No.1 Zhanlanguan Road, Beijing 100044, People's Republic of China
Key Laboratory of Regional Energy and Environmental Systems Optimization, Ministry of Education, North China Electric Power University, No. 2 Beinong Road, Beijing 102206, People's Republic of China
Institute for Energy, Environment and Sustainable Communities, University of Regina, 120, 2 Research Drive, Regina, SK S4S 7H9, Canada

Yongwei Gong
Key Laboratory of Urban Stormwater System and Water Environment, (Beijing University of Civil Engineering and Architecture), Ministry of Education, No.1 Zhanlanguan Road, Beijing 100044, People's Republic of China

Chunjiang An
Institute for Energy, Environment and Sustainable Communities, University of Regina, 120, 2 Research Drive, Regina, SK S4S 7H9, Canada

Ming Lin
Key Laboratory of Regional Energy and Environmental Systems Optimization, Ministry of Education, North China Electric Power University, No. 2 Beinong Road, Beijing 102206, People's Republic of China

Hui Yu
Department of Chemical Engineering, University of New Brunswick, Fredericton, NB, E3B 5A3, Canada MOE Key Laboratory of Regional Energy Systems Optimization, S&C Academy of Energy and Environmental Research, North China Electric Power University, Beijing 102206, China

Huining Xiao
Department of Chemical Engineering, University of New Brunswick, Fredericton, NB, E3B 5A3, Canada

Dunling Wang
Ministry of Agriculture, Government of Saskatchewan, Regina, SK, S4S 0B1, Canada

Shan Zhao
Faculty of Engineering and Applied Science, University of Regina, Regina, Saskatchewan S4S 0A2, Canada

Yang Zhou
Faculty of Engineering and Applied Science, University of Regina, Regina, Saskatchewan S4S 0A2, Canada

Mengyuan Wang
Faculty of Engineering and Applied Science, University of Regina, Regina, Saskatchewan S4S 0A2, Canada

Xiaying Xin
Faculty of Engineering and Applied Science, University of Regina, Regina, Saskatchewan S4S 0A2, Canada

Fang Chen
Faculty of Engineering and Applied Science, University of Regina, Regina, Saskatchewan S4S 0A2, Canada

Aaron N Wiegand
University of the Sunshine Coast, Locked Bag 4, Maroochydore DC, QLD 4558, Australia

Christopher Walker
University of the Sunshine Coast, Locked Bag 4, Maroochydore DC, QLD 4558, Australia

Peter F Duncan
School of Ocean Sciences, Bangor University, Anglesey, LL59 5AB, United Kingdom

Anne Roiko
University of the Sunshine Coast, Locked Bag 4, Maroochydore DC, QLD 4558, Australia

Neil Tindale
University of the Sunshine Coast, Locked Bag 4, Maroochydore DC, QLD 4558, Australia

Jin Si
Bartlett School of Graduate Studies, University College London (UCL), London WC1H-0NN, UK

Fuzhan Nasiri
Bartlett School of Graduate Studies, University College London (UCL), London WC1H-0NN, UK

Peng Han
The Key Laboratory of Water and Sediment Science, Department of Environmental Engineering, Peking University, Beijing 100871, China

Tianhong Li
The Key Laboratory of Water and Sediment Science, Department of Environmental Engineering, Peking University, Beijing 100871, China

Amer Ibrahim Al-Omari
Department of Mathematics, Faculty of Science, Al al-Bayt University, Mafraq 25113, Jordan

Abdul Haq
Department of Statistics, Quaid-i-Azam University, Islamabad 45320, Pakistan

Julian Scott Yeomans
Operations Management & Information Systems Area, Schulich School of Business, York University, 4700 Keele Street, Toronto, Ontario M3J 1P3, Canada

Matt Davison
Departments of Applied Mathematics and Statistical & Actuarial Sciences and the Richard Ivey School of Business, The University of Western Ontario, London, ON, Canada

Ozgur Gurtuna
Turquoise Technology Solutions Inc, Montreal, Canada

Claude Masse
Environment Canada, Montreal, Canada

Brian Mills
Adaptation and Impacts Research, Environment Canada, Waterloo, Canada

Liang Jing
Faculty of Engineering and Applied Science, Memorial University of Newfoundland, St. John's, NL A1B 3X5, Canada

Bing Chen
Faculty of Engineering and Applied Science, Memorial University of Newfoundland, St. John's, NL A1B 3X5, Canada

Baiyu Zhang
Faculty of Engineering and Applied Science, Memorial University of Newfoundland, St. John's, NL A1B 3X5, Canada

Hongxuan Peng
Faculty of Engineering and Applied Science, Memorial University of Newfoundland, St. John's, NL A1B 3X5, Canada

www.ingramcontent.com/pod-product-compliance
Lightning Source LLC
Chambersburg PA
CBHW050438200326
41458CB00014B/4994